Optics and Lasers in Biomedicine and Culture

Springer
Berlin
Heidelberg
New York
Barcelona
Hong Kong
London
Milan
Paris
Singapore
Tokyo

Series of the International Society on Optics Within Life Sciences

Series Editor:
Gert von Bally *Laboratory of Biophysics, Institute of Experimental Audiology, University of Münster, Robert-Koch-Str. 45, 48129 Münster, Germany*

Volume V

OWLS V

International Society on
Optics Within Life Sciences

International Commission for Optics

Foundation for Research & Technology – Hellas

C. Fotakis · T. G. Papazoglou
C. Kalpouzos (Eds.)

Optics and Lasers in Biomedicine and Culture

Contributions to the Fifth International Conference on Optics Within Life Sciences OWLS V
Crete, 13–16 October 1998

With 276 Figures

Springer

Editors:

Costas Fotakis
Theodore G. Papazoglou
Costas Kalpouzos

FORTH-IESL
PO Box 1527
GR 71110 Heraklion
Greece

Library of Congress Cataloging-in-Publication Data

International Conference on Optics within Life Sciences (5th: 1998; Crete, Greece)
Biomedicine and culture in the area of modern optics and lasers: contributions to the Fifth International Conference on Optics and Life Sciences OWLS V, Crete, 13–16 October 1998 / C. Fotakis, T. Papazoglou, C. Kalpouzos (eds.).
p. cm. – (Series of the International Society on Optics within Life Sciences; 5)
Includes bibliographical references.
ISBN 3540666486 (hardcover: acid.free paper)
1. Imaging systems in medicine–Congresses. 2. Medical Sciences–Congresses. 3. Imaging systems–Congresses. I. Fotakis, C. (Costas). II. Papazoglou, T. III. Kalpouzos, C. IV. Title. V. Series.
R857.O6I575 1998 610'.28–dc21 99-053478

ISBN 3-540-66648-6 Springer-Verlag Berlin Heidelberg New York

Printed in Germany

Typesetting: Camera ready by authors/editors
Cover design: design & production GmbH, Heidelberg

Printed on acid-free paper SPIN 10735738 57/3144/xo - 5 4 3 2 1 0

Preface

The International Society on Optics Within Life Sciences (OWLS) was founded in 1990 as a non-profit scientific organization based on individual membership. Its objective is to forward interdisciplinary information and communication of specialists in optics as well as in medicine, biology, environment and cultural heritage. To serve this purpose a series of *International Conferences on Optics Within Life Sciences* (OWLS I – V) has been established. The latest conference entitled ***Biomedicine and Culture in the Era of Modern Optics and Lasers*** took place on October 13-16, 1998 in Aghia Pelagia, Crete, Greece. The event was supported by the European Union as a Euroconference, by the International Commission for Optics (ICO), by the Japanese Medical Optical Equipment Industrial Association (JMOIA) and by the Karl Storz GmbH & Co. KG, which has financed the publication of this book by Springer Verlag.

Following the previous OWLS conferences devoted to diverse applications of optics in life sciences, this OWLS V Conference focused on recent achievements in applying lasers and optics in biomedicine and the preservation of cultural heritage. Particular attention was paid to laser diagnostics in medicine, the interaction of laser radiation with biological tissue, aspects of the preservation of cultural heritage, and the development of new systems for these studies.

The editors wish to thank Ms. M. Kokolaki for her assistance in preparing the manuscript of this volume. We also acknowledge the financial support of Karl Storz Endoskope GmbH & Co. KG.

C. Fotakis, Heraklion, Greece
C. Kalpouzos, Heraklion, Greece
T. Papazoglou, Heraklion, Greece

Contents

Optics

In Search of the Earliest Known Lenses (Dating Back 4500 Years) 3
Jay M. Enoch

Study on Reproduction of Sound from Old Wax Phonograph Cylinders Using the Laser 14
T. Asakura, J. Uozumi, T. Iwai and T. Nakamura

Dynamic Holographic Image Projection: The Key to Optical Interfacing 20
N.A. Vainos, S. Mailis, G. Siganakis, A. Bonarou, V. Tornari, G. Betzos and P. Mitkas

A Novel Electronic-Speckle-Pattern-Interferometer (ESPI) for Dynamic-Holographic-Endoscopy 29
B. Kemper, A. Merker, S. Lai and G. von Bally

The Endoscopic Application of the Deconvolution Filter Set Method 33
M. Hattori and S. Komatsu

A Coherent Optical Technique Useful in Biomechanics: Holographic Interferometry 37
P.M. Boone

Relative and Absolute Coordinates Measurement by Phase-Stepping Laser Interferometry 50
V. Sainov, J. Harizanova, G. Stoilov, P. Boone

Reproduction of Sounds from an Old Russian Wax Phonograph Cylinder by Various Optical Methods 54
T. Nakamura and T. Asakura

Optical Reproduction of Sounds from Negative Phonograph Cylinders 58
J. Uozumi, T. Ushizaka and T. Asakura

Applications of Video Holographic Interferometry in Textile Conservation 62
P.S. Fowles, A. Lord, Nguyen Huu Thanh and P.M. Boone

Second-Harmonic Generation Imaging 66
R. Gauderon, P.B. Lukins and C.J.R. Sheppard

Non-Steady-State Photocurrents in Indium-Oxide Thin-Film Holographic Recorders 70
S. Mailis, A. Ikiades, N.A. Vainos, V.V. Kulikov and I.A. Sokolov

Bending Loss in Mid-Infrared Waveguides and Fibers 75
A.A. Serafetinides, K.R. Rickwood, E.T. Fabrikesi, G. Chourdakis, N. Anastassopoulou, Y. Matsuura, Yi-Wei Shi, M. Miyagi and N. Croitoru

Twin-Image Elimination in Digital In-line Holography 79
S. Lai, B. Kemper and G. von Bally

Examination of Jade Conditions by Laser Technology 83
M.H. Khin, Y. Emori, H. Ohzu

Laser and Optics in Art Conservation

Application of UV-Lasers in Historical Glass Cleaning 89
K. Dickmann, J. Hildenhagen, F. Fekrsanati, H. Römich, C. Troll and U. Drewello

Humanoid Machine Vision Systems and Underwater Archeology 95
P. Greguss

Near-Ultraviolet Pulsed Laser Interaction with Contaminants and Pigments on Parchment: Spectroscopic Diagnostics for Laser Cleaning Safety 100
W. Kautek, S. Pentzien, P. Rudolph, J. Krüger, C. Maywald-Pitellos, H. Bansa, H. Grösswang, E. König

A Preliminary Study into the Suitability of Femtosecond Lasers for the Removal of Adhesive from Canvas Paintings 108
J. Shepard, C.R.T. Young, D. Parsons-Karavassilis, K. Dowling

UV-laser Ablation of Polymerized Resin Layers and Possible Oxidation Processes in Oil-Based Painting Media 115
V. Zafiropulos, A. Galyfianaki, S. Boyatzis, A. Fostiridou, E. Ioakimoglou

Vibration Monitoring by TV-Holography — a Diagnostic Tool in the Conservation of Historical Murals 123
T. Fricke-Begemann, G. Gülker, K.D. Hinsch, H. Joost

Algorithms for Pigment Identification 127
M. Breitman

Investigations of Handwritten Manuscripts by Means of Optical Correlation Methods 131
M. Senoner, S. Krueger, G. Wernicke, N. Demoli and H. Gruber

Optical Fibers for the Cultural Heritage II: The Monitoring of Lighting in Museum Environments 135
R. Falciai, A.g. Mignani, C. Trono, B. Tiribilli

Fast and Precise Determination of Painted Artwork Composition by Laser Induced Plasma Spectroscopy 139
A. Ciucci, V. Palleschi, S. Rastelli, A. Salvetti, E. Tognoni, R. Fantoni and I. Borgia

Application of Factor Analysis and Multivariate Curve Resolution Techniques for the Separation and Identification of Raman Spectra from Pigments on Artworks 143
L. Coma, M.J. Manzaneda and S. Ruiz-Moreno

Automatic Acquisition and Evaluation of Optically Achieved Range Data of Medical and Archaeological Samples 147
D. Dirksen, G. von Bally and F. Bollmann

Surface Modification of Wood by Laser Irradiation 151
H. Wust, M. Panzner, G. Wiedemann, K. Henneberg, L. Pöppel, T. Wittke

Decoration of Glass by Surface and Sub-surface Laser Engraving 155
A. Lenk, T. Witke

Experimental Studies on Black Crusted Sandstone Cleaning by Various UV Wavelengths 159
S. Klein, V. Zafiropulos, T. Stratoudaki, J. Hildenhagen, K. Dickmann and Th.. Lehmkuhl

Laser Induced Breakdown Spectroscopy in the Analysis of Pigments In Painted Artworks. A Database of Pigments and Spectra 163
T. Stratoudaki, D. Xenakis, V. Zafiropulos and D. Anglos

Optical Fibers for the Cultural Heritage I: Picture Varnishes as Thermosensitive Fiber Cladding 169
A.G. Mignani, M. Bacci, C. Trono

Non-invasive Measurements of Damage of Frescoes Paintings and Icons by Laser Scanning Vibrometer: A Comparison of Different Exciters Used with Artificial Samples 174
P. Castellini, E. Esposito, N. Paone, E.P. Tomasini

FTIR Imaging Spectroscopy for Organic Surface Analysis of Embedded Paint Cross-Sections 179
R.M.A. Heeren, J. van der Weerd and J.J. Boon

Advanced Workstation for Controlled Laser Cleaning of Paintings 183
J.H. Scholten, J.M. Teule, V. Zafiropulos, R.M.A. Heeren

Illumination System for the Theodosius' Plate 188
E. Bernabeu, J. Alda, A. García-Botella, J.A. Gómez-Pedrero, E. Olivera

Research and Development of Raman Spectroscopy with Optical Fibre. Application to Pigments Identification 192
F.J. Sierra, J.M. Yúfera, S. Ruiz-Moreno, M.J. Soneira and C. Sandalinas

The Potential Uses of Lasers and Layer Manufacture in Conservation 197
P.S. Fowles

Identification of Pigments by Raman Microscopy: Relevance to the Authentication or Otherwise of Egyptian Papyri 201
L. Burgio and R.J.H. Clark

Progress in the Use of Excimer Lasers to Clean Easel Paintings 203
A.E. Hill, A. Athanassiou, T. Fourrier, J. Anderson and C. Whitehead

Discrimination of Photomechanical Effects in the Laser Cleaning of Artworks by Means of Holographic Interferometry 208
V. Tornari, V. Zafiropulos, N.A. Vainos, D. Fantidou and C. Fotakis

Cleaning of Ceramics Using Lasers of Different Wavelength 213
T. Stratoudaki, A. Manousaki, V. Zafiropulos, N. Huet, S. Pétremont, A. Vinçotte

Non-Destructive Analysis of Two Post-Byzantine Icons by Use
of the Multi Spectral Imaging System (MU.S.I.S 2007) 218
O. Theodoropoulou, G. Tsairis

Lasers in Medicine

TMLR: Potential Mechanisms of Action and Design Parameters
for Excimer-Based Clinical Systems 225
R.E.N. Shehada, T. Papaioannou and W.S. Grundfest

Lasers in Modern Cataract-Surgery 238
E. Alzner and G. Grabner

Correlation of Thermal and Mechanical Effects of the Holmium Laser
for Various Clinical Applications 244
M.C.M. Grimbergen, R.M. Verdaasdonk and C.F.P. van Swol

Intra-operative Fluorescence Imaging of ALA-induced PpIX Using
the Double Ratio Imaging Technique in Malignant Brain Tumours 250
A.J.L. Jongen, R.E. Feller, J.G. Wolbers and H.J.C.M. Sterenborg

Photon Interactions with Ceramic Bone Implants 254
St. Szarska, K. Sarnowska

Easy Human Visual MTF Measurements 260
N. Nameda, M. Masuya, Y. Shimizu and E. Hatae

Measurement of the Anterior Segment of the Eye
Using an Improved Slit Lamp 264
L.H. Liu and H. Ohzu

Optics in Photomedicine and Photobiology

STM of Light-Sensitive Biological Systems 269
P.B. Lukins and T. Oates

Optical Inspection of Surface Quality of Pharmaceutical Compacts 273
*K.-E. Peiponen, R. Silvennoinen, V. Hyvärinen, P. Raatikainen and
P. Paronen*

Single Shot, Laser Plasma X-Ray Contact Microscopy of *Chlamydomonas* 277
A.C. Cefalas, P. Argitis, E. Sarantopoulou, Z. Kollia, T.W. Ford, A. Marranca, A.D. Stead, C.N. Danson, J. Knott, D. Neely

Image Formation in Optical Coherence Tomography and Microscopy 281
C.J.R. Sheppard and M. Roy

In vivo Measurement of the Optical Properties of Human Tissues in the Wavelength Range 610-1010 nm 285
R. Cubeddu, A. Pifferi, P. Taroni, A. Torricelli and G. Valentini

Depth Estimation of an Absorbent Embedded in a Dense Medium Using Diffused Wave Reflectometry 289
T. Iwai, K. Tabata, G. Kimura and T. Asakura

In vitro Optical Characterization of Female Breast Tissue with Near Infrared fsec Laser Pulses 294
G. Zacharakis, V. Sakkalis, G. Filippidis, A. Zolindaki, E. Koumantakis, T.G. Papazoglou

Bio-Speckle Phenomena for Blood Flow Measurements: Speckle Fluctuations and Doppler Effects 297
Y. Aizu and T. Asakura

Layered Gel-Based Phantoms Mimicking Fluorescence of Cervical Tissue 301
S. Chernova, A. Pravdin, Y. Sinichkin, V. Tuchin, S. Vari

Interferometrical Microscope for Investigations of Biological Objects 307
D. Tontcheva, I. Sainova, N. Metchkarov, V. Sainov

Realization of a Double-Exposure Interferometer with Photorefractive Crystals for Biomedical Applications 312
M. Weber, F. Rickermann and G. von Bally

Comparative Studies of Laser Induced Fluorescence and Intravascular Ultrasound for the Human Coronary Artery Diagnosis of Atherosclerosis 316
A. Manolopoulos, A. Vasileiou, V. Kokkinos, G. Athanasopoulos, E. Agapitos, N. Kavantzas and D. Yova

Temporal and Spectral Narrowing of Sub-picosecond Laser-Induced Fluorescence of Polymeric Gain Media 324
G. Zacharakis, G. Heliotis, G. Filippidis, T.G. Papazoglou

Laser Induced Fluorescence in Atherosclerotic Plaque
with Different Excitation Wavelengths 328
*M. Makropoulou, H. Drakaki, N. Anastassopoulou, Y.S. Raptis,
A.A. Serafetinides, A. Paphti, B. Arapoglou and P. Demakakos*

In vitro Laser-Induced Fluorescence Measurements
of Human and Lamb Heart Tissue 332
*G. Filippidis, G. Zacharakis, G.E. Kochiadakis, S.I. Chrysostomakis,
P.E. Vardas, T.G. Papazoglou*

Eye Model Using a CCD Camera for Observing
the Images Constructed by IOLs 336
K. Ohnuma, Y. Shiokawa, N. Hirayama and Q. Hua

Optical Wavefront Sensing Using a 2-Dimensional Diffraction Grating:
Application to Biological Microscopy 340
H. Ohba and S. Komatsu

Measurement of Chlorophyll Distribution
in the Leaf by Laser Induced Fluorescence 344
K. Takahash, Y. Emori

High-Resolution Color Holography for Archaeological
and Medical Applications 349
F. Dreesen, H. Delere and G. von Bally

Protoporphyrin IX Kinetics in Rat Peritoneal Organs and Tumor After
Systemic 5-Aminolevulinic Acid Administration 353
M.C.G. Aalders, H.J.C.M. Sterenborg, F.A. Stewart, N. van der Vange

Optics

In Search of the Earliest Known Lenses (Dating Back 4500 Years)

Jay M. Enoch
School of Optometry , University of California at Berkeley
Berkeley, California 94720-2020. E-mail: jmenoch@socrates.berkeley.edu

Abstract. Early lens history is controversial. First lenses arose among the artisan/artist communities rather than among the scientific community. These lenses were not used for visual corrections. The author has sought to locate lenses still in their original context, that is, with the lens and the object to be viewed through the lens still intact in their original relationship ("frozen in space and time"). Included among these items are remarkably fine lenses from the IVth and Vth Dynasties of Egypt (ca. 4500 years ago), and a lens from the Little Palace of Knossos (Minoan, ca. 3500 years ago) in Greece. Among other early examples, surprisingly, lenses may have existed in Peru during the first centuries after the birth of Christ. Clearly skills were often lost and rediscovered.

1. Introduction

Defining the earliest lenses remains an enigma[1-11]. Visiting a number of distinguished art or archeological museums reveals numerous lens-like objects in cabinets, or contained in pieces of sculpture or other works of art. However, were these lenses, *per se*, or were they decorative pieces, jewelry, architectural embellishments, symbols of purity, burning glasses, a form of a timepiece, etc.? Did they take origin from the jewelers or seal-makers art, or were these glassblowers products, either in lens form or perhaps as empty vessels filled with fluid (e.g., Seneca)[12]? We must distinguish between the various uses of lenses and mirrors as optical elements. These need not have been used for visual corrections[13-15] or magnifiers. To further confound issues, in these discussions one often encounters ancient beliefs about vision, philosophy, and superstitions.

Early lenses derived from artisans/artists, not the scientific community. Hence, we do not encounter the orderly traditions associated with the development of science. In the pertinent literature, there are debates which consider questions such as: Were lenses present and used, versus, were demanding visual tasks accomplished by young and myopic individuals?[1-5]

A number of items from ancient times can be defined which would have benefitted from lens use, e.g., seals of various types[1-6,16-18], fine cameos[19], coins[1,2,4-6,8,10,11], fine pieces of jewelry (various)[1,2,4-6,8,9,11,18-20], miniature paintings[1,4,5,21], tiny texts such as micro-cuneiform writings[2,4-6,8], the Dead Sea scrolls and accompanying documents (e.g., phylacteries)[5,6], secret codes inscribed on coins (e.g., early Christians),[a] etc.

Few individuals involved in this literature have had training in optics or vision science, and there are marked deficiencies when considering lenses used in conjunction with ambient lighting, or effects of contrast, or utilization of shadows (as in carving or reading/in`terpreting intaglio works), or the use of a fluid to smooth an imperfect surface (see refs. [4-6,9]). Lens-like items have rarely been found in indoor work sites. Drops of water or oil were also used as lenses[4-6,21], and pinhole and slit devices have been employed by different peoples[4,5,8,11] etc. How then does one proceed? One way to address the issue of "did lenses exist and when" is to find lenses in their original context with the object employed by the artisan in place[3-6]. The artisan had to want the observer to view this item through the lens. A number of such items have been located at several sites. At least, the artisan must have appreciated magnification effects, and probably aberrations imparted by the early lenses.

Some artifacts prove not to be lenses, although, at first glance, they seem to be lenses, e.g., (1) During the Renaissance, crystalline lens-like objects were often contained in crosses - these crystals could have been fine lenses, but were symbols of purity (this point was emphasized by Turner)[4-6,22]; (2) holes drilled in crystals, transparent jewel beads, or seals which were used for stringing - they clearly provide obvious magnification of the string, fiber or wire employed. However, the same holes were drilled in non-transparent items of a similar size and use (but, perhaps the obvious optical effects gave people ideas!)[16-18,20,23,24]; (3) I have examined a rock crystal lens-like element in a fine Egyptian amulet at the British Museum; it had a fine hair-line crack which allowed dirt to penetrate over millennia, and this collected to form a figure-like "object" behind the convex-plano transparent jewel[25].

Lenses did exist early and at different locations. These fine items were all contained in distinguished works of art; they certainly were not first pieces created by these artisans/ apprentices. Many lenses considered here were convex/flat (flat="plano") in shape. Exceptions were: (1) a number of the Egyptian lenses had a small concave curve ground in the central "pupillary aperture" on the plano rear surface of the convex/plano lens; (2) Peruvian items mentioned (which I have examined only partially), and (3) a convex/concave (meniscus-shaped) Roman item located in Bologna. No doubt lenses were discovered and lost several times in history. An excellent example of such loss is the extraordinary group of lenses having origin in the Egyptian IV/V Dynasties[26-33].

2. Attributes of a Lens

Earlier, I presented definitions of a lens[3,5,6,10]. Each lens surface acts as a refracting device; i.e., it changes the vergence of an incident wavefront of radiant energy in the visual spectrum by a definable amount. The lens must be reasonably transparent and homogeneous and should have generally regular surfaces, at least one of which must be curved. Surface irregularities must be modest. There should be a principal axis for the two lens surfaces, and the lens must be able to form a reasonable focus. This implies the presence of primary and secondary focal points. I am careful not to consider sphericity of the surface, nor correction of aberrations. These are not reasonable expectations in early lenses. The lens must be able to form a real or virtual image of the object, and there must be a recognizable correspondence of points and lines in image space to points and lines in object space.

I do not consider use of single lens units which act as Galilean telescopes (i.e., each has converging and diverging surfaces resulting in magnification but not a change in resultant wavefront vergence for a distant object), e.g., the product of a glassblower (the walls of a near sphere). These were not used as lenses, *per se*, in early times. However, such near-spheres could be filled with fluid, as described by Seneca[12], creating a converging lens for magnification or alleviation of presbyopia.

3. Examples of Lenses Located

3.1 Egyptian, IV/V Dynasties, "Schematic Eyes" in Statues

Professor Jean Royer and Dr. Robert Heitz[26,27] called my attention to a remarkable group of statues having origin during the IV/ V Egyptian Dynasties (2600-2350 B.C., ca. 4500 years ago, see Table I)[26-33]. The eyes in these statues were exceptional![28,29] Not only were fine lenses used as corneal and anterior chamber elements in eyes of these statues, but these constructs *may have been* components of entire schematic eyes. High quality, very clear rock crystal corneal lenses were used. They had a convex front surface, and a plano rear surface with (in several) a central steep concave area in the pupillary zone[26-29]. They had a painted iris located on the flat rear lens surface, and an aperture for the eye pupil[29].

Table 1. Egyptian Statues Containing Eyes with Lenses IV/V Dynasties (2600-2350 B.C.)

Location	Identification of Statue	Notes
a. Louvre Museum	"*Le scribe accroupi*" The seated scribe (Figs. 1, 2)	2600-2350 B.C.[26-33]
b. Louvre Museum	"*Le haut fonctionnaire Kai*" The major functionary, Kai	2600-2350 B.C.[26-28]
c. Cairo Museum	"The Headman of the Village, The Sheikh El-Beled" (Fig. 3)	Wood, covered with plaster, painted, eyes inset. ca. 2450 B.C. [27-30,33] [b]
d. Cairo Museum	Prince Ra.Hotep, and his wife Nofert"	High Priest of Heliopolis and General. Nofert was a royal family member. 2723-2563 B.C. [27, 28, 30, 31] Limestone, painted
e. Cairo Museum	"A seated scribe"	Limestone, painted [27, 29, 30, 31]
f. Cairo Museum	"A statue of a man with arm extended"	Wood, probably covered with plaster [28,30,31]

Were the concave pupillary aperture zones open or were they painted? A number of these lenses reveal the presence of resins within at least part of the pupillary aperture-was this material (if glue) extruded from the surrounding surface areas or was it painted on?[28,29] If the aperture was open, the pseudo-cornea plus concave rear surface might be a primitive form of meniscus lens(?). There was a "vitreous chamber" behind the iris plane composed of curved sheets of copper (which, if the aperture was open, might have yielded a reddish pupillary reflex when these sculptures were first constructed [prior to corrosion]). In a number eye-units, the ensheathed ocular unit had a sclera-conjunctiva-like structure (red-veined marble, over-painted with added "fine red blood vessels") and separate eyelids.

When one views these schematic eyes(?)/lenses, it appears as if the eyes follow the observer[26-29]. The sense of movement may be imparted by the prismatic action of the converging+diverging lens as one rotates about the lens/iris unit (the hemispheric pupillary aperture seems to translate). Shadows from the partially overlying edge of the lid over the conjunctival element (under certain lighting conditions) may have added to the effect.

Fig 1. "Le scribe accroupi" (the seated scribe); Louvre, Paris (dated 2600-2350 B.C.). Reproduced courtesy of Time-Life Books, Inc., Alexandria, Virginia.

Construction was achieved by groups of specialized artisans working together as a team[30]. "Eye units" were separately constructed and these were inserted *en bloc* into the statues. Examples of these separate constructs exist today in both The Louvre[26,27,29] and at The Egyptian Museum in Cairo[27,28,31]. In these eye constructs, at least the artisan/sculptors had to appreciate the optical properties of these remarkable assemblies[28,30]. After an approximate 200 year "construction window" for these *objets d'art*, the understanding of ocular anatomy revealed by these artisan/sculptors was evidently lost[27], and except for brief and sporadic use[28], the knowledge represented here did not surface again for more than two thousand years. However, as we will see, lenses, *per se*, surfaced again in ca. 1500 B.C. in Greece. These Egyptian lenses are exceptional in quality and application, and are found in outstanding works of art[6,26-29]. They could not have been first lenses.

Two of these Egyptian statues are found in the Louvre. These are, "*Le scribe accroupi*" (Figs. 1, 2 a,b), and "*Le haut fonctionnaire Kai.*" Both were found at Sakkara/Saqqara, and were made of painted limestone ("*calcaire*")[26,27,29]. Other outstanding examples (e.g., Fig. 3) are to be found at the Cairo Museum (Table 1)[27,28,31].

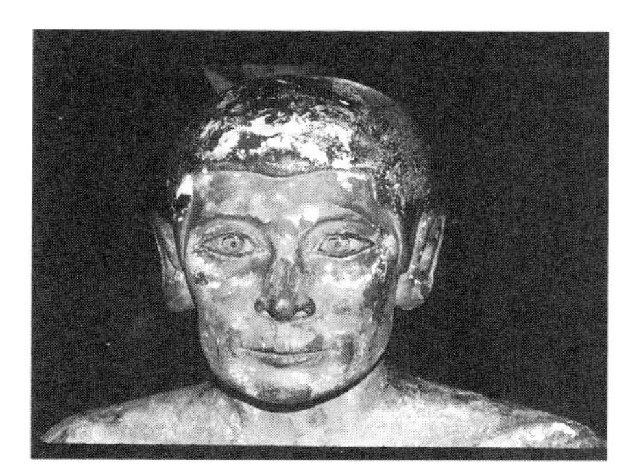

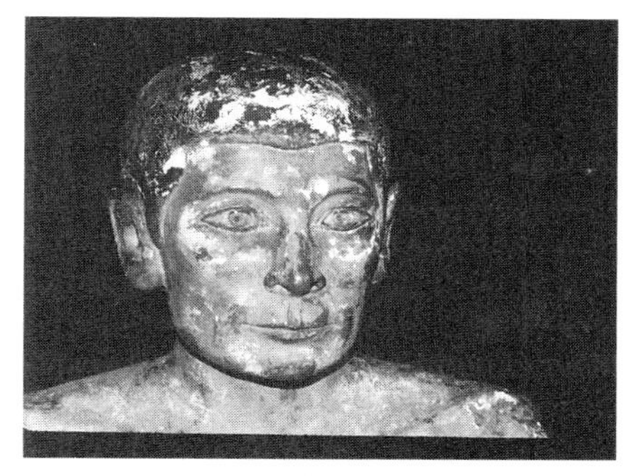

Fig 2. "Le scribe accroupi:" (a) front view (photographed by the author); (b) photographed, after the author moved to the side. Note the apparent following of the observer by the eyes of the statue (observe eye pupils, the eyelids and their shadows).

3.2 The Bull's Head Rhyton, the Archaeological Museum, Heraklion, Crete

This item was found at the Little Palace of Knossos (Minoan, New Palace Period), and had origin between 1500 and 1550 B.C.[16,17,24] I have described this fine Greek piece previously[3-6], and considered in some detail the significance/meaning(s) of the human face in silhouette which was located at the rear of the convex/plano "corneal lens" in the then remaining right eye of the bull's head[3-6,16,24,34]. This painting was rather delicate and beautiful, and quite distinct (off-white in color). The lens is composed of rock crystal. In his book on Minoan and Mycenaean civilization and art, Higgins[17] noted that the rear side of the remaining right lens was painted (he also identified the "cornea" as a lens). Also, Davaras[34] points out that only one of the eyes is preserved, and it is made of rock crystal painted on the underside. I viewed this outstanding sculpture during a visit to the Museum in the 1980s. Photographs were not allowed. Available publications did not magnify well[16]. Figs. 4,5 are more recent photographs.

Visiting Heraklion in October 1997, the face in silhouette was no longer visible (Fig. 4-6). Had I been mistaken? No, I located 6 photographs of the face at the

Museum store[24,34]. Today, behind the corneal lens, one observes a "blood red" pupil set in a black iris (duplicated now in a left eye in the sculpture). I asked the guards present about the face; they stated that they had not been able to observe the face for ca. two years. I could not determine what had happened to the painted face in silhouette.

Fig. 3. "The Sheikh El-Beled:" The Egyptian Museum, Cairo, Egypt (dated ca. 2450 B.C.). See Table I and Note 2. Copyright: Mr. Roger Wood, Deal, Kent, England.

I reaffirmed that the corneal "lens" was indeed a converging (plus) lens by using a method termed "clinically" as "hand neutralization." [3-6] The same method has been used in examinations of all of lenses described here, and these are positive or converging lenses (except the central areas of the Egyptian lenses) .

3.3 Other Lenses

Glass Lenses with Gold Inserts on the Plano Side, Greek and Roman: Converging lenses were inserted in bezels of Greek and Roman rings, and applied to other *objets d'art*. They combine convex-plano glass lenses (the glass in the Greek lenses had a light green cast) and generally incorporate raised gold or other material insets on the plano side viewed through the lens[1,4-6,9,11,20] The Greek items have origin from ca. 400 B.C. One of the Roman medallions at the Archaeological Museum in Bologna[1] is broken in half. It is a meniscus lens with a convex front and a concave rear surface on which is placed the figure to be viewed through the lens[14,15].

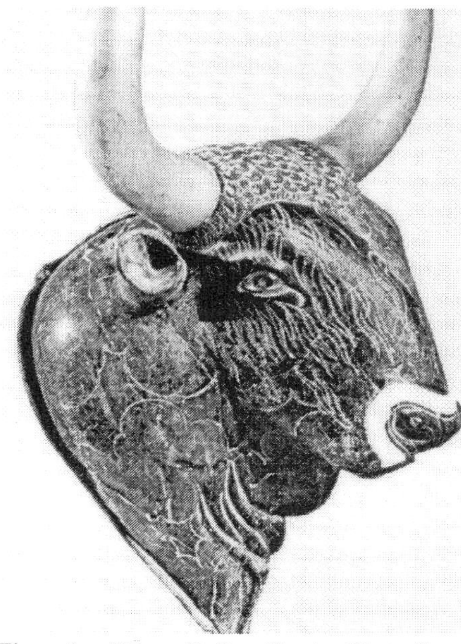

Fig. 4. "The Bulls Head Rhyton", Archaeological Museum, Heraklion, Crete (dated 1550-1500 B.C.). Reproduced courtesy, Ekdotike Athenon S.A., Athens, Greece.

Fig. 5. "The Bulls Head Rhyton": Detail of the finely painted head on the rear of the "corneal" lens in the right eye. Reproduced courtesy Ekdotike Athenon S.A., Athens, Greece.

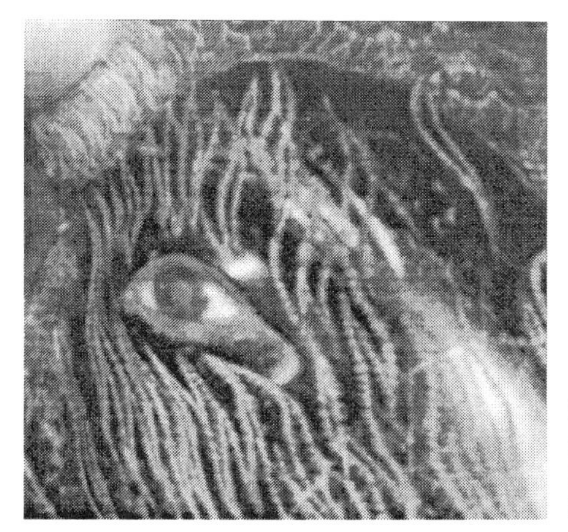

Fig. 6 "The Bull's Head Rhyton": Detail of the right eye of this sculpture photographed by the author in October, 1997. Note change in pupillary appearance!

Statues Containing Eyes with Lenses Used in Eye Inserts: Greek, Hellenistic-Roman, and Roman statues are found with "eyes" containing a convex-plano lens with a painted iris on the rear surface. Those I have seen are dated from ca. 400 B.C.-ca. 14 A.D. and probably later. In many instances these were bronze statues. The sculptor had to appreciate the magnification properties of the lens element when he viewed the iris element painted on the rear surface. Prime examples of such work are found in the British Museum, the Louvre, the National Archaeological Museum in Athens, etc[35].

The Museo Raphael Larco Herrera, Lima, Peru: This museum contains very interesting items dating to about the time of Christ (0-800 A.D.)[23]. I intend to visit this site soon. At an exhibit in San Francisco (1996/7), there was displayed a remarkable necklace wholly composed of pierced round rock crystal beads with an aubergine-shaped rock crystal pendant of enormous size and of weight. The pendant was remarkably clear (there were some imperfections), and one could see every stitch of a very fine velvet fabric viewed through the pendant. Also, the substantial magnification of the "string" joining the transparent rock crystal beads was clearly evident. The Museum Director, Andrιs Alvarez-Calderσn Larco, called my attention to two additional sizeable items with a central rock-crystal circular zone surrounded by a mother-of-pearl collar. These items are located at the Museum in Lima. These objects were NOT found in context, but represent the possible development of the necessary arts/skills needed for lens development in the New World prior to European intervention.

The Flinders Petrie Lenses at the British Museum: Two lenses were found by Flinders Petrie, in Tanis, Egypt, in the home of an artist. These are dated ca. 147 A.D.[1,4-6]. These items, convex/plano, are located at the British Museum; they are about 2.5 inches in diameter with a focal length of ca. 3.5 inches. I examined, in detail, one of these large, convex/plano lenses (badly damaged). This artifact had sufficient magnification to have been used as a lens. It was ground and polished. On the curved surface there is a flat spot, and there is considerable astigmatism, but it was possible to magnify an object through this item. This lens was NOT found in context, but the artist would have benefitted from use of this lens.

The Talisman of Charlemagne, 9th Century A.D.: This pendant is located at the Palace de Tau, Reims, France[4,36]. This talisman was composed of two ground lenses (convex/plano) claimed to be sapphires with slivers of the true-cross displayed in cruciform shape between them. The latter are tied at their junction. F. Ward[36] notes that these lenses were probably not true sapphires, but rather colored glass; he also argues that one of these two lenses might be white glass rather than sapphire blue.

The Layard lens at the British Museum (900-700 B.C.):[1-5,10] This convex-plano lens was discovery at Nimrud in 1854. If one looks through discrete areas *outside* of a central and complex imperfection zone, it is possible to obtain modest magnification. This lens-like object is particularly important because it raised the issue, were lenses present and used?

The Berkeley Lens: A remarkably clear, well polished, but highly astigmatic Turkish lens. It was purchased with the ancient Berkeley Turkish seal collection, and is another probable example of an early lens, although its provenance is not well established[4-6].

As noted, the first reported use of a glass sphere-like container filled with water was reported by the Roman of Spanish descent, Seneca[12]. He used this water-filled glass globe as a presbyopic reading aid (he died about 65 A.D.).

Reports of the Emperor Nero's emerald are too speculative to comment upon[8]. Narinder Singh Kapany[21] notes that ancient Persian and Indian artists created lenses by using a drop of oil on a glass or crystal plate. These were used to paint miniature objects. The droplet served as a magnifier.

R. Temple of London has been studying Medieval lenses of Scandinavian origin[36]. They date somewhat after the Talisman of Charlemagne[35].

This survey is certainly not intended to be exhaustive[1,2,4-6,8,11].

4. Summary

Did lenses exist in the ancient world? Yes. Were they in general use? I do not know. Was the art of lens construction passed on outside of limited family or artisan groups or guilds? I do not know. If so, sharing of knowledge was limited. Were the early lenses described here used for visual corrections? No. Were other early lenses used for visual corrections? I do not know. Might early lenses have been used as magnifiers? Possibly, certainly magnification was appreciated by those constructing these fine pieces of art, and magnification would have benefitted a number of different artisan groups, etc. Were myopic and/or young individuals used for certain fine visual tasks at a near distance? Probably. Were concave surfaces ground? Probably, the capability did exist[4-6,13-15]. These seem not to have been used for lenses in early applications - although the purpose of the concave depressions ground in front of the pupillary zones of the Old Kingdom Egyptian lenses (IVth and Vth Dynasties) has not yet been determined. I hope to pursue this issue.

I infer that the first lenses were made by jewelers/grinders of gems/artisan groups working with sculptors, or seal makers. They most probably were working with rock crystal or rose quartz (these quartz crystals have a hardness of Grade 7 on the Mohs scale - a remarkable finding, in and of itself).

5. Acknowledgments

The author wishes to express his appreciation to Dr. R.B. Parkinson, Assistant Keeper, Department of Egyptian Antiquities, British Museum, London. And very special thanks are offered to both Dr. Robert Heitz, Haguenau, Alsace, France, and to Professor Jean Royer, U. Besanηon, Chβtillon-Le-Duc, France; as well as to other individuals who kindly aided me.

References

1 Sines G., Sakellarakis Y. (1987) Lenses in antiquity. Am J Archaeology 91:191- 196
2 Gorelick L., Gwinnett A.J. (1981) Close work without magnifying lenses? Expedition 23 (Winter):27-34
3 Enoch J.M. (1987) It is proposed that the cornea of the eye of the Bull's Head Rhyton from the Little Palace of Knossos (artifact dated 1550-1500 BC, Heraklion, Crete, Greece) is a true lens. In Fiorentini A., Guyton D.L., Siegel I.M. (Eds.) Advances in Diagnostic Visual Optics. Springer-Verlag, Heidelberg, 15-18
4 Enoch J.M. (1996) Early lens use: Lenses found in context with their original objects. Optom Vis Sci 73:707-715
5 Enoch J.M. (1998) The enigma of early lens use: What is a lens? How do we know an apparent lens was used as a lens? Technology and Culture 39(2,April): Cover +273-291
6 Enoch J.M. (1998) Ancient lenses in art and sculpture and the objects viewed through them. Paper EI-3299-46. In Rogowitz B., Pappas T. (Eds.), Human Vision and Electronic Imaging III, San Jose, CA, Jan.26-Jan.29, 1998. Proceedings SPIE 3299:424-430
7 Lindberg D.C. (1976) Theories of Vision from Al-Kindi to Kepler, Chicago History of Science and Medicine Series. U. Chicago Press, Chicago
8 Krug A. (1987) Nero's Augenglas Realia: Zu Einer Anekdote. Archıologie et Mıdecine, Actes du Colloque 23-25 Octobre, 1986, Centre de Recherches Archıologiques du CNRS, Musıe Archıologique d'Antibes, 7th Rencontres Internationales d'Archıologie et d'Histoire, Antibes, France, October 1986, Editions A.P.D.C.A., Juan les Pins, France, 459-475
9 Sines G. (1992) Precision of engraving of Etruscan and archaic Greek gems. Archeomaterials 6 (1,Winter):53-68
10 Lewis B. (1997) Did ancient celators use magnifying lenses? The Celator 11(11):40-41
11 Plantzos D. (1997) Crystals and lenses in the Graeco-Roman world. Am J Archaeology 101:451-464
12 Seneca, "Natural Questions", QNat 1.6.5-7
13 Rosen E. (1956) The invention of eyeglasses. J History of Medicine Allied Sciences 11:13-53 & 183-218
14 Ilardi V. (1993) Renaissance Florence: The optical capital of the world. J European Economic History 22(3,Winter): 507-541
15 Ilardi V. (1976) Eyeglasses and concave lenses in fifteenth-century Florence and Milan: New documents. Renaissance Quarterly 29(3):341-359
16 Sakellarakis J.A. (1995) Herakleion Museum, Ekdotike Athenon, S.A., Athens, Greece. Bull's Head Rhyton, 34,35 & cover; beads /items strung, 18,74; seals, many 17 Higgins R. (1981) Minoan and Mycenaen art. Revised Edition, Thames and Hudson, London, England. Bull's head rhyton, 161,162 and cover; seals, several
18 Demakopoulou K. (Ed.) (1996) The Aidonia Treasure: Seals and Jewellery of the Aegean Late Bronze Age. National Archaeological Museum, Athens, Exhibit May30-Sept.1, 1996, Ministry of Culture, Athens, Greece. Seals (various); transparent beads/items strung, 64,85,86,113,114
19 Nicosia F., Tondo L., Virtuoso D., et al. (1996) I Cammei: Die Medici e Dei Lorena. Museo Archeologico di Firenze, Florence, Italy
20 Williams D., Ogden J. (1994) Greek Gold: Jewellery of the Classical World. Trustees of the British Museum, British Museum Press, London, England 30-31, 220-221
21 Kapany N.S. (1997), Palo Alto, CA., Personal communication re: lens like materials used to prepare Persian/Indian miniatures

22 Turner G.L'E. (1985) Animadversions on the origins of the microscope. In North J.D., Roche, J.J. (Eds.) The Light of Nature, chap. 12. Essays in the History and Philosophy of Science Presented to A. C. Crombie, Dordrecht, The Netherlands 193-207; + personal communication on lens-like objects which are not lenses, 1995
23 Alvarez-Calderon Larco A., Executive Director, Museo Arqueologico Rafael Larco Herrera, Lima, Peru (1997) + Personal communication 1997 re: Unique lens-related objects in this museum; rock-crystal pendant; transparent beads/items strung
24 Papapostolou J.A. (1981) Crete, Clio Editions, Athens, Greece. Bull's Head Rhyton (the right eye shows a human face in silhouette), Fig. 110, follows 134; Transparent beads/items strung, stringing magnified, Fig. 91 on 126, Fig.139 on 161
25 Andrews C. (1994) Amulets of Ancient Egypt. University of Texas Press and British Museum Press, printed by The Bath Press, Avon, England. Fig. 43c (follows 40, but page not numbered)
26 Bouquillon A., Quırı G. (1997) Le regard du scribe. Pour La Science #232 (February):27
27 Royer J. (1996-1997) Les ocularistes de la statuaire ıgyptienne (presented at Troyes, April 28, 1997). Bulletin Socitıı Francophone d'Histoire de l'Ophtalmologie 4:49-52
28 Lucas A., Harris J.R. (1962) Ancient Egyptian Materials and Industries, 4th edn. Edward Arnold, London, England. There are several meaningful and relevant chapters (inlaid eyes, resins, glazes, glass, precious and semi-precious stones, etc.).
29 Ziegler C. (1997) Les Statues Igyptiennes de L'Ancien Empire. Catalogue, Musıe Du Louvre, Dıpartment des Antiquitıs Igyptiennes, Iditions de la Rıunion des Musıes Nationaux, Paris, France. References and details relative to pertinent materials in the Louvre (Egyptian IV & V Dynasties +)
30 Strouhal E. (1992) Life of the Ancient Egyptians. University of Oklahoma Press, Norman, OK
31 Lambelet E. (1981) Orbis Terrae Aegiptiae: Museum Aegiptium, 3rd edn. Lehnert, Landrock and Co., Cairo, Egypt, pp. 33, 39, 1981.
32 Richman R. (Ed.) (1987) The Age of the God-Kings: Time Frame 3000-1500 B.C. Time-Life Books, Alexandria, Virginia 65,114
33 Baines J. Malek J. (1996) Atlas of Ancient Egypt. Andromeda Oxford, Ltd., Abingdon, England 8-9, 36, 147, 149
34 Davaras C. (No date listed, not paginated) The Palace of Knossos. Editions Hannibal, Athens 309, Greece. The Bull's Head Rhyton: see 27 (counting pages after the frontspiece and the page listing the publisher). Bull's Head Rhyton: A clear view of the right eye with the painted face visible on the rear side of the "corneal" lens is seen in a "fold-in" portion of the cover and elsewhere
35 Caleca A., Gioseffi D., Mellini G.L., Collobi, L.R. (1967) British Museum London. Great Museums of the World Series. Newsweek, Inc., A.Mondadori Editore, Milan, Italy, New York; Simon and Schuster, Inc., New York 115
36 Ward F. (1991) Rubies and sapphires. National Geographic Magazine 180 (4,October):100-124 (see 119-121 and figure on 120)
36 Temple, R. (1997) London, England, personal communication re: early Finnish lenses

[a] These references are contained in the biblical archaeology literature. Their author was Prof. J. Vardaman, Mississippi State University, Mississippi 39762.

[b] Apparently workmen observed this statue (Fig. 3), which resembled the headman of their village; and named it after him, "Sheikh El-Beled." It is a statue of Ka-Aper, e.g., see Humpheries A. (4/1998) Cairo. Lonely Planet, Hawthorn, Victoria, Australia 87

Study on Reproduction of Sound from Old Wax Phonograph Cylinders Using the Laser

Toshimitsu Asakura[1], Jun Uozumi[2], Toshiaki Iwai[2], and Takashi Nakamura[3]

[1] Faculty of Engineering, Hokkai-Gakuen University
Sapporo, Hokkaido 064-0926, Japan
[2] Research Institute for Electronic Science, Hokkaido University
Sapporo, Hokkaido 060-0812, Japan
[3] Kushiro National College of Technology
Nishi 2-32-1,Otanoshike, Kushiro, Hokkaido 084-0916, Japan

Abstract. Various methods using the laser for reproduction of sound from old wax phonograph cylinders have been developed by the Asakura's group during the past 15 years. Not only the information recorded in the wax cylinders but also the cylinders themselves are a cultural inheritance. Since the developed optical methods are noncontacting and nondestructive, they are powerful tools for repaired or partly broken wax cylinders, and may well be employed in the future for reproducing valuable sounds from old wax cylinders. The principles of the method using the laser are introduced together with their special characteristics and the physical properties of reproduced sounds.

1 Introduction

As well known, Thomas Edison in the United States, a very famous scientist and engineer, invented a recording machine of human voice which was called "Phonograph." After the 10 years' improvement, the phonograph using wax cylinders became very popular. These wax cylinder phonographs were distributed all over the world during about 40 years from 1887 to 1932. In the United States, the wax cylinder phonographs were used mainly for the purpose of amusement. In Europe, on the other hand, these phonographs were used for the various purposes of recording not only the voices of famous people but also the famous music and songs. In addition, these phonographs were used for the academic purpose of recording the various languages of, especially, minority races.

Using the phonograph over the years from 1902 to 1905, B. Piłsudski (1866–1918), a polish anthropologist, recorded the talks and songs of the Ainu people in Sakhalin and Hokkaido on wax cylinders to study their culture. In 1977, Piłsudski's 65 wax cylinders were discovered in Poland and brought to the Research Institute for Electronic Science (RIES), Hokkaido University in 1983 for the purpose of reproduction and investigation of the sounds recorded on them[1].

On the other hand, using the phonograph over the years from 1920 to 1935, Takashi Kitazato (1870–1960), a language professor at Osaka University, recorded the talks and songs of many people in Japan, Taiwan, Philippine, Malaya, Singapore and Indonesia to investigate an origin of the Japanese language. In 1985, Kitazato's 240 wax cylinders were discovered in Kyoto and also brought to the RIES for reproduction of their sounds.

Since the discovery of Piłsudski's and Kitazato's cylinders, many wax cylinders have been brought to the RIES mainly from Europe for the purpose of reproduction of the sounds. The most interesting wax cylinder existing in Europe is of Johannes Brahms (1822-97) whose piano music was recorded.

The stylus of the Edison type phonograph gives a heavy pressure of approximately 20 g to the grooves of the wax cylinder in the reproduction process and, therefore, there is a great risk of damaging the wax cylinder. This is a big problem because not only the sounds recorded on the wax cylinder but also the cylinder itself are part of our cultural inheritance.

Therefore, a reproduction system using a very light pressure stylus was developed in our laboratory. Using the stylus method, we successfully reproduced 60 % sounds from wax cylinders having good conditions. However, there were 40 % more cylinders that had not yet been reproduced. The stylus method cannot be used because the cylinders are cracked or in pieces. These broken wax cylinders were repaired. Figure 1 shows an example for the conditions of the wax cylinder before and after repair. Since the valuable sounds of the talks and songs lost so long ago were recorded on these cylinders, reproduction was most important. To reproduce the sounds from the repaired cylinders, we developed a laser-beam method which is noncontacting and nondestructive[1]. In this paper, we would like to introduce the laser-beam method which is very useful for reproducing sounds from the repaired wax cylinders.

2 Principle of the Laser-beam Reflection Method

Figure 2 shows the principle of a laser-beam method. The laser beam is incident onto the grooves cut on the surface of the wax cylinder and reflected with the angle obeying the reflection law. The reflected beam reaches the detecting plane, placed perpendicularly to the optical axis. The intersection point

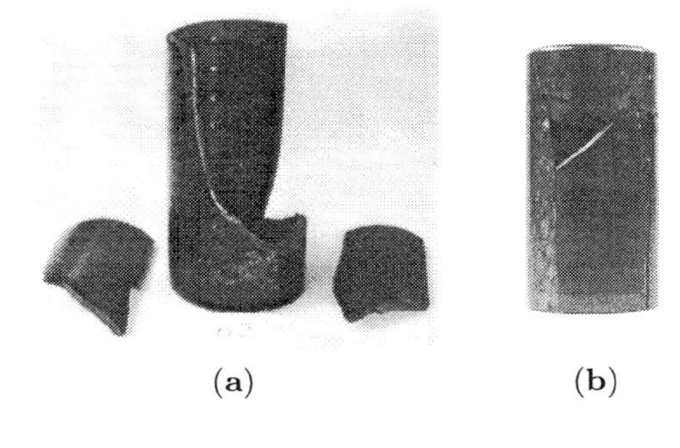

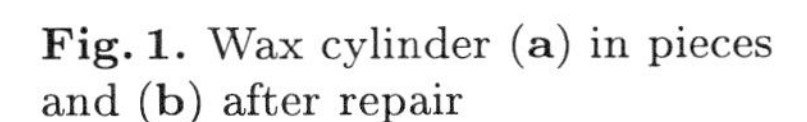

Fig. 1. Wax cylinder (**a**) in pieces and (**b**) after repair

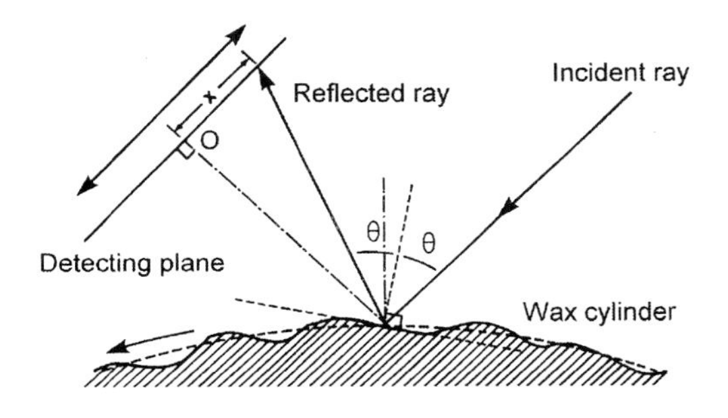

Fig. 2. Schematic 1-D diagram of the laser-beam reflection method

of the reflected beam is separated from the origin by a distance proportional to the reflection angle. When the wax cylinder is rotated, the intersection point moves temporally on the detecting plane. The temporal variation of the intersection position is detected by a position sensitive device (PSD) as a sound signal.

3 Properties and Problems of Reproduced Sounds

The laser-beam method is based on geometrical optics. However, the actual reflection phenomenon from the grooves does not exactly obey the law of geometrical optics because of the finite diameter of the illuminating laser beam. We have found the several problems in the development of this method:

1. Fidelity of the reproduced sound
2. Noise characteristics
3. Existence of echo
4. Occurrence of tracking error

We solved these problems from the quantitative investigation of the reproduced sound signals.

To study the problems of fidelity and noise characteristics, we investigated the long-time frequency spectra of the sound signal produced by using the stylus method and the laser-beam method. In the stylus method, we used the Edison type phonograph. In the laser-beam method, the illuminating spot diameter is 80 μm. As a result, we found that, in the case of the stylus method, the sound is in the frequency range from 250 Hz to 6 kHz. Especially, the resonant frequencies are at 400 Hz, 2 kHz and 4 kHz. In the laser-beam method, the sound intensity with the low resonant at 400 Hz is strong but the sound intensity with the high frequency is weak. The lack of high frequency components makes the consonant indistinct.

The existence of noise inherent in the laser beam method was also examined. The low frequency noise under 300 Hz is very strong. The low frequency noise gives rise to a great deal of degradation on the articulation. On the other hand, the high frequency noise over 1 kHz masks the reproduced sounds and becomes more obstructive to hearing the reproduced sounds.

To investigate the cause of the noise in the laser beam method, we studied the reflected spot at the detecting plane and found a random granular intensity pattern existing together with the reflected beam spot. This granular pattern may be produced from interference of the laser light reflected from the micro-structure distributed over the surface of the wax cylinder.

We also investigated the fidelity of the reproduced sound by changing the diameter of the illuminating laser beam. The diameter was controlled by adjusting the distance z of the illuminating point from the waist of the beam. Figure 3 shows the variation of long-time frequency spectra of the reproduced sounds with the spot diameters. With an increase of the beam diameter, the high frequency components are greatly decreased. This may be due to the smoothing effect for the time-varying directions of beams reflected from the groove within the illuminating beam spot. The lack of high frequency components makes the consonant indistinct. From this investigation, we conclude that the most suitable spot diameter is 30–130 μm from the viewpoint of the fidelity of the reproduced sound.

Figure 3 also shows that the high frequency noise suddenly decreases with an increase of the illuminating beam spot diameter. On the other hand, the low frequency noise exists independently of a variation of the spot diameter. Therefore, the high frequency noise can be effectively suppressed by using the illuminating beam with the large diameter. On the other hand, the noise signals in the low frequency region have a constant intensity independently of the beam diameter. This low frequency noise below 300 Hz may be sup-

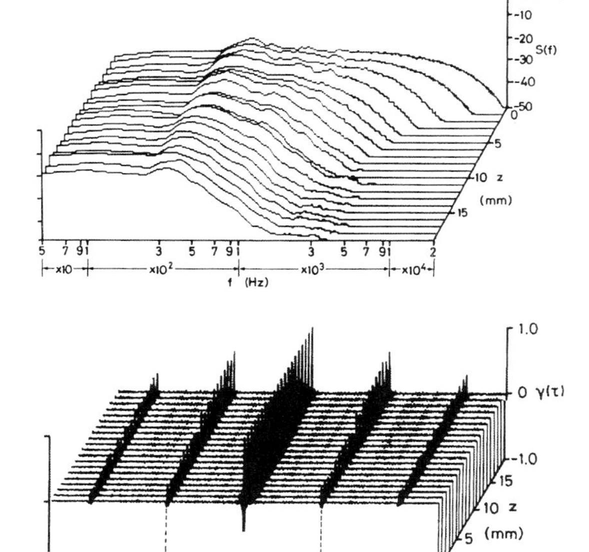

Fig. 3. Long-time frequency spectra obtained from sound signals reproduced by the laser-beam reflection method as a function of the distance z from the beam waist

Fig. 4. Autocorrelation functions of reproduced sound signals as a function of the distance z from the beam waist

pressed by using a high-pass filter since the sound information in this region is not originally recorded. From the viewpoint of noise reduction, the noise is suppressed by using the laser beam with the spot diameter over 80 μm.

Another problem is to reduce an echo. The echo is overlapped on the reproduced sounds with an increase of the illuminating beam diameter. Figure 4 shows the autocorrelation functions of time-varying sound signals as a function of the distance z. The maximum peaks at $\tau = 0$ sec results from the sound signals reproduced from the grooves illuminated by the laser beam. The second peaks at $\tau = 0.4$ sec come from the echo and their magnitudes correspond to the intensity. It is seen that the intensity of the echo increases with an increase of the beam diameter.

The laser beam having a larger spot diameter than the width of grooves illuminates the neighborhood of the grooves on both sides and, therefore, the echo is produced together with the main sounds. We found that the intensity of the echo is under 30 % of the main sound intensity for the spot diameter under 100 μm. Under this condition, the existence of the echo does not disturb the hearing of the reproduced sounds.

4 Tracking Error

There is the possibility that the illuminating beam gets out of the grooves, producing the tracking error, because it does not directly trace the grooves of the wax cylinder like in the case of the stylus method. The tracking error results mainly from the position error of the driving part in the reproduction system. As shown in Fig. 5, if the tracking error happens, the incident beam is reflected to the y-direction. In this case, the intensity of the reproduced sounds decreases suddenly because the intersection point of the reflected beam gets out of the one-dimensional PSD.

To avoid the tracking error, we used the two-dimensional (2D) PSD and the lens driver of the compact disk player. The 2D-PSD allows independent detection of the x- and y-coordinates of the beam spot position. The time-varying values of the x- and y-coordinates of the reflected beam become the sound and the tracking error signal, respectively. The tracking error signal is

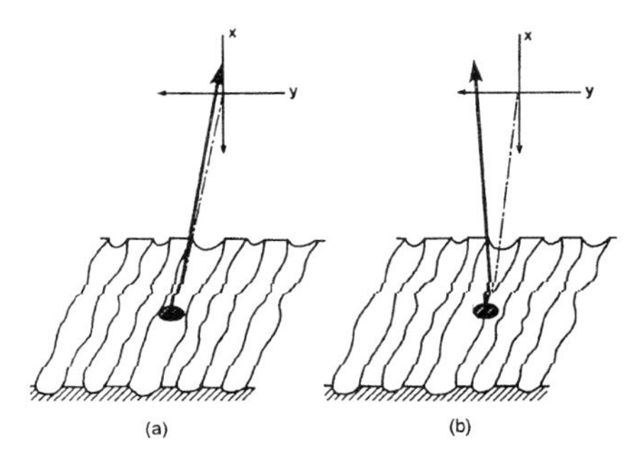

Fig. 5. Optical rays reflected from the grooves (**a**) without and (**b**) with the tracking error

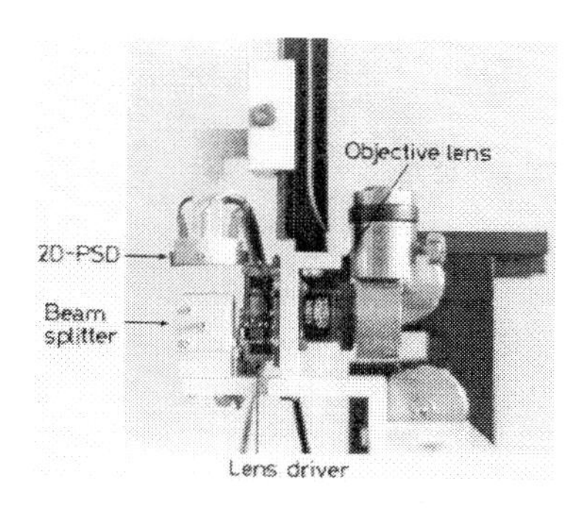

Fig. 6. Optical reproduction system with 2D-PSD and the lens driver to compensate the tracking error

fed to the lens driver to move the lens. By moving the lens, the illuminating spot moves to the center of the groove to keep the normal illumination.

Figure 6 shows the reproduction system with the 2D-PSD and the lens driver. By using this system, we obtained the stable reproduction of the sound with the constant intensity.

5 Further Development of the Method

The laser-beam reflection method introduced above has been developed further to improve its performance and to be applied to various types of old records. To solve the noise problem, an incoherent light source was tried, by which the granular pattern in the beam spot at the detection plane is suppressed. To improve the tracking, modifications of the mechanical part were conducted. A contact method on the basis of light-weight optical fiber was also proposed as an alternate way for improving the tracking[2].

A method using the diffraction property of the laser beam was developed to reproduce sound from old disk records in which the sound is recorded as lateral winding of the sound grooves[3]. More recently, the laser-beam reflection method was applied to negative cylinders which are replicas of wax cylinders[4]. The system for long cylinders are also developed[5].

Finally, it is emphasized that the developed optical methods are basically noncontacting and nondestructive. These methods are powerful tools for various types of old records, and may well be employed in the future for reproducing valuable sounds from them.

References

1. Iwai, T., Asakura, T., Ifukube, T., Kawashima, T. (1986) Appl. Opt. **25**, 597–604
2. Nakamura, T., Ushizaka, T., Uozumi, J., Asakura, T. (1997) Proc. SPIE **3190**, 304–313
3. Uozumi, J., and Asakura, T. (1988) Appl. Opt. **27**, 2671–2676
4. Uozumi, J., Ushizaka, T., Asakura, T. (1998) Proc. OWLS V
5. Nakamura, T., Asakura, T. (1998) Proc. OWLS V

Dynamic Holographic Image Projection: The Key to Optical Interfacing

N. A. Vainos, S. Mailis[1], G. Siganakis, A. Bonarou and V. Tornari,
Foundation for Research and Technology-Hellas (FORTH), P.O. Box 1527, Heraklion 71 110, Greece. E-mail: vainos@iesl.forth.gr
G. Betzos and P. Mitkas
Colorado State University, Dept. of Electrical and Computer Engineering, Fort Collins, CO 80523, USA, E-mail: mitkas@engr.colostate.edu
[1]Present address: Optoelectronics Research Centre, University of Southampton, Southampton, SO17 1BJ, UK

Abstract. Holographic information systems may stimulate further novel applications in the future, provided their inherent abilities to deal with three-dimensional image fields are explored. The concept of holographic interfacing among machine layers and the user is addressed here in the frame of optical storage, processing and delivery. Preliminary results on the amplified projection of three-dimensional, holographically reconstructed images indicate that dynamic holographic projectors can facilitate efficient three-dimensional data transfer. The fruitful exploitation of tandem photorefractive amplification and phase-conjugation stages, potentially operating in a combined pulsed/ continuous-wave mode, may open up new avenues in cineholographic archiving, large scene dynamic holographic projection and multimedia.

1. Introduction

Conventional holography utilising sensitive permanent recorders such as, for example, silver halide materials is now well established. Holographic methods are currently applied successfully in art display and non-destructive testing operations[1]. The cinematic recording and display of three-dimensional (3D) object scenes had been addressed in the past, but technological limitations still exist[2]. The potential large-scene cineholographic projection may lead, however, the way to novel concepts relating to multimedia applications in the future. The huge information capacity implied in this case, even after applying information redundancy techniques, necessitates the use of a new class of holographic systems. Those would comprise high information capacity recorders and systems capable of delivering high-intensity fields for 3D imaging purposes. Aiming to such holographic operations, object scene radiometry as well as object safety issues, particularly associated with the recording of living beings or light sensitive items such as works of art, are trading off materials recording sensitivity, long-term information storage and dynamic processing aspects. Limitations concerning the currently available average laser power also add further impediments.

Owing to recent advances, the above concepts are now naturally directly linked to holographic information storage[3]. This technology is currently representing a major candidate in information systems and a plethora of novel and diverse applications are foreseen[4]. A rapidly accessed holographic memory will provide the "ideal medium" for the recording and reproduction of 3D-image data, however, further developments are required to achieve this goal.

Individual aspects of the forthcoming holographic storage technology, from the storage materials, to information encoding and addressing schemes, to system architecture, have been addressed[5]. Storage materials such as photopolymers[6] and inorganic photorefractives (PR)[7,8] are very attractive but are not yet offering an ideal performance. Their drawbacks mainly relate to the contrasted recording-sensitivity and storage-time issues and considerable research is directed towards viable alternatives, including photorefractive polymers[9] and photosensitive inorganic oxides[10-12]. System architecture also plays an important role. Hologram fixing[8], memory refreshing[13], non-volatile recording[14] and low-intensity reconstruction by signal amplification[15] are methods aiming to alleviate existing disabilities. However, the widely applied hybrid-optoelectronics approach deals exclusively with two-dimensional image formats, which are provided by spatial light modulation input devices and detected by two-dimensional photodetector or CCD arrays. Data is stored in the recorder by means of hologram multiplexing[16-18] and a major effort to perfect the process, increase the information capacity and decrease error rates is currently under way.

The information capacity of holographic memories along with the high bandwidth and associative processing capabilities make them ideal for multimedia database applications[19-21]. In this case, optoelectronics-based data transfer may be the preferred scheme[22]. Nevertheless, this latter method is unable to cope with the considerable phase content of 3D-image fields addressed here. An all-optical interfacing approach is thus required[23], which has not received a proper attention so far.

Faithful reconstruction of a generic 3D image requires handling of a huge amount of data and the current state of technology prohibits the application of hybrid optoelectronics. Conventional optical holography represents, on the other hand, an established 3D-image projection scheme. Problems relating to sensitivity limitations had been identified in the past and the use of photorefractive or other non-linear optical amplifiers for increasing the signal beam intensity for subsequent hologram recording has been proposed[24]. It was only recently proved, however, that PR materials are definitely capable for holographic 3D-image projection operations under reduced spatial coherence. The dynamic nature of these materials enables them to perform important real-time phase conjugation and perspective inversion thus yielding 3D-images with a relatively large field depth[25] and an appreciable parallax[26]. A combined amplification and phase conjugation operation becomes, therefore, a beneficial prerequisite. It further produces simultaneous adaptive all-optical linking and signal phase-modulation operations[27], which corroborate all-optical image-and-voice-transmission. In this context, the concepts of all-optical holographic projection[28,29], may now be revived in the frame of holographic memories. Such schemes incorporated in a holographic memory system provide novel optical interfacing means, promote further the current potential and lead the way forward to a new class of holographic multimedia applications.

2. Dynamic Holographic Projectors and Interfacing

The combination of permanent and dynamic holography becomes significant, facilitating the accurate all-optical information transfer between the "natural world", represented by the object and the observer and the "virtual world" represented by the memory. Coherent amplification enhances the signal being here a diffuse 3D-image field. To preserve the original form, size and perspective, phase-conjugation operations are becoming vital. Such phase conjugation operations also provide accurate hologram reconstruction, as well as efficient coupling amongst storage media and/or the media and the user, by facilitating adaptive transmission operations[27].

To demonstrate the interfacing concepts a series of experiments have been carried out towards the typical integrated arrangement depicted schematically in Figure 1. A HeCd laser source delivering about 40mW at

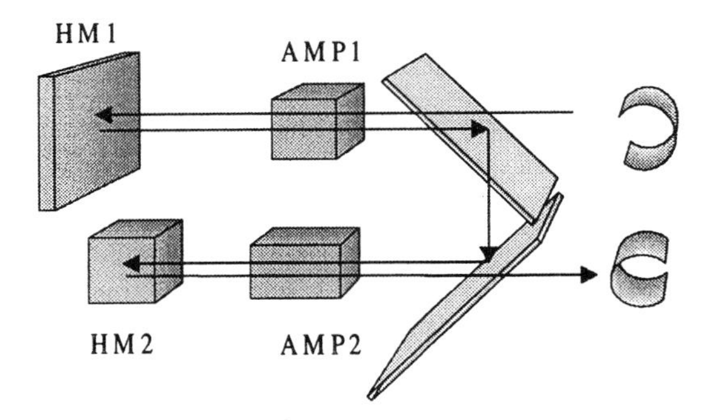

Fig. 1. Dynamic holographic projection system.

442nm has been used in these experiments, in conjunction with a versatile optical arrangement. The full set up and including reference beams are omitted in figure for clarity. Permanent holographic storage is performed by means of silver-halide transmission phase-holography. These static holograms represent "read-only" holographic memory banks (HM1). In addition, poled single-domain $BaTiO_3$ and SBN PR crystals are used to provide image-amplification (AMP1,2), as well as short-term storage and phase conjugation (HM2) through nonlinear wave mixing processes.

The operation of these materials at the particular conditions used here has been investigated before[23]. Noise growth imposes several limitations and requires special attention. Neither noise reduction measures[30] have been taken here, nor all the available dynamic range of the materials has been exploited at the low intensity levels used. The spatial coherence was, in addition, heavily reduced. Nevertheless, the required operations were demonstrated quite reliably.

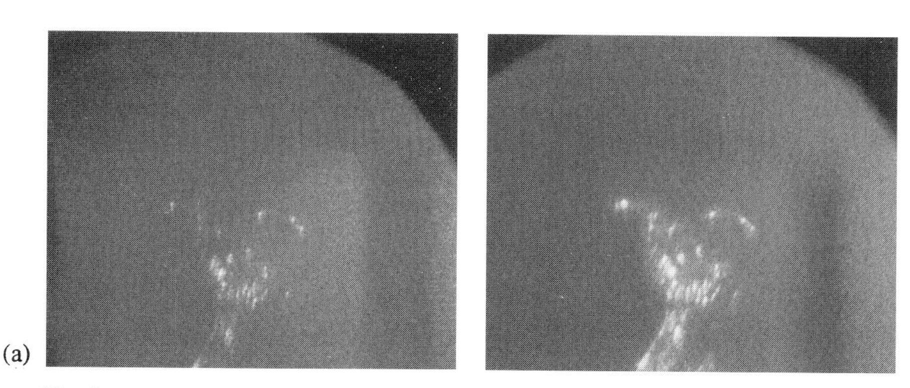

Fig. 2. (a) Original 3D holographic reconstruction and (b) amplified projection.

In the typical arrangement outlined in Figure 1, a very weak image field, light scattered off an object, is directed through the PR amplifier AMP1. Following stabilisation of the PR process, the amplified image field is directed to expose the silver halide emulsion forming a "read-only" holographic record. This arrangement exhibits several advantages and it is of particular relevance to low light level holography. Such cases may either concern holographic recording of light sensitive object-scenes, not excluding living organisms, or may be implied by the large object size and the limited laser power resources. This operation has been successfully demonstrated in the low light level holographic interferometry of sensitive items[31] for structural monitoring of Byzantine icons, or the enhancement of a digital holographic memory operation[23].

A complementary operation, which is also incorporated in Figure 1, concerns the dynamic projection of a holographic image, which is of greater interest here. The three-dimensional record of a miniature statue (*Idefix*: Character of Cosignny and Uderzo in the series Asterix©, Dargaud Editeur, Paris 1966.) has been generated for convenience, in a silver-halide holographic plate. Recording was performed in the object space of an imaging lens, far from the conjugate plane, by using diffuse illumination. To observe holographic fidelity, a low modulation index and low intensities have been used, resulting in a relatively low diffraction efficiency of about 10^{-3}. Both virtual and real reconstructed images have been used however, the real, but orthoscopic image reconstructed with an almost 7:1 reduction ratio to the original object was considered to avoid perspective problems. In Figure 2(a), a photograph of the original hologram reconstruction passing through the crystal is depicted. This reconstructed field was input into a $BaTiO_3$ PR amplifier (AMP1) away from the conjugate planes. The field in this region was diffuse with a preferential direction implying some increase of the degree of spatial coherence. The measured average signal intensity including losses was in the range of $5\mu W/cm^2$. This weak field was input in the amplifier, AMP1. The amplifier was pumped at ~ $5mW/cm^2$. Not a fully optimal geometry has been used although a nominal central grating spacing of about $0.5\mu m$ was utilised. No special noise reduction was considered at this stage, but care has been taken to avoid noise in the field of view of the holographic projector, although sacrificing amplification gain. The average intensity of amplified scattered noise (in the absence of signal) measured in the useful field of the projector was typically in the range of $2\mu W/cm^2$. A typical result of amplification of this three-dimensional scene is depicted in Figure 2(b). On average, a ×15 amplification over the available field has been observed, which, although not large, demonstrates the principle of the operation. The depth and parallax of the original reconstruction are fully preserved, the latter being naturally affected by the imaging arrangements with respect to the object. A compromise has been made here for simplicity and low total laser power.

A further phase conjugation stage is appropriate, as it is included in the lower arrangement of Fig.1 by means of AMP2 and HM2. To improve the signal level the use of higher intensities is not appropriate due to gain saturation and noise growth effects. In this case two or more PR amplifiers (represented by the combination AMP1 and AMP2 in Fig. 1) must be used for the realisation of an efficient operation[32,29]. A typical arrangement of this kind and corresponding result appear in Figure 3, in which two tandem $BaTiO_3$ amplifiers are pumped by an argon ion laser beams at relatively low signal-to-pump beam intensity ratios maintaining operation in the linear gain region. A ×100 increase of intensity is observed due to the first amplifier (region A) and an about ×20 increase due the second (region B) which yet appears to be far from saturation. A very large total gain of more than 2000 is thus achieved with good linearity. The advantages of such operation, as contrasted to the single amplifier case, are several and relate to power management, image contrast preservation and final signal-to-noise considerations.

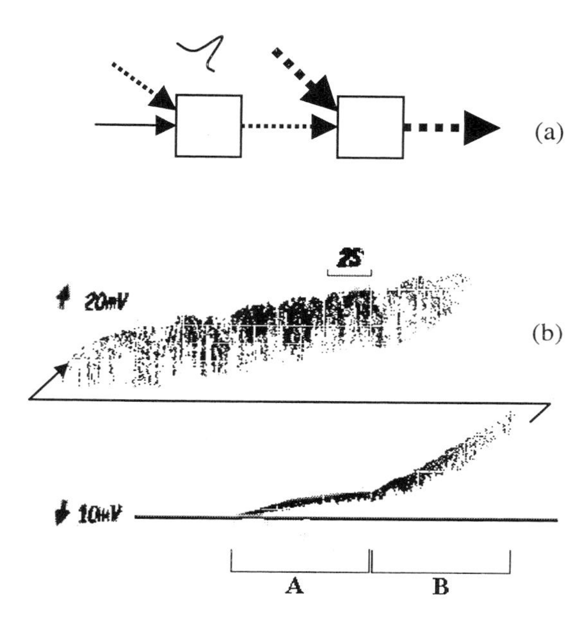

Fig. 3. (a) Tandem amplifiers may operate in a pulsed-cw mode and (b) representative result of tandem amplification by two $BaTiO_3$ amplifiers

To highlight the potential of the above schemes, we consider the case of pulsed holographic recording as indicated in Fig. 3(a). In its more relevant to the present operation form, a weak continuous wave (cw) signal may be amplified in a pulsed fashion[33], to provide appropriately a sequence of frames of three-dimensional images at the desirable frame rate. This operation addresses directly the pulsed and efficient interfacing discussed here. In this case the pumps of the devices shown in Figure 3(a) are pulsed and synchronised. A light-sensitive moving object scene is cw illuminated at low light levels, while the system is delivering high intensity frames of the object at a high repetition rate, for subsequent storage into the memory or for transfer from the reconstructed memory to the observer or a secondary memory bank. The latter may be further assisted by dynamic phase conjugation providing space adaptation as well as perspective and distortion correction.

3. Conclusions

The advantageous combination of static and dynamic holographic media promotes the development of all-optical interfacing systems. The primary issues addressed in this work concern the incorporation of coherent optical amplification and dynamic phase conjugation in the otherwise conventional holographic recording scheme. The preliminary results presented encourage the utilisation of photorefractive amplifiers in the projection of three-dimensional, holographically reconstructed images and further work towards integration is in progress.

Dynamic holographic projectors may provide the means for alleviating existing problems in large scene holographic recording and non-destructive testing. Such an approach may lift radiometric and safety limitations involved in three-dimensional holographic recording of living-organisms and, generally, light-sensitive items including works of art. Furthermore, the proposed schemes, combined with phase conjugation operations, are linked to current advances in holographic memories, additionally facilitating holographic reconstruction from sensitive recorders and addressing the concept of large-scene cineholographic recording and projection. Even though holographic memories are currently directed to hybrid optoelectronics solutions, the all-optical interfacing proposed here offers unique means for transferring the phase content of 3D-image data, which will potentially lead to the future generation of cineholographic multimedia.

4. Acknowledgements

Work supported by the EU funded Ultraviolet Laser Facility at FORTH. Support by the NATO Collaborative Research Grant CRG960767 is also gratefully acknowledged. Original aspects of this work had been earlier explored at Rutherford Appleton Laboratory, UK, and the past collaboration with M.C. Gower is gratefully acknowledged.

References

1. See for example: *Proc. Int. Symposium on 3D Image Technology and Arts*, (Asian Tech. Info. Program; Tokyo, 1992).
2. P. Smiglieski, in *Industrial Laser Interferometry*, Proc.SPIE, **746**, 29 (1987).
3. E. Chuang and D. Psaltis, Appl. Opt., **136**, 8445 (1997).
4. D. Psaltis and F. Mok, Scientific American, **273**, 70 (1995).
5. J. H. Hong, I. Mcmichael, T.Y Chang, W. Christian, E. G. Paek, Opt. Eng. **34**, 2193 (1995).
6. K. Curtis and D. Psaltis, Appl. Opt. **33**, 5396 (1994).
7. S. I Stepanov, Reports on Progress in Physics, **57**, 39 (1994).
8. J.P. Huignard and P. Gunter, Eds, *Photorefractive Materials and Applications I & II* (Springer Verlag; Berlin, 1989-1991).
9. S. M. Silence, D.M. Burland and W.E. Moerner, in *Photorefractive Effects and Materials*, D. Nolte, Ed. (Kluwer Academic Publishers; Boston, 1995).
10. K. Hirao, J. Non-Crystal. Solids, **196**, 16 (1996).
11. S. Mailis, L. Boutsikaris, N. A. Vainos, C. Xirouhaki, G. Vasiliou, N. Garawal, G. Kyriakidis and H. Fritzsche, Appl. Phys. Lett. **69**, 2459 (1996).
12. C. Grivas, D.S. Gill, S. Mailis, L. Boutsikaris and N.A. Vainos, Appl. Phys. **A66**, 201(1998).
13. J.J.P. Drolet, E. Chuang, G. Barbastathis and D. Psaltis, Opt. Lett., **22**, 552 (1997).
14. D. Psaltis, F. Mok, H.Y.S. Li, Opt. Lett., **19**, 210 (1994).
15. H. Ranjbenbach, S. Bann, and J.P. Huignard, Opt. Lett., **17**, 1712(1992).
16. G. Barbastathis and D. Psaltis, Opt. Lett. **21**, 432 (1996).
17. C. Alves, G. Pauliat and G. Roosen, Opt. Lett., **19**, 1894 (1994).
18. C. Denz, T. Dellwig, J. Lembcke and T. Tschudi, Opt. Lett., **21**, 278 (1996).
19. B. J. Goertzen and P. Mitkas, Opt. Eng., **35**, 1847 (1996).
20. P. A. Mitkas, G. Betzos, and L. J. Irakliotis, IEEE Computer, **31**, 45 (1998).
21. G. A. Betzos, K. G. Richling, and P. A. Mitkas, in *Proc. 4th Int. Workshop on Multimedia DBMS*, pp. 190-197. (IEEE Comp. Soc. Press; Dayton, Ohio, August 5-7, 1998).
22. S. Cambell, Y.H. Szang and P.Y. Yeh., Opt. Commun., **123**, 27 (1996).
23. P. Mitkas, G.A. Betzos, S. Mailis and N.A. Vainos, Proc. SPIE, **3388**, (in print) (1998-9).
24. J.-P. Huignard US Patent 4,458,981
25. N.A. Vainos and M.C. Gower, J. Opt. Soc. Am., **B 8**, 2355 (1991).
26. B.P.Ketchel, G. Woods, R.J. Anderson and G. J. Salamo, Appl. Phys. Lett., **71**, 7(1997).
27. S. Mailis and N.A. Vainos, Appl. Opt., **32**, 7285 (1993).
28. N. A. Vainos and M.C. Gower, in Proc. Ninth Quantum Electronics Conference QE-9 (Inst. Of Physics; Oxford UK, 1989).
29. N.A. Vainos and M.C. Gower, in Proc. Topical meeting on Photorefractive Materials, Effects and Devices II (Opt. Soc. Am.-Soc. Franc. Opt.; Aussois, France, 1990).
30. S. Breugnot, H. Rajbenbach, M. Defour and J.-P. Huignard, Opt. Lett., **20**, 447(1995).

31. V. Tornari, S. Mailis, L Boutsikaris and N.A. Vainos, in *Academie-Verlag Series in Optical Metrology*, W. Juptner and W. Osten, Eds, v **3**, p.228 (Acad.-Verlag; Berlin 1997).
32. S. Breugnot, H. Rajbenbach, M. Defour and J.-P. Huignard, Opt. Lett., **20**, 1568(1995).
33. N. A. Vainos, S. Mailis and M. C. Gower, Appl. Phys. Lett., 60, 1529 (1992).

A Novel Electronic-Speckle-Pattern Interferometer (ESPI) for Dynamic Holographic Endoscopy

B. Kemper[1], A. Merker[1], S. Lai[2], and G. von Bally[1]

[1] Laboratory of Biophysics, University of Münster, Robert–Koch–Straße 45, D–48129 Münster, Germany
[2] presently on leave from Zhengzhou University of Technology, P. R. China

Abstract. A novel Electronic-Speckle-Pattern Interferometer based on an endoscope camera system is presented which can be applied to hand-held examinations of technical objects and for in-vivo minimal invasive medical diagnostics. Optical fibers and endoscopic optics permit a very compact and flexible setup of this ESPI system. An amplitude modulated cw-laser is used for illumination. A CCD-camera in combination with a fast frame-grabber system allows dynamic "on line"-image subtractions up to a frequency of 25 Hz with high fringe contrast. First results of measurements with this system from investigations of technical objects and biological objects (in-vivo) are obtained. With this method, the operator's tactile contact with the treated tissue which is lost in endoscopic minimal invasive therapy can be replaced by visual information (endoscopic taction).

1 Introduction

The combination of holographic metrology with endoscopic imaging allows the development of a special class of instruments for nondestructive quantitative diagnostics within body cavities. This method includes the analysis of structure, form, deformation, and vibration of the object. This aspect opens up new perspectives in minimal invasive diagnostics, especially in medical applications. The development of digital imaging, digital holographic interferometry [1], Electronic-Speckle-Pattern Interferometry (ESPI), and microoptics makes new techniques of fringe processing by endoscopy possible. Contrary to earlier attempts in holographic endoscopy where single interferograms with Q-switched lasers were recorded [2], modern diagnostic techniques require process analysis with a video frame rate of 25 Hz. Here, an endoscopic ESPI camera system with an external (proximal) holographic camera using standard endoscopic optics is presented, which can be applied hand-held for in-vivo minimal invasive diagnostics. The system replaces the operator's tactile sense by visual information. Therefore "on line" information about in-vivo motions of biological objects and deformations of technical objects are gained by producing speckle interferograms with the double exposure image subtraction method in video real time using a cw-laser.

2 The Endoscopic ESPI Camera System

Fig. 1 shows the schematic setup of the endoscopic ESPI system. The light source is an Argon-ion-laser using the green 514.5 nm line. The laser beam is split and coupled into mono-mode optical fibers for object illumination and reference wave guidance. Both beams are expanded by a special optical micro lens system. The reference beam directly illuminates the CCD-chip of a progressive-scan camera (Sony XC8500). For imaging the object a Hopkins-optic (Storz Otoskop 1215 A, Storz Laparaskop 26031 A) is connected to an adapter, which includes a beam splitter for joining the object and reference beams together. Fig. 2 shows a photo of the system assembled as a prototype at the Laboratory of Biophysics in cooperation with Karl Storz GmbH & Co., Tuttlingen, Germany. The endoscopic ESPI camera system is linked to a fast image processing system which allows the subtraction of two video frames at 25 Hz. For double exposure image subtraction of two subsequent images

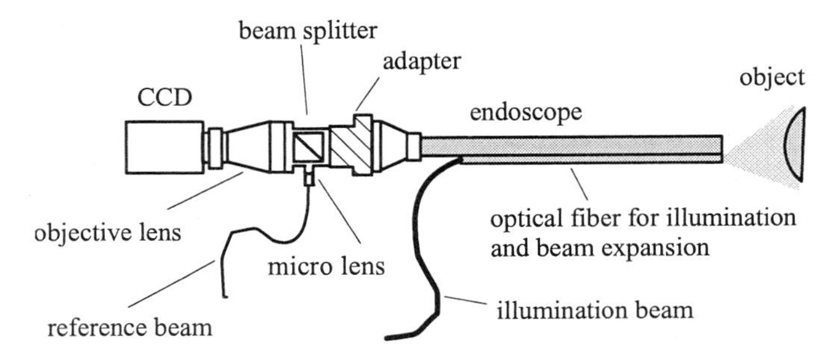

Fig. 1. Schematic setup of the endoscopic ESPI camera system.

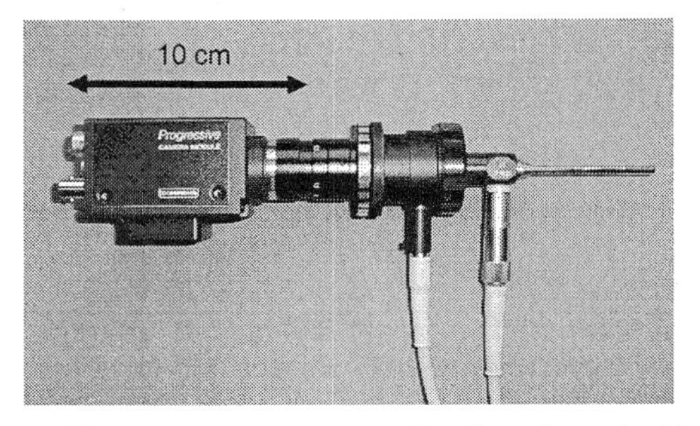

Fig. 2. Endoscopic ESPI camera system as developed at the Laboratory of Biophysics in cooperation with Karl Storz GmbH & Co., Tuttlingen, Germany (here attached to an otoscope with a Hopkins-optic-system; the illuminating beam and the reference beam are guided via mono-mode optical fibers).

in video real time, an exposure scheme is used which is illustrated in Fig. 3. The light pulses have to be placed variably within the camera frames to allow different time delays between two exposures. Also, the exposure times themselves have to be variable. Therefore, a double pulse generator, which is

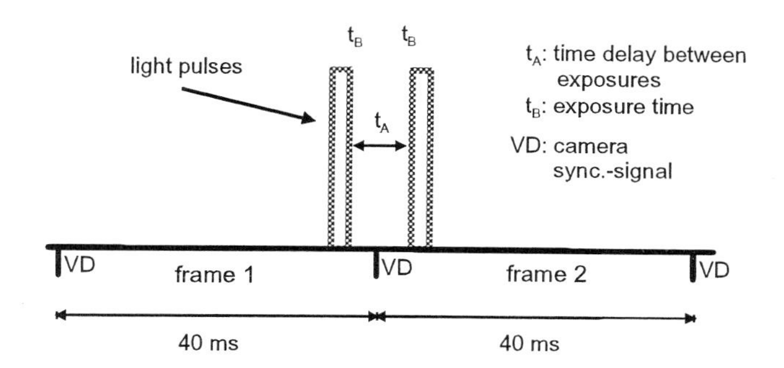

Fig. 3. Double pulse exposure in two subsequent video frames.

synchronized to the vertical delay signal (VD) of the CCD-camera, controls an acousto-optic modulator for light modulation. The image frame rate of the speckle interferograms is only limited by the camera frame acquisition rate. Using a standard grayscale camera working in interlaced mode, 25 Hz are reached. In progressive-scan mode (full frame mode), 12.5 images per second in full frame resolution with high fringe contrast are achieved.

3 Investigation of Technical and Biological Objects In-Vivo

Figures 4 (a) and (b) show the results of tests performed on technical objects. The images are extracts of a recorded series (progressive-scan mode, image repetition rate 12.5 Hz). The test object (metal plate) was tilted and pointlike deformed (out-of-plane deformation). The result is a moderate fringe density per frame with high fringe contrast for parallel and circular fringe patterns. Figures 4 (c) and (d) present first in-vivo investigations of a human hand. The fingertip as well as the fingernail were stimulated by a needle. The elasticity of the specimen (fingertip: soft, fingernail: hard) can be clearly differentiated by observing the different fringe patterns. This aspect is an important new result for minimal invasive diagnostics. It could enhance conventional endoscopic examinations by replacing the operator's tactile sense with visual information.

4 Summary

A new compact and flexible endoscopic ESPI camera system has been developed, which can be used for hand-held examinations in minimal invasive

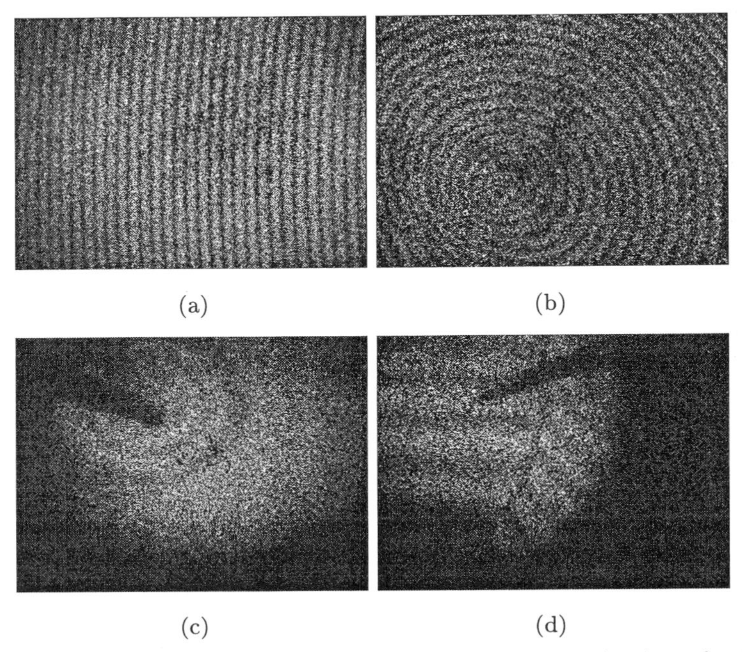

Fig. 4. Investigation of a technical object and in-vivo investigation of a human hand: (a),(b) tilt and pointlike deformation of a metal plate; (c),(d) fingertip and fingernail stimulated by a needle.

diagnostics. The speckle interferograms processed with the double pulse exposure image subtraction technique are obtained with high fringe contrast in video real time by using a progressive-scan CCD-camera. First promising investigations on technical objects and biological objects (in-vivo) have been carried out. The method developed here can replace the operator's tactile capability by supplying visual information (endoscopic taction). Thus, it has the potential to form an important extension of industrial intracavity inspections and minimal invasive medical diagnostics.

References

1. Coquoz, C. et al.: Holography with a flexible microendoscope, Series on Optics Within Life Sciences, Elsevier, Amsterdam, Vol. 3 (1994) 235–238.
2. von Bally, G. et al.: Techniques for endoscopic holography, Series on Optics Within Life Sciences, Elsevier, Amsterdam, Vol. 1 (1993) 13–21.

This work was supported by the German Bundesministerium für Bildung und Forschung (BMBF) and Karl Storz GmbH & Co., Tuttlingen, Germany.

The Endoscopic Application of the Deconvolution Filter Set Method

Masayuki Hattori[1] and Shinichi Komatsu[1,2]
[1] Department of Applied Physics, Waseda University, 3-4-1 Okubo, Shinjuku, Tokyo 169-8555, Japan
[2] Materials Research Laboratory for Bioscience and Photonics, 3-4-1 Okubo, Shinjuku, Tokyo 169-8555, Japan

Abstract. We propose the combination of the endoscopic system with the deconvolution fileter set method, a digital image recovery and defocus amount evaluation technic for the defocussed images. This combination extends the focus range of the small, fixed-focus simple optical systems, and is expected to reduce the operating time by omitting the focus adjustment on site. In this paper, experimental results of the focus extension and object distance measurement are shown.

1 Introduction

We have proposed a digital image recovering technic called deconvolution filter set method, by which the clear image is recovered from one defocussed frame, even if the focus parameter used in the exposure is not designated.

The effectiveness of the deconvolution filter set method is investigated by the computer simulation [1], and its capability of restoring the realistic optical image is confirmed in our latest researches [2]. In these researches, defocussed image restoration, object distance measurement, and the focus detection is successfully performed from single frame, real defocussed images. In this paper, we briefly summarize deconvolution filter set method and propose its new application to the enodscopic system.

2 Deconvolution filter set method

We briefly summarize the deconvolution filter set method, which is capable of restoring the images exposed in an unknown focussing condition.

It is well-known that the defocussed image img(r) is mathematically repreted as the convolution * of the object irradiance distribution obj(r) with the point spread function psf(r) caused by the imperfect focus adjustment.

$$\begin{aligned} \mathrm{img}(\mathrm{r}) &= \int \mathrm{psf}\left(\mathbf{r}-\mathbf{r}'\right) \times \mathrm{obj}\left(\mathbf{r}'\right) \mathrm{d}^2\mathbf{r}' \\ &= \mathrm{obj}\left(\mathbf{r}\right) * \mathrm{psf}\left(\mathbf{r}\right) \end{aligned} \quad (1)$$

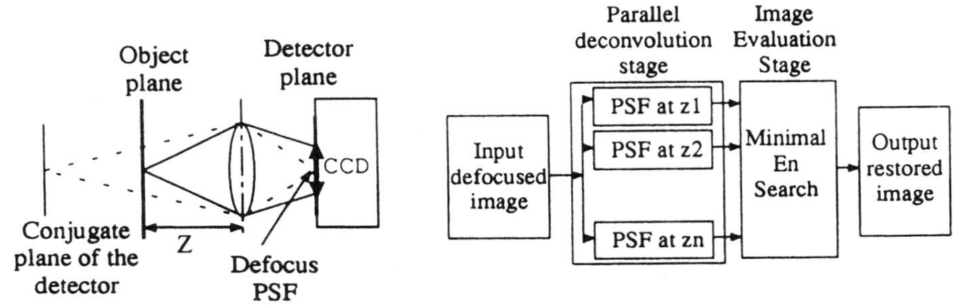

Fig. 1 Defocus PSF

Fig. 2 Deconvolution Filter Set Method

The point spread funtion psf (r) of the fixed-focus imaging optics mainly depends upon the object distance z in Fig. 1

In the deconvolution filter set method Fig. 2, the defocussed point spread function is measured in advance as the image of the point light source at the various possible distances z. Then, input defocussed image is restored by the most probable psf at z, using digital deconvolution filter Eq. (2) in the Fourier domain.

$$\widehat{\mathrm{obj}}(\mathbf{x}) = \mathrm{F}^{-1}\left\{\mathrm{W}(\mathbf{u}) \times \widehat{\mathrm{IMG}}(\mathbf{x})\right\}$$
$$\mathrm{W}(\mathbf{u}) = \frac{\mathrm{PSF}^{*}(\mathbf{u})}{\left|\mathrm{PSF}(\mathbf{u})\right|^{2} + \Gamma(\mathbf{u})} \tag{2}$$

$\Gamma(\mathbf{u})$ is noise to signal ratio and $\mathrm{F}^{-1}\{\ \}$ denotes inverse Fourier transformation. In our current method, the most probable psf at $\hat{z}$ is chosen, as the best matching psf that minimize the following negative error ratio defined by Eq. 93) for the out put of the Wiener optimal deconvolution filter in Eq. (2).

$$\mathrm{E}_{\mathrm{neg}}\left(\widehat{\mathrm{obj}}(\mathbf{r})\right) = \sqrt{\frac{\sum_{r;\widehat{\mathrm{obj}}(r)<0} \left|\widehat{\mathrm{obj}}(\mathbf{r})\right|^{2}}{\sum_{r;\mathrm{all}} \left|\widehat{\mathrm{obj}}(\mathbf{r})\right|^{2}}} \tag{3}$$

Thus, the restored image is obtained as the most clear deconvolution with the least negative error ration E_n. And the object distance is maejoured as the distance z corresponding to the most probable psf.

In the adjustable-focus system [2], this method can be used to search for the best focus parameter.

3 Endoscopic application

Originally, the deconvolution filter set provides the means for the restoration of defocussed images and distance measurement from single frame defocussed images. This application to the fixed-focus imaging optics may extend the focus range without focus adjustment operation. These are the great advantage for endoscopic systems because the focal range of the fixed-focus single lens objective on a small endoscopic camera head can be expanded, and on sight operation time may be reduced because promptly taken image will be sufficient for obtaining clearly detailed images.

4 Experimental setup

We investigated the performance of the filter set method in a small optical camera head with a fixed-focus, single lens objective. The conjugate plane of the image detecotor is fixed at 20 cm ahead of the objective. The NTSC video signal output is connected to the computer. A red LED used for the illuminating light source. The PSFs are measured in advance as the defocussed images of a pin hole ($\varphi = 1\text{mm}$) taken at various distances and stored in the computer. The object distance is about 10 cm, that is only a half distance of the conjugate plane of the detector.

5 Results

The experimental results of the image restoration are shown in Fig. 2. The vanished letters in the out-of-focus images in Figs. 2 (a) and 2(b) are recovered as shown in Figs. 2 (c) and (d). It is recognized that the focal range of the system is expanded by the filter set method.

Fig. 3 is the experimental result of object distance detection from a single defocussed image in Fig. 2 (a). Error metric E_n is minimal at the true object distance, i.e., $\hat{z} = 10\text{cm}$. The result suggests that the focus range of the endoscopic system with the fixed-focus small objective may be extended by the present method.

6 Conclusion

We introduced the digital image processing method called deconvolution filter set method which is effective especially for the small fixed-focus imaging optics to extend the focus range as well as to measure the object distance.

Its application to the endoscopic system is also proposed, and some fundamental experimental results are shown.

This study was partly supported by the High-Tech Research Center Project by the Ministry of Education, Science, Sports and Culture.

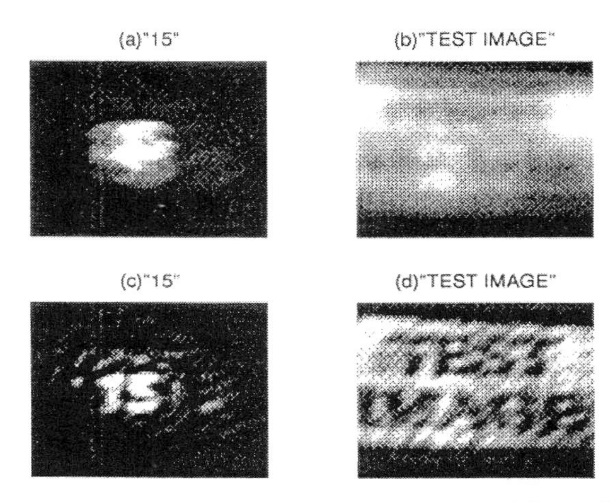

Fig. 3. Focus range extension results. (a) and (b) are the raw defocused images of close objects, (c) and (d) are the restored outputs.

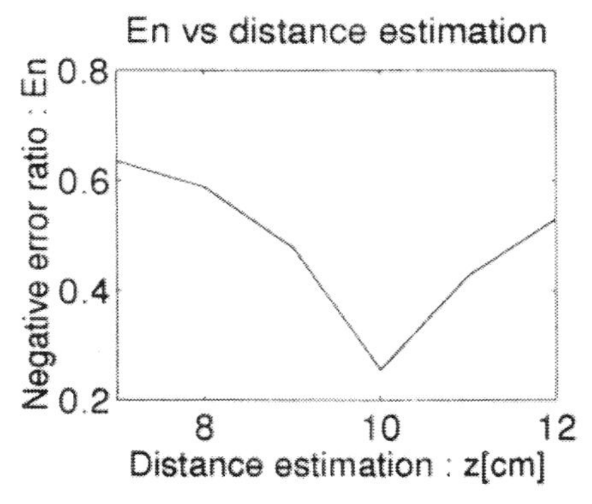

Fig. 4 Result of the object distance estimation. Negative error ratio is minimal at the true object distance (10 cm)

References

1 Sakima, N. et al. (1994) Filter bank method for determining defocus amount of lens system. ICO Topical Meeting Digest. 242

2 Hattori, M., Komatsu, S. (1998) Defocus Evaluation by the Deconvolution Filter Set Method. Jpn. J. Opt. (KOGAKU). 27, 160-156.

A Coherent Optical Technique Useful in Biomechanics: Holographic Interferometry.

Pierre M. Boone
Division of Experimental Mechanics, Department of Mechanical Production and Construction University Gent, Sint Pietersnieuwstraat 41, B-9000 Gent, Belgium
e-mail: Pierre.Boone@rug.ac.be

Abstract: The use of pulsed ruby lasers in recording holographic interferograms of specific surface displacements of the human body in vivo is demonstrated. Technical peculiarities of the technique are discussed and some practical examples shown.

1 Introduction

The application of a pulsed laser in holographic interferometry is certainly not new. Particularly at NPL, remarkable work has been done since the late sixties by Gates, Hall and Ross.[1,2]. From then on, the number of people working in this field is continuously growing.

It took some decades after the first laser action before the pulsed ruby laser became a reliable took, with predictable and reproducible pulse energies and intervals. However, only few publications in the biomechanical field can be found. This brought us to the idea of using our equipment – initially designed for investigations in applied mechanics – for experiments on human beings, principally to obtain a general idea of the specific technical difficulties characterizing this method when used in traumatology and orthopedics, and what can be expected from the pictures.

It must be kept in mind that the experiments shown here were not part of true medical investigations, but rather give some examples of potential applications.

2 The Equipment

2.1 General

It is supposed that the underlying fundamental principles of ruby lasers are known so one can concentrate on specific particularities of our equipment. We used two different lasers; the first was a JK type 2000 holocamera[3] comprising a holographic ruby oscillator, coupled to two amplifiers, to provide 2J TEM_{00} output in one or two pulses (see Figure 1), and additional optics for generating the reference beam. The second laser was a HLS 2 Lumonics laser; its characteristics are discussed further.

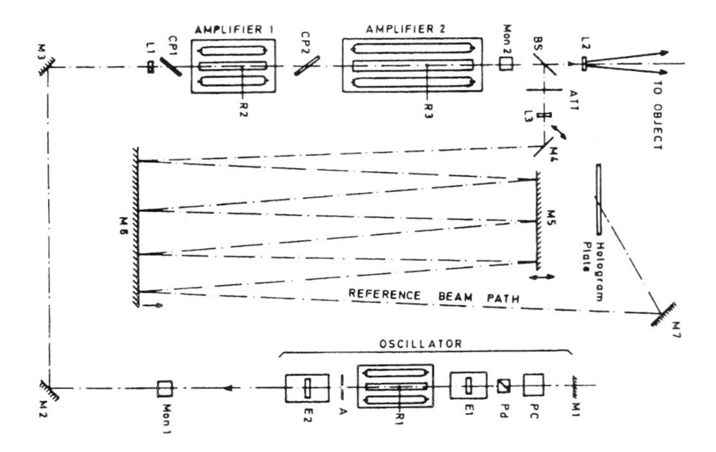

Fig. 1. Schematic layout of holocamera JK 2000.

The oscillator incorporates a 4x0,25 inch ruby rod (R1) with its flash tubes, intracavity etalon (E1), output etalon (E2), Pockels cell (PC) with Glan polariser (Poll), Q switch and a mode selecting aperture (A).

The beam emerging from the oscillator is coupled via two 45° mirrors (M2, M3) and a negative lens (L1) into the amplifier chain, comprising a 4x0,25 inch first (R2) and an 8x3/8 inch second amplifier rod (R3), both with their appropriate flash tubes. Angle correction plates (CP1, CP2) compensate for the 1° inclined input face of each amplifier rod.

Two energy monitors (Mon 1, Mon 2) are provided, allowing on line monitoring of both oscillator and amplifier output energies when coupled to a storage scope. The reference beam is generated by the beam splitter B5. Its path passes via an attenuator (ATT) and a diverging lens (L3), and its length can be altered by changing the angle of M4 and the distances of M5 and M6 as to correspond to the object beam path length. The hologram plate holder is attached to the frame. The holocamera is mounted on a inclinable table and mounted on rollers for easy displacement.

2.2 Working principles

The Q-switch assembly comprises a Pockels cell (PC, Figure 1) and uses a KDP crystal. When a high bias voltage (2,5 to 3 KV) is applied to the crystal, 90° rotation of polarisation of the laser light occurs. When this rotated wave passes through the Glan-polariser it is completely extinguished. The result is comparable to that of a closed shutter. When a negative voltage pulse is applied, effectively removing the bias for about 500 ns, laser action occurs.

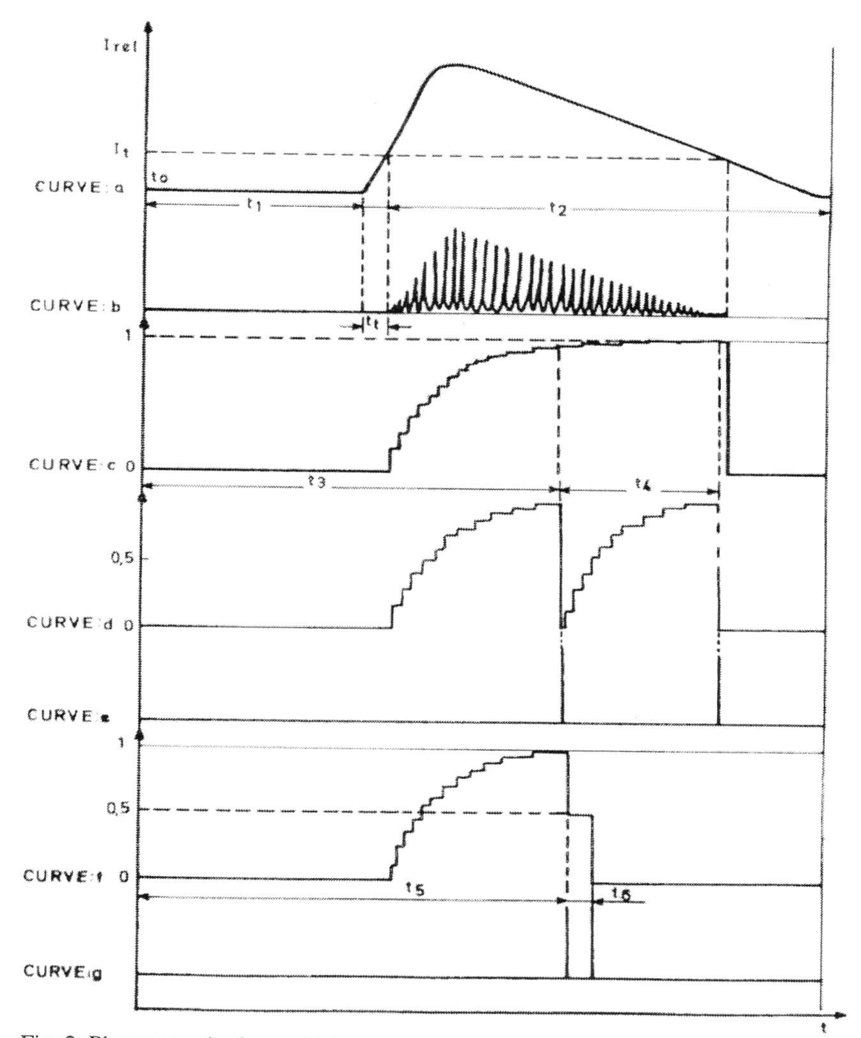

Fig. 2. Phenomena in the oscillator stage.

Figure 2 shows the phenomena in the oscillator: Curve a shows the relative flash tube intensity in function of time (the abcissa for all curves), t_0 corresponding to the closing of the trigger contact (possible jitter is neglected). There is a certain delay time t_1 before the flash intensity starts to rise, caused by the impedance of the electrical circuit. The flash burns a constant time, typically 500μs. When the flash intensity reaches the threshold value I_t, laser action starts randomly (curve b), with intensity peaks proportional to the flash intensity.

Integrating, we obtain the energy curve c if the Pockels cell remains in the "open" position. If it is "closed", energy is stored in the rod, also following curve c, but no laser output will be generated. At the moment the energy reaches half its maximal value the Pockels cell is opened and the stored energy released as a laser flash.

However, the pump flash is still active and a second build-up of energy can take place, whereafter a second laser flash can be generated (curve d and e).

The foregoing applies only for long interval times between pulses (100 to 700μs). If this time becomes too short to store the energy for the second pulse, we must close the shutter again when half the stored energy is used, leaving the other half available. This situation is represented in curve f and g and is typical for interval times between 1 and 100 μs. The duration of a single pulse is typically 20 to 30 ns.

If the output of the energy monitor Mon 1 is coupled to a storage oscilloscope with adjustable trigger delay via integrating electronics, one can see a trace as shown in Figure 3 in double pulse operation. The vertical parts correspond to a calibrated energy, the time interval to the horizontal part between two pulses.

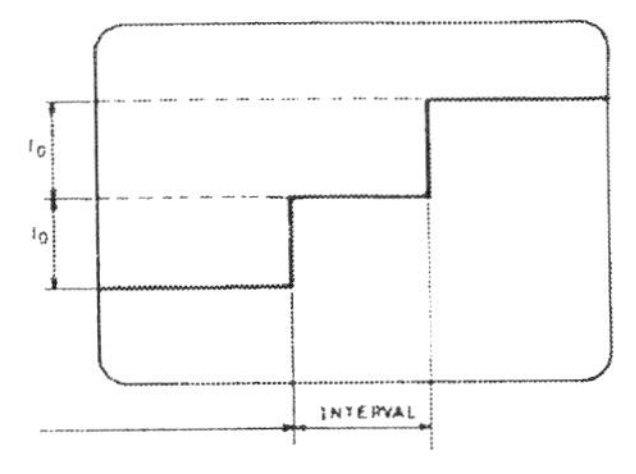

Fig. 3. Double pulse operation.

The action of the amplifiers is similar to that of the oscillator. The flash tube delay times however are chosen so that the maximal energy becomes available at the moment between both pulses. In practice, the output is monitored via Mon 2 and the delays optimized for maximum output. The energy of the flash pulses can be adjusted by changing the voltage over the amplifier flash tubes. The output lens L2 is interchangeable and must be chosen in function of the area to be illuminated.

2.3 Set-Up for Transmission Holograms

The use of the internal reference was abandoned in most experiments, as it was originally designed for illuminating structures of around 2x2m, requiring large distances to the holographic plate. We needed much shorter distances as our subjects were rather small (head, hand, arm and chest). Besides, a large set-up flexibility is wanted for this field of applications. The easiest set-up is shown on Figure 4. Only one diverging beam is used, H is the hologram plate, S is a screen to avoid back illumination of H, and M is a mirror. By moving M laterally a suitable beam ratio can be obtained. The whole set-up is made on a ordinary table.

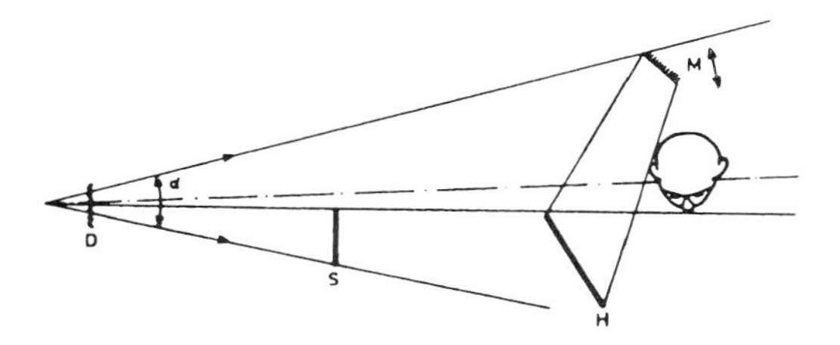

Fig. 4. Setup for recording transmission holograms.

The uniformity of the intensity of this first laser leaved much to be desired, probably due to thermal effects, impurities and/or parasitic reflections. External spatial filtering is very complicated, as the energy density in the focal point of lenses easily surpasses breakdown of air, and was not attempted.

Thanks to extensive work done at NPL[4], we were able to produce a suitable scatter plate (D, Figure 4) giving minimal attenuation of intensity and change in divergence.

To avoid backreflections in the laser, the scatter plate must be slightly inclined, and to ensure sufficiently fine speckle on the subject, it must be mounted some distance away from the output lens. These parameters must be determined first in a suitable set-up with a He-Ne laser (α and β, Figure 5). The divergence angle α of the pulsed laser must also be known. A glass master diffusor with equivalent divergence angle was produced by attacking window glass with the vapour of ammonium chloride.

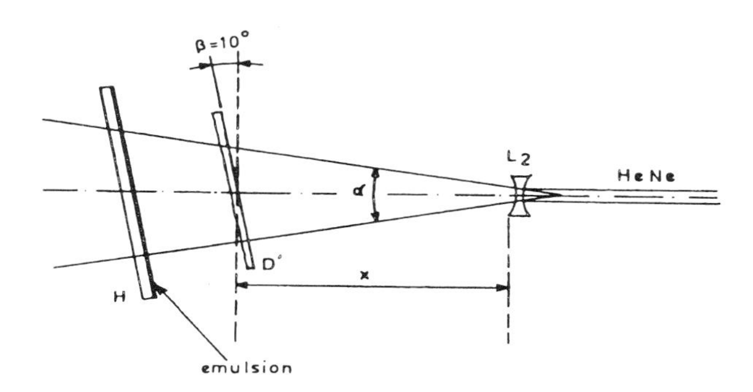

Fig. 5. Recording of a scatter plate.

The final scatter plate was made by exposing a holographic plate H (Agfa Gevaert 8E75HD), with a He-Ne laser beam of divergence α through the master diffusor D'. Several plates were exposed with increasing energy to obtain densities between 0,4 and 1,2 when developed in a 1:1 mixture of 5 gr. Pyrogallol/1 water and 60 gr. Na_2CO_3/I water.

Afterwards, bleaching is performed in a solution of 1,5 cc sulfuric acid and 3 gr Ammonium-bichromate/l water. The same processing was used for the transmission holographic interferograms.

After processing, the plate giving minimal attenuation was selected and mounted in the pulsed laser beam, in the same position as it had in the He-Ne laser beam.

The equipment is provided with a low power He-Ne laser (0,3 mW) for internal alignment purposes; it can hardly be used for external alignment. For easy positioning of the external optical parts in our set-up, we used a white light projector mounted close to the output beam axis, with a divergence comparable to that of the laser beam.

With the aid of an additional integrating photocell, the intensities of reflected objected beam and reference beam were equalised. The total energy impinging on the plate was measured and the amplifier gains were adjusted until the required energy level was reached, typically about 20μJ/cm^2 for a prebleach density of 1.6.

3 Initial Experiments

3.1 Impact on a safety helmet

Figure 6 shows holographic interferograms of a shock wave propagating through a safety helmet and the face of the wearer, due to an impact of a 300 gr. Hammer dropped from a height of 10 cm. The laser was triggered by a current flowing via the hammer to a brass strip stuck on the helmet, at the moment of impact. The delay between contact and first pulse is about 1.3 ms and cannot be adjusted as it is function of the electronics (see later); the interval between pulses can be regulated with an accuracy of ±1%.

Figure 6a is taken at 5 μs interval and shows the propagation of the shock wave in the helmet. Figure 6b, at 50μs interval, shows severe increase in helmet displacement and deformation concentration around the ventilation holes and the attachment points of the inner structure; some displacement around the eyes is already visible. Finally, in Figure 6c, taken at 260μs interval, the helmet displacement is too large to leave fringes visible on the helmet, while the shock wave passing over the face becomes clear.

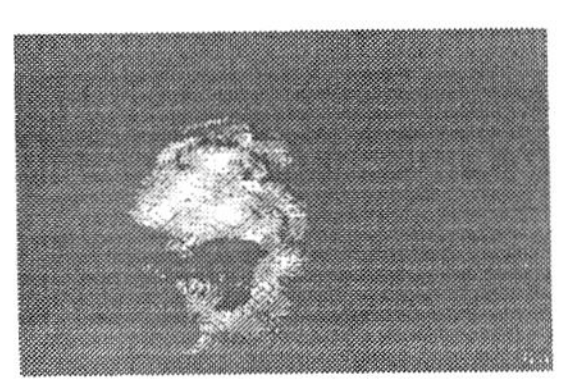
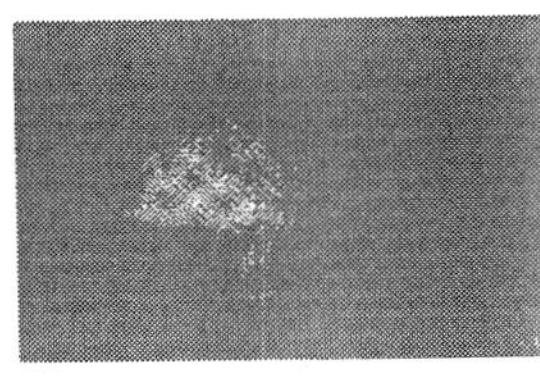
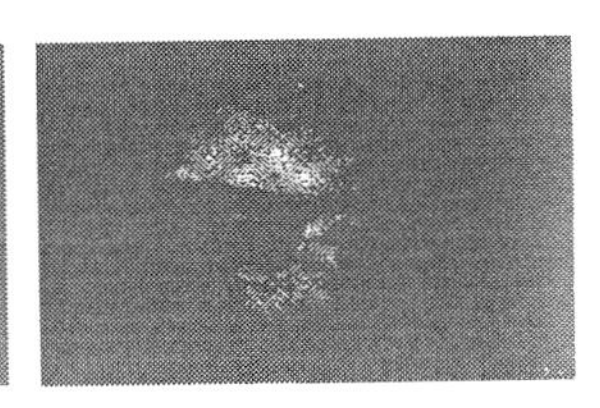

Fig. 6. Hammer impacts on a safety helmet.

From these pictures one can conclude, from the engineering viewpoint, that the ventilation holes create spots of strain concentration in the helmet and that principally the region of the eyes are submitted to large movements long before the shock wave enters the lower skull.

3.2 Firing an air gun

An air gun can be a cheap tool to create reliable impact loads in a mechanical laboratory[13].

The triggering of the laser occurs when the pellet breaks a 0.05 mm thick wire mounted over the barrel mouth. Figure 7a shows the displacement of the hand (interval 1 μs). Note the rather homogeneous displacement of the fist and the slight displacement of forefinger and wrist. In Figure 7b (interval 94 μs) one can see the displacement of the weapon substructure due to the release of the compression spring. The hand has moved further although the displacement of the wrist is still low.

The Figure 7c (interval 600 μs) we see the deformation of the substructure due to the percussion of the piston in the chamber. The displacement of the hand is too large to be visible.

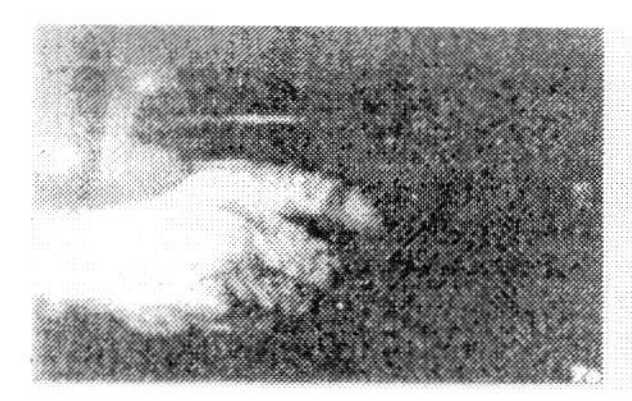
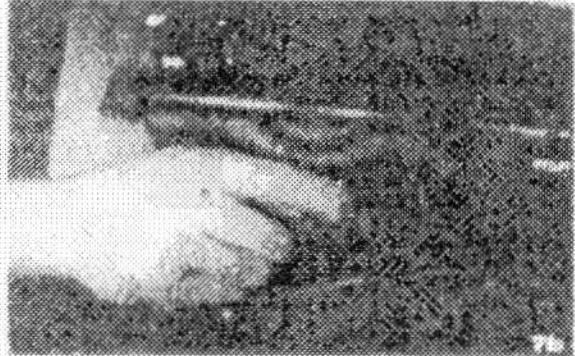
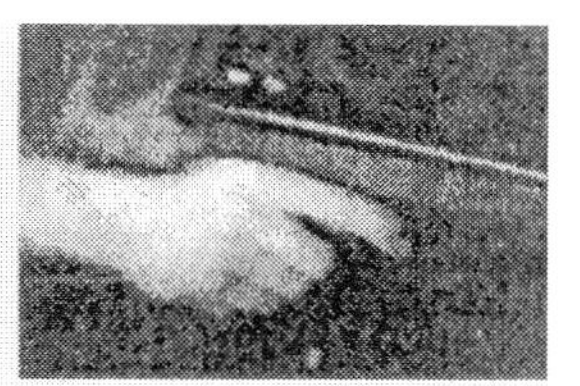

Fig. 7. Firing an air pistol.

3.3 Behaviour of the arm in a tennis strike

To obtain adequate information about displacements in holographic interferometry, the loads applied to the bodies under study can be very small. That is why large and quick movements of the human body – as generally involved in sports – can easily be simulated.

In the case of a tennis strike for instance – shown in Figure 8 – the arm was held steady and a pellet was fired with the air gun onto an aluminium disc fixed excentrically with respect to the "sweet spot".

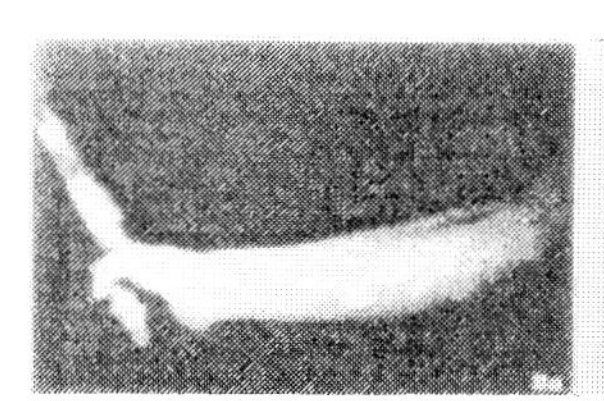
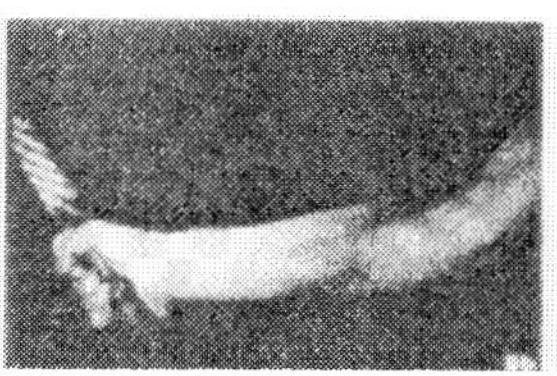
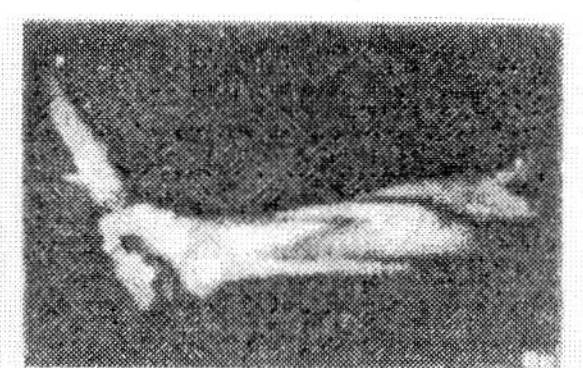

Fig. 8. Displacements during a tennis strike.

The mechanical energy released on this way is 4J, to be compared to the 40J in reality[5]. The trigger system is described in foregoing paragraph. Figure 8a is taken with an interval of 10μs; only one fringe is visible in the racked shaft while in Figure 8b taken at 110μs interval – there is bending and rotation of the shaft, slight deformation of the ball of the tumb and starting rotation of the arm near the elbow. Figure 8c taken at 515 μs interval, shows large bending and rotation of the shaft, turning to pure rotation of the grip and severe rotation of the arm. Special attention can be paid to the fringe discontinuity near the elbow.

3.4 Displacement of abdomen and thorax

If the movements of the body parts under investigation are slow, electric trigger systems are no longer necessary, as the laser can be fired by hand. This is the case while investigating respiratory movements. The beginning of the expiration if shown in Figure 9a, taken with an interval of 52 μs. At maximal inspiration rate the pattern of Figure 9b occurs, with equal time interval.

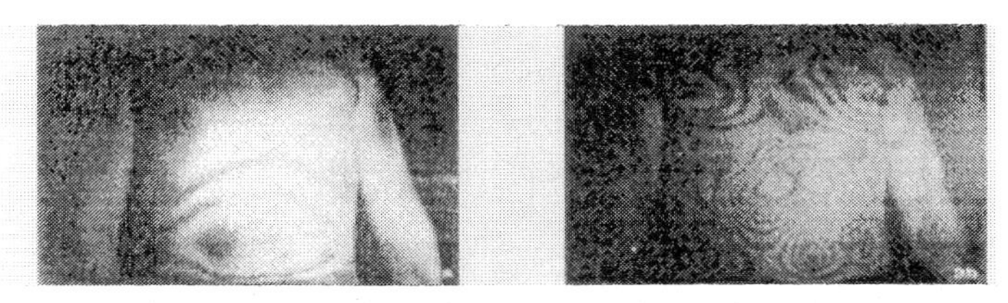

Fig. 9. Displacements of chest due to respiration.

3.5 Image quality

It can be seen from the pictures that when using a scatter plate, the fringe sharpness is not so high. The scatter plate reduces the spatial coherence of the laser beam. This is of less importance for the object beam, but when the reference beam has reduced spatial coherence, the sharpness of the hologram and hence of the fringes decreases.

4 Second Series Of Experiments

In the second series we used a more recent generation of ruby lasers: the HSL2, also made by the Rugby group, that was succesively named JK, Lumonics, Asahi, and now Innolas.

It is rated 1 Joule per (double) pulse, but can easily give the double for the intervals longer than 100 μs. It has the advantage of having only one amplifier stage, and is equipped with a diamond spatial filter between oscillator and amplifier. In this way, the output is quite uniform and allows one to work without scatter plate. In this way, both transmission and reflection (Denisyuk) setups can be used, and a good quality of the resulting images can be obtained in both setups. A number of photographs taken for transmission holograms will be presented in this paper; some examples of reflection holograms were shown in the exhibition accompanying the OWLS 5 meeting.

4.1 The Denisyuk Setup

The very simple Denisyuk[14] setup is shown in fig. 10.

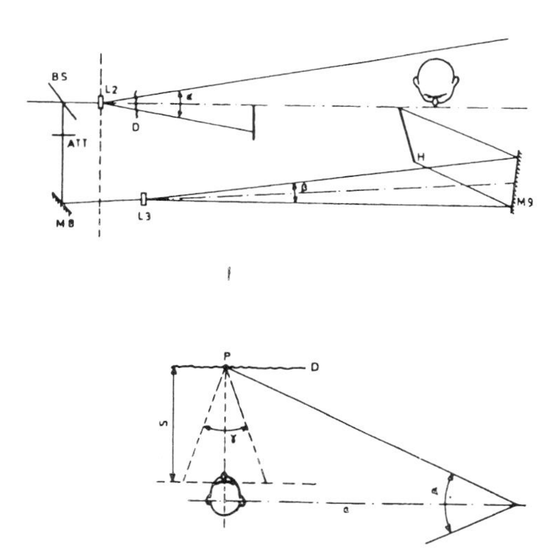

Fig. 10. Denisyuk setup.

It only comprises three components: the laser, a beam diverging lens (normally biconcave, again for avoiding air breakdown) on the hologram. The light falling directly on the emulsion forms the reference beam; light passing through the emulsion illuminates the object, whose reflection forms the signal beam.

The setup is thus very economical with light, as the beam is effectively used twice. Drawbacks are that the hologram has to be larger than the object (if the object is close to the hologram, what normally is the case, to achieve the best possible I_{ref}/I_{obj} ratio) and the extreme resolution that has to be achieved by the photographic material (typically 4 to 5000 linepairs/mm). The "photographic" scheme is also more demanding, involving controlling a number of parameters[15].

The experiments shown here were:

- Movements of the face and the throat when whispering the letter "a" (fig.11).
- Biting on a quite hard nut (fig. 12).

- Again impacts on a safety helmet (fig. 13 a,b,c), but of another type than the one shown before.
- Movements of the hand muscles when pinching on a excercising spring (fig. 14).
- Trials to locate a subsurface ecchymosis by visualising the difference in hardness when the muscles are loaded mechanically (fig. 15).

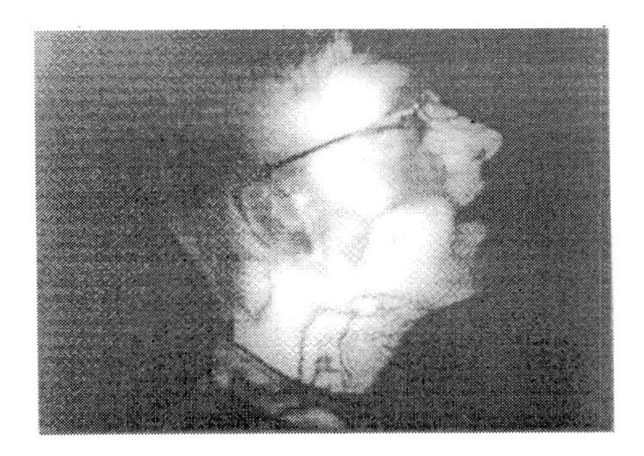

Fig. 11

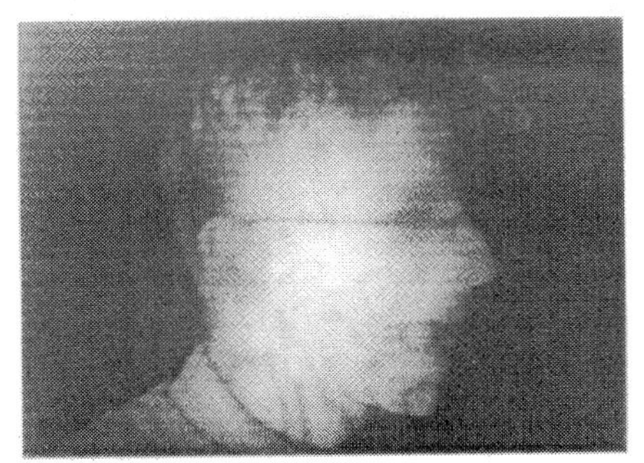

Fig. 12

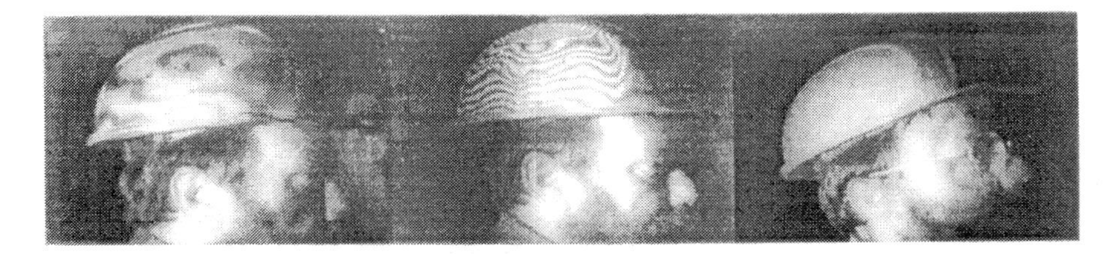

Fig. 13a, b, c

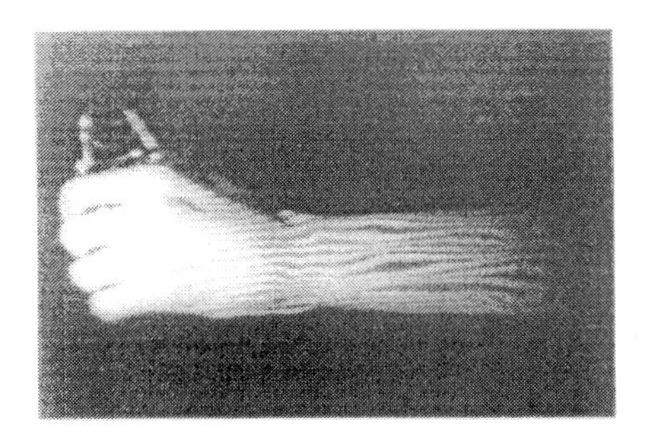

Fig. 14

Fig. 15

5 Discussion

5.1 General remark

The interpretation of the fringe patterns given here are only based on mechanical considerations. It is obvious that if medical research is to be derived, full benefit of the technique is only possible if teamwork between engineers and medical doctors is organised.

5.2 Qualitative measurements

Everyone working with holographic interferometry knows that derivation of quantitative data from a pattern is a tedious and time-consuming task, generally requiring the use of a computer and a sophisticated optical set-up. Whenever allowable one should therefore rely on qualitative interpretation, or try to use ESPI (video variant of holographic interferometry) that allows automated image analysis.

When the set-up is kept constant, comparison with similar bodies will give no problems, especially in the medical field where comparison is an usual interpretation method.

Looking at the symmetry of the pattern will often give good results but one should remember that the fringes do not always form on the object but sometimes in the space in front or behind it. This gives always asymetrical fringes except when the displacement vector is normal to the middle plane of the test object.

It sould be emphasized that qualitative interpretation is mainly based on experience, and that the study of some practical examples can largely help to start [6,7,8,9].

5.3 Time lapse

As mentioned in the paragraph on the working principles, there exists a time lapse between the trigger pulse and the first laser flash of between 800 μs and 1,5 ms, depending on the preset pulse interval. With the HSL2 system, it is consistently about 1,3ms. It must be kept in mind while interpreting displacement patterns at what moment the recorded action has occurred, with respect to the moment of triggering. That time lapse can easily be measured with the aid of a storage oscilloscope.

To make things clear we will use the test on the safety helmet as an example. In this case the time lapse (t3 or t5 on Figure 2) turned out to be 1,36 ms; and moment of impact and triggering were equal. This means that the first picture (Figure 6a) is the situation 1,36 ms after impact superimposed to one 5μs later.

If measurements have to be done closer to the moment of impact, other trigger techniques are required, e.g. the electromagnetic release of a weight dropping on the helmet could be used. The release signal could be converted to the trigger signal and retardation arranged by changing the height or by a suitable delay line system. For equal impact energy the weight should have to be adapted to the height.

This example shows the necessity for inventive considerations on the trigger system, that must be adapted for every particular application.

6 Conclusions

If safety precautions are taken, double pulsed holographic interferometry can easely be applied on human beings in vivo. The information collected allows easy derivation of qualitative displacement data and even quantitative values can be derived, be it wish more sophisticated set-ups and data acquisition. Extensive use of the technique demands for multidisciplinary team work.

7 Acknowledgment

The work reported here was supported by the Belgian National Fondation for Collective Fundamental Research through grants 2001275 and 204486.

The author also gratefully acknowledges the personal grant placed at his disposal by the National Science Fondation to study reflection hologram optimization.

References

1. Ross I.N., Gates J.W.C., Hall R.G.N., 1972, "Practical lasers for photographic and holographic recording". Proc. 10th International Conference on high speed Photography, Ni
2. Hall R.G.N., Gates J.W.C., Ross I.N., 1972, "Short interval double-pulse holography of reflection objects". Proc. 10th International Conference on high speed Photography, Ni
3. J.K. Lasers LTD. "System 2000". Somers Road, Rugby Warwickshire, England. Tel. Rugby (P788) 70321; Tlx:311540.
4. Gates J.W.C., 1968. "Holography with scatter plates". J. Sci Instr. Series 2, 1989
5. Brannigan M., Adali S., 1981, "Mathematical modelling and simulation of a tennis racket". Med. And Science in Sports and Exercises 13 (1).
6. Schumann W., Dubois M., 1979, "Holographic Interferometry". Springer-Verlag, Berlin New-York, 194.
7. Von Bally G., 1979, "Holography in Medecine and Biology". Springer-Verlag, Berlin New-York, 269.
8. Wedendal P.R., Bjelkhagen H.I., 1974, "Dental holographic interferometry in vivo utilizing a ruby laser system". Acta Odont. Scand. 32, 131.
9. Boone P.M., 1981, "Application Medicales de l'Holographie". Cahiers de Medecine du Travail, vol. XVIII (1), 47.
10. Sliney D.H., Freasier B.C., 1973, "The Evaluation of optical radiation research". Applied Optics, 12 (1).
11. British Standards, 1972, "Protection of Personnel against Hazards from Laser Radiation" BS 4803.

12. Tozer B.A., 1979, "The calculation of maximum permissible exposure levels for laser radiations". J. Phys. E. 12, 922.

13. De Caluwe, M., Boone, P.M., "Double pulsed holography on human beings in vivo", Proceedings of the Max Born Centenary Conference, SPIE, vol. 369, 1982, p. 596-605

14. Denisyuk, Y., "Photographic Reconstructions of the Optical Properties of an object in its own scattered field", Sovjet Physics, Doklendy 7, (1962), p. 543-545.

15. P.M. Boone, "Sone problems associated with processing Agfa Gevaert 8E75HD sheet film for reflection holography". SPIE. vol. 600, 1985, p. 172-177.

Relative and Absolute Coordinates Measurement by Phase-Stepping Laser Interferometry

Ventseslav Sainov, Jana Harizanova, George Stoilov, Pierre Boone*
Central Laboratory of Optical Storage and Processing of Information
Bulgarian Academy of Sciences, G. Bonchev Str., P.O.Box 95, Sofia 1113, Bulgaria
E-mail: vsainov@optics.bas.bg

*University of Gent, Sint Pietersnieuwstraat 41, B 9000, Gent, Belgium
E-mail: Pierre.Boone@rug.ac.be

Abstract. Relative and absolute coordinates measurements of real objects by phase-stepping laser interferometry are presented. The method proposed is suitable for remote non-destructive testing and accurate measurement of real object such as machine's parts, modules, constructions, as well as archaeological monuments and in biomedical investigations for spinal deformations measurements of human body, etc. Theoretical analysis for calculation of the coordinates from the measured phase difference in illumination of the objects with different phase-shifted interference patterns as well as uncertainty of the measurements are proposed. A good agreement with experimentally obtained results is achieved.

1. Introduction

Relative and absolute coordinate's measurements are a pressing task for remote non-destructive testing of real objects in working conditions. [1-3] The most popular at the moment are digital triangular methods as well as laser scanning and confocal system for 3D micro/macro investigation. The present work deals with theoretical and experimentally obtained result for relative and absolute coordinate determination by phase-stepping laser interferometry with projection of interference patterns.

2. Methodology and Theoretical Approach of the Experiment

The experimental setting shown on Fig.1 is used for the purposes of the present work. The vertical interference fringes, generated with a collimate laser light and adjustable Michelson's interferometer (one mirror is mounted on a phase-shifting device) are projected on the plane $(x',y',0)$ with a period d which can vary in broad limits - $d = m\lambda/2\sin(\theta/2)$, where: λ - wavelength, θ - angle between interfering beams, m – magnification of the projection lens.

The opticoelectronic feedback is realized by beam splitter, incorporated into the system, reflecting part of the interference pattern to the vertical slits of large – format PIN-photodiodes. Important is the three-point suspension of the phase-shifting mirror and low-voltage control (0-12 V), allowing automated tracking and control of the density and orientation of the projected interference fringes. The CCD camera (resolution 604 x 576 pixels) operates in measurement mode in which exposure time can vary from 50ms to 10s.

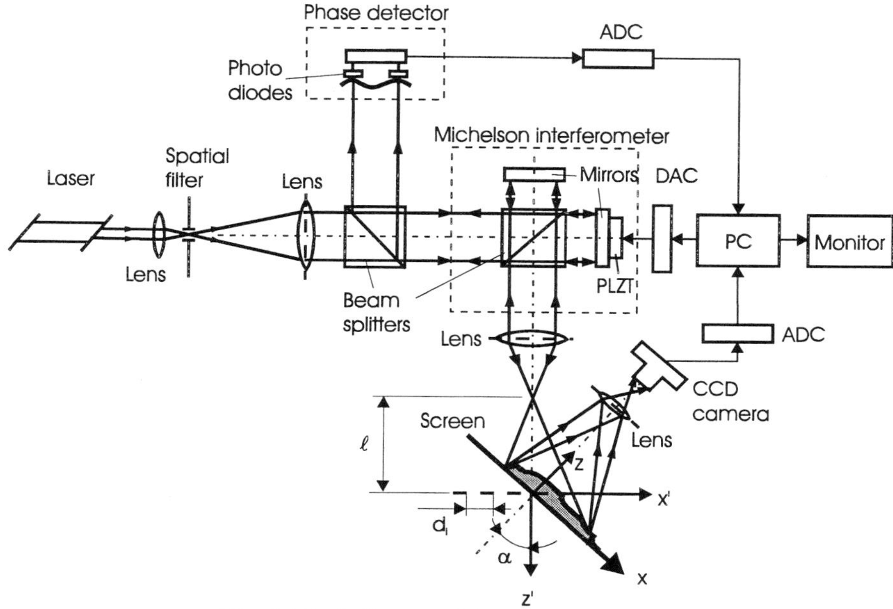

Fig. 1. Block diagram of the measurement system

The parasitic phase noise, due to vibration, turbulence and other inhomogenietes is time-averaged and filtrated during the capture of the interference pattern.

The relative coordinates are determined from the phase difference in the interference patterns of illumination of a flat screen in the plane *(x, y, 0)* and of the object placed in the same plane. According to [1,2] for the z-coordinates we have:

$$z(x,y)=\frac{N_{x,y}d\,(l+x\sin\alpha)^2}{N_{x,y}d\cos\alpha(l+x\sin\alpha)-l(l\sin\alpha+x)} \tag{2.1}$$

where the $N_{x,y}$ are numbers as expressed below.

The *absolute coordinates* of the measured object can be obtained from the phase difference in the interference patterns with different period d_i $(i=1,2)$ at successive illumination of the object surface. Taking into account that the distance from exit pupil of the projection lens to the object is much greater than its longitudinal size L $(l>>L>>z)$, the geometrical distortion in the observation field is very small. If the phase in the plane $(x', y', 0)$ is $\varphi_i = 2\pi x'/d_i$, after rotation by angle $(180 + \alpha)$ around y-axis, for non-checked coordinate system (x, y, z) we have:

$$|\varphi_i(x,y,z)|=\frac{2\pi}{d_i(x,y)}\bullet\frac{lx\cos\alpha+lz\sin\alpha}{l-z\cos\alpha+x\sin\alpha} \tag{2.2}$$

Substituting :

$$N_{x,y} = \frac{[\varphi_2(x,y,z) - \varphi_1(x,y,z)]}{2\pi} \tag{2.3}$$

for the *z(x,y)* coordinate we obtain:

$$z(x,y) = \frac{N_{x,y}(l + x\sin\alpha) - \dfrac{d_2 - d_1}{d_1 d_2} lx\cos\alpha}{N_{x,y}\cos\alpha + \dfrac{d_2 - d_1}{d_1 d_2} l\cos\alpha} \tag{2.4}$$

For the fixed values of *d, l,* and α, the accuracy in measurement mainly depends of the phase difference, i.e. by the accuracy of determining $N_{x,y}$. According to (2.4), the accuracy of the z-coordinate's estimation is determined by following expression:

$$|\delta z(x,y)| = \frac{\dfrac{d_2 - d_1}{d_1 d_2} l(l\sin^2\alpha + x)\delta N_{x,y}}{\left[N_{x,y}\cos\alpha + \dfrac{d_2 - d_1}{d_1 d_2} l\sin\alpha\right]^2} \tag{2.5}$$

For example at the basic values of $l = 7$ m, $\alpha = 52°$, $d_1 = 2$ mm, $d_2 = 1$mm and $\delta N_{x,y} \sim 0.02$, the systematic error for z-coordinate determining does not exceed 0.25mm. The influence of the other parameters (l, d, α) is assessed in detail in [3] and could be neglected.

3. Experimental Results

The experimentally obtained results for 3D measurement of "Madara Rider" relief on a Bulgarian coin at illumination with He-Ne laser are presented on the Fig.2.

4. Conclusion

The proposed method can be successfully applied for measurement of relative and absolute coordinates of real objects in a wide dynamic range. Especially for investigation of biomedical and archeological objects it's of great importance. From the derived equations for the coordinates determination, the normal displacement is accessed. This is essential for fault detection, early diagnostics and non-harmful examination of human body. For example the developed approach and designed system could be applied for examination of scoliosis and other spinal deformations and for accurate testing of archaeological monuments. The planed experiments for measurements mentioned above an especially for the famous rock bas-relief "Madara rider" in Bulgaria (sized 2.6 m x 3.2 m) are bases for the further developments and scientific collaboration.

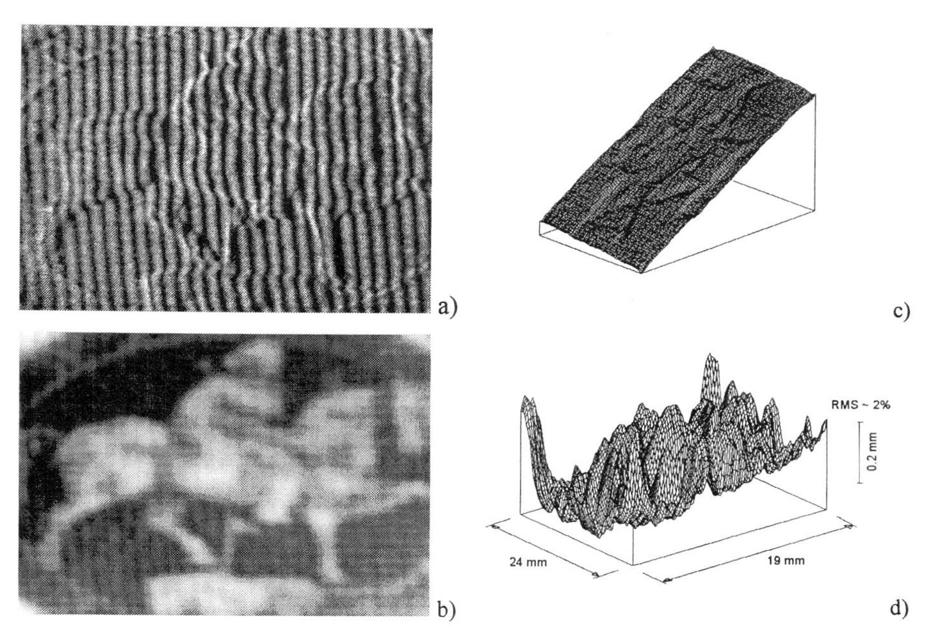

Fig. 2. Interferometric measurement of "Madara Rider" relief from a Bulgarian coin a) Interference pattern projected on the object; b) Phase map modulus 2 π c) 3D image of the tilted object; d) 3D image in the normal (z) direction.

Acknowledgements

The present work is funded from the contract for bilateral scientific collaboration.

References

1. V.Sainov Accuracy and dynamic range in shape measurement of a large-format objects, Proc.2-nd Int. Workshop on Authom. Process. of Fringe Patt., Akad. Verlag, Berlin, v.19, 182-187 (1993)
2. P. M. Boone, V.Sainov, J. Ven der Linden In-plane and out-plane displacement measurement in a wide dynamic range with a modified ESPI system, SPIE, v. 2545, 44, (1995)
3. V.Sainov, G.Stoilov, D. Tonchev, T. Dragostinov Shape and normal displacement measurement of real objects in a wide dynamic range, Akad. Verlag, Balatonfured, v.2, 52-60 (1996)

Reproduction of Sounds from an Old Russian Wax Phonograph Cylinder by Various Optical Methods

Takashi Nakamura[1] and Toshimitsu Asakura[2]

[1] Kushiro National College of Technology, Kushiro, Hokkaido, 084 JAPAN
[2] Hokkai-Gakuen University, Sapporo, Hokkaido, 060 JAPAN

Abstract. The reproduction of sounds from old phonographic wax cylinders using a laser-beam reflection method was successfully developed at Hokkaido University about ten years ago. This development has made great contributions to the fields of folklore and philology because a lot of cylinders with culturally important contents remain in the world still now. That method was, however, limited to be applied to normal wax cylinders. Furthermore, an unavoidable noise problem caused by laser speckles occurs in this method. To overcome these demerits, active efforts have been made and, consequently, various methods for the reproduction of sounds are attempted. In this paper, the reproduction of sounds from an old Russian wax cylinder using various methods is introduced.

1 Introduction

After the invention of the tinfoil recording by Thomas Edison in 1877, a great progress has been made in the audio technology. Especially, the optical technology contributed a great deal to that technology such as CDs and MDs. It is evident that the analog audio has been overcome by the digital technology in audio markets. It is, however, recognizable that there are needs for the analog audio. In some cases, such needs come from a nostalgia or a maniac interest to an old technology. In other cases, there are academic needs for the analog audio because analog recordings include important cultural materials. The phonograph based on the tinfoil recording was invented by Edison in 1877. It was innovated until 1888 to a phonograph using wax cylinders and was used for both recordings and reproduction of sounds until about 1927. The most important characteristic of the innovated phonograph had the ability of recordings. Simply by replacing a stylus for reproduction with that for recording, it was easy to make recordings on a cylinder. Therefore, the phonographic wax cylinder system was used as a portable recording machine in the world. As a result, numerous kinds of recordings were made by field researchers at that time, and some of those recordings have remained until now and become important cultural inheritances.

The reproduction of sounds from old phonographic wax cylinders are easily realized by using an Edison-type phonograph. Although this phonograph is a proper way of the reproduction, it may cause damages on sound grooves of wax cylinders because of its heavy stylus pressure. The laser-beam reflection method was first developed at Hokkaido University [1] for the purpose of non-destructive and non-contacting reproduction of sounds from Piłsudski's wax cylinders discovered at Mickiewicz University in Poland. This new method has made a great contribution to both philology and folklore of the Ainu people and originated various works on the optical reproduction of sounds from old phonographic wax cylinders and old disk records [2,3].

In this paper, the wax cylinders and the Edison-type phonograph are briefly introduced first. Next, the principle of the laser-beam reflection method is described as a basis for the optical reproduction of sounds from old wax cylinders. Furthermore, various attempts for the reproduction are shown. Finally, conclusive remarks are noted.

2 Wax Cylinders and Edison-Type Phonograph

The dimension of the standard type of wax cylinders are ≈2 inches diameter and ≈4 inches long. Sounds are cut on 100 grooves per inch. The playing speed is 90rpm for speech and 120–160rpm for music. Therefore, the playing time is about 4 minutes for speech and 2 minutes for music. There are cylinders of dense sound grooves(200 per inch) and of the longer dimension(150mm long) which are supposed to have been produced in order to cope with disk records.

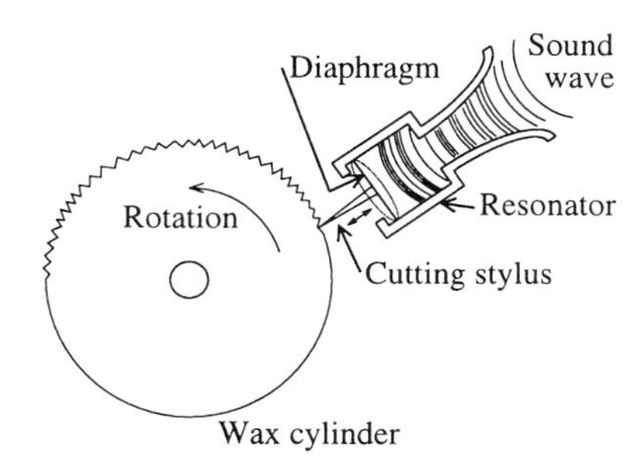

Fig. 1. Schematic diagram of the procedure for recording sounds on the wax cylinder

By using the Edison-type phonograph with a cutting stylus, the recordings were made on wax cylinders with a direct cutting method. Figure 1 shows this situation. Most of the phonographs did not provide an electric amplification system. Therefore, the cutting was performed only with the sound pressure transmitted to the stylus through the diaphragm and results in small variations of the cutting depth on sound grooves. The Edison-type phonograph had spread rapidly and widely all over the world after its invention. Although the major use of it was for amusements especially in the United States, its properties of portability and easy recordings made a great contribution to philology and folklore. The field workers at that time brought a phonograph with them and recorded various materials. In spite of the capability of easy recordings, the Edison-type phonograph gradually declined and was finally replaced by disk records until 1929.

3 Laser-Beam Reflection Method

Figure 2 shows an intersection of a wax cylinder and the principle of the laser-beam reflection method. Sound signals are recorded as vertical variations of the grooves. A laser beam is incident on the bottom of a groove of the wax cylinder and is reflected according to the law of reflection. Apparently, the direction of the reflected ray is proportional to the variation of the depth of the groove. Hence, by placing a position sensitive device (PSD) in the detecting plane, sound signals recorded on the cylinder are reproduced as the output signal of the PSD.

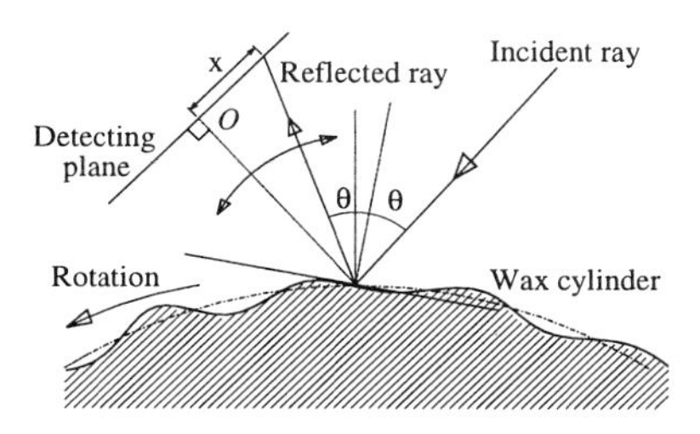

Fig. 2. Intersection of the wax cylinder and the schematic 2–D diagram of the laser-beam reflection method

The use of a laser source provides an ease of optical configuration. Especially, the focusing of a laser beam onto the bottom of a sound groove is easily establishable. On the contrary, some difficulties arise with the use of a laser. The major difficulty is a problem concerned to speckles. The surface roughness of sound grooves causes a speckling effect in the detecting plane and this results in an unnecessary noise for the reproduction of sounds because a PSD's output is proportional to the mean intensity distribution on its sensitive area. In addition to the noise problem described above, a tracking problem also arises inherently in the case of a non-contacting method. The larger radius of an illuminating laser beam spot causes an overillumination over the grooves of the wax cylinder and results in an echo problem for reproduced sounds. An adjustment error of the beam spot directly affects the direction of the reflected beam and produces an unrequired beam shift out of the detecting position.

4 Russian Wax Cylinder

A series of Russian wax cylinders were found at the museum of Petersburg and were considered to be the recordings of Chaliapin, a famous Russian Opera singer in the late 19th. But no notes concerned to the contents of the cylinder were found on the package. The dimensions of the cylinder are different from those of the standard type. The diameter is the same as the standard type, but the length is $\approx$6 inches, by which the playing time becomes longer. Figure 3 shows a photograph of the Russian wax cylinder and its package together with a standard wax cylinder on the right side. The phonographs which are available still now for this type of cylinders are scarcely found and a prototype of the laser-beam reflection method is not available for the Russian cylinders because of the mechanical restriction. Therefore, a new system using the laser-beam reflection method which is applicable to various types of cylinders has been constructed.

Fig. 3. Photograph of the Russian wax cylinder and its package. The small cylinder shown on the right side is a standard type

By using the new system, the reproduction of sounds from the Russian wax cylinder has been successfully performed. There are 5 parts of the recordings which seem to be the readings of poetry. Various methods, including the use of an incoherent light source, the very low pressure contacting technique and the other attempts, have been applied to increase the quality of reproduced sounds. The main purpose is to establish a non-destructive and non-contacting method for the faithful reproduction of sounds from old phonographic wax cylinders. Each method successfully reproduced the sounds from the wax cylinder and the reproduced sounds seem to be good.

5 Conclusion

A new system which is applicable to various types of wax cylinders has been successfully developed. By using this system, various methods for the reproduction of sounds from old wax cylinders are investigated. Especially, the reproduction of sounds from the Russian wax cylinder with the longer dimension was made with this system. Although each method has its advantages and disadvantages, these methods will contribute to the non-destructive screening test of cylinders which have remained in obscurity in the world.

References

1. T. Iwai, T. Asakura, T. Ifukube, and T. Kawashima (1986) Reproduction of sounds from old wax phonograph cylinders using the laser-beam reflection method. Appl. Opt. **25**, 597–604
2. J. Uozumi and T. Asakura (1988) Reproduction of sound from old disks by the laser diffraction method. Appl. Opt. **27**, 2671–2676
3. T. Nakamura, T. Ushizaka, J. Uozumi, and T. Asakura (1997) Optical reproduction of sounds from old phonographic wax cylinders. Proc. SPIE **3190**, 304–313

Optical Reproduction of Sounds from Negative Phonograph Cylinders

Jun Uozumi[1], Tsuyoshi Ushizaka[1], and Toshimitsu Asakura[2]

[1] Research Institute for Electronic Science, Hokkaido University
Sapporo, Hokkaido 060-0812, Japan
[2] Faculty of Engineering, Hokkai-Gakuen University
Sapporo, Hokkaido 064-0926, Japan

Abstract. The laser-beam reflection method that has been developed for the reproduction of phonograph wax cylinders was applied to metallic negative cylinders called galvanos, for which the traditional phonograph cannot be used. To this end, a compact optical head consisting of a laser diode, a position sensitive device, and illumination and detection optics was developed together with drive and control units. Some valuable sounds were reproduced successfully by this system.

1 Introduction

The invention of the phonograph by Thomas Edison has provided a first tool for recording sounds, and numbers of valuable recordings were made from the end of the 19th century to the beginning of the 20th. Many such recordings remain as phonograph wax cylinders, which can be reproduced by traditional phonographs as long as they do not have serious damages. To reproduce wax cylinders with serious damages, the laser-beam reflection method has been developed and applied successfully to various wax cylinders [1–4].

Recently, the authors come to know that many phonograph cylinders are preserved in Germany in the form of metallic negative cylinders in which folk music of various countries over the world are recorded. The negative cylinder was made by plating a wax cylinder with copper and, then, by melting down the original wax cylinder, being also referred to as *galvanos*. In this process, the sound is transferred into convex portions on its inside surface. Because of this, any stylus method is ineffective and those negatives have been left for a long time without investigations of the valuable sounds in them.

It is found, however, that the laser-beam reflection method developed for wax cylinders is effective also to the negatives. To apply the method to the negatives, some modifications were made and a new instrument was constructed. In this paper, we describe some features of the negatives and the developed *optical galvano player*.

2 Negative cylinders

Three negative cylinders were employed for the development of the instrument (Fig. 1(**a**)). They seem to be produced from typical wax cylinders and

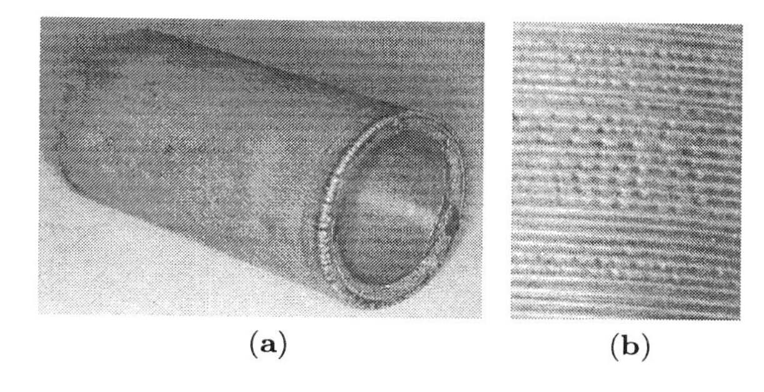

Fig. 1. (**a**) Negative cylinder. (**b**) Inside surface of the negative cylinder showing the height variation of the sound bank which is a replica of the sound groove of the original wax cylinder

have nearly the same dimension of 56, 54 and 110 mm in the outer and inner radii and length, respectively. The typical rotation rate was 144 and 160 rpm, and a sound of 2–3 min is encoded in the inside surface of each cylinder as a spiral *bank* with a pitch of 1/100 inch (Fig. 1(**b**)).

3 Principle of the reproduction

The principle of the laser-beam reflection method is shown schematically in Fig. 2. In case of the negative, the sound is encoded as a height variation of the surface of the sound bank. Suppose a laser beam incident on the center

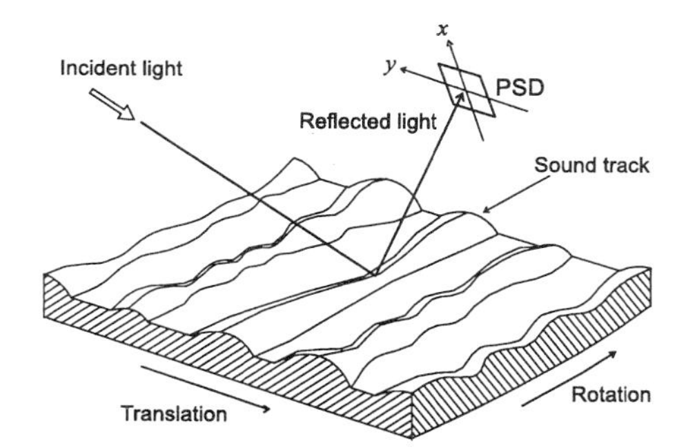

Fig. 2. Principle of the laser-beam reflection method. As the cylinder rotates, the reflection direction varies in the x-direction due to the height variation of the sound bank (sound track). The deviation of the illumination spot from the center of the sound bank gives the variation in y-direction

of the sound bank. The beam is reflected according to the reflection law and impinges a detection plane (xy-plane) perpendicular to the optical axis. With a rotation of the cylinder, the illuminating position in the xy-plane moves along the x-axis. In fact, the x-coordinate of this illuminating position is proportional to the tangent of the sound bank at the illumination spot and gives the sound signal. On the other hand, if the illumination spot deviates from the center of the bank, the reflected beam digresses in the y-direction, which gives a signal for the tracking error and can be used to compensate the deviation of the illuminating spot by adjusting the translation speed of the cylinder. These two signals in the x- and y-directions can be detected independently by setting a two-dimensional position sensitive device (PSD).

4 Optical system

To apply the above principle to the negative cylinder, a compact optical system that can be inserted into the cylinder is required. A schematical diagram and a photograph of the optical head we have constructed are shown in Figs. 3 and 4, respectively. A laser beam from a laser diode LD with the wavelength of 670 nm is guided by mirrors M_4, M_3, M_2 and M_1, and a beam splitter HM to a lens L_1, which converges the beam to the inside surface of the cylinder. The beam reflected from the cylinder surface goes back through the same part of the system up to HM and, then, is led by the lens L_2 to the PSD. A pinhole P is placed at the focal plane of L_2, where the image of the cylinder surface is formed, and passes only the image of the illuminating spot, thus rejecting effectively the stray light due to multiple scattering and reflection inside the cylinder.

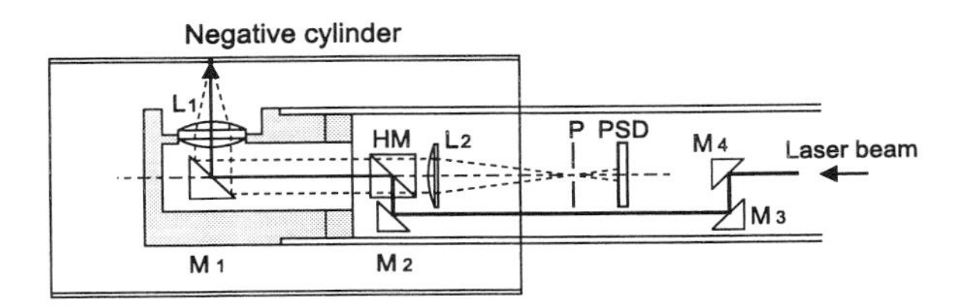

Fig. 3. Schematic diagram of the optical head. The lens L_1, mirrors M_1 and M_2, and beam splitter HM are common to the illumination and detection optics

Fig. 4. Photograph of the optical head with the cover removed for display. The mirror M_1 is not seen in this photograph. P indicates the position of the pinhole which is unmounted in this image

5 Reproduction system

A schematic diagram of the developed system which may be called *optical galvano player* is shown in Fig. 5. The system consists of the drive and control units, the optical head being mounted on the former. A photograph of the drive unit is shown in Fig. 6. This system has two tracking modes: the auto-tacking mode and the constant translation mode. The former utilize the tracking-error signal to keep the tracking of the laser beam along the sound bank, while in the latter the cylinder is translated with a constant speed determined by the rotation rate of the cylinder. The rotation rate is adjustable in the range of 140–160 rpm.

Using this system, we reproduced successfully the sounds from some negatives including performances of Japanese musical instruments, *shamisen* and

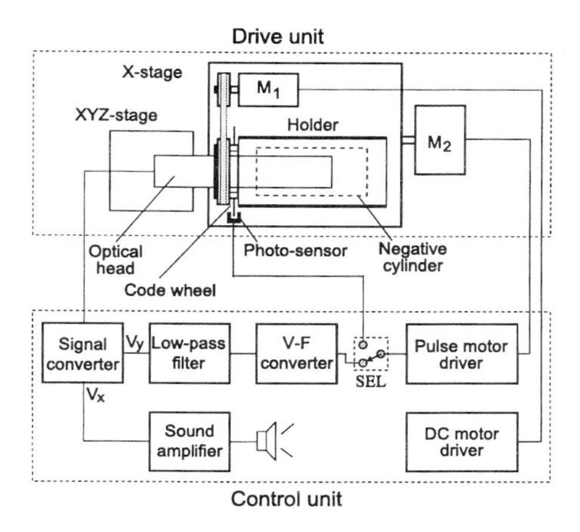

Fig. 5. Diagram of the optical galvano player. The system consists of the drive and control units. The DC motor M_1 rotates the holder while the pulse motor M_2 drives x-stage. The signal converter processes the output of the PSD and yields the sound and tracking error signals, V_x and V_y. V_x is amplified and drives the speaker, while V_y is sent to the pulse motor driver via the low-pass filter and V-F converter (auto-tacking mode). The pulse motor can also be controlled by the signal from the photo-sensor (constant translation mode)

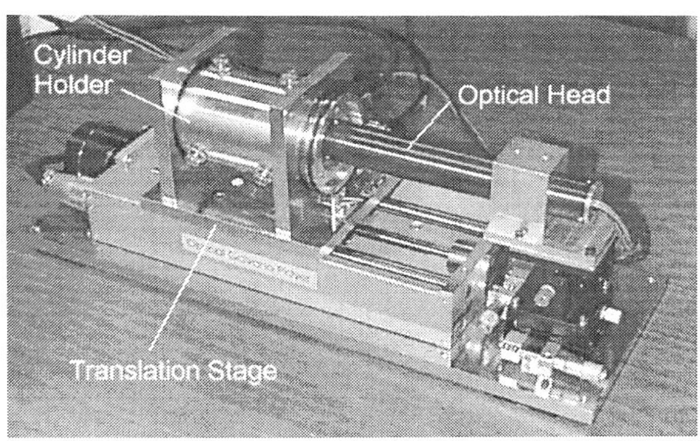

Fig. 6. Drive unit of the optical galvano player. The optical head is inserted into the negative held inside the cylinder holder which is mounted on the translation stage (x-stage)

koto, which were recorded in Berlin in 1901. The reproduced sounds have much better quality than existing wax cylinders. This is partly because wax cylinders currently available have been played many times and worn out considerably while the negatives preserve the initial quality of their original wax cylinders. The present instrument is expected to be used for revealing valuable sound information that has been left unknown for long time.

References

1. Iwai, T., Asakura, T., Ifukube, T., Kawashima, T. (1986) Reproduction of sound from old wax phonograph cylinders using the laser-beam reflection method. Appl. Opt. **25**, 597–604
2. Nakamura, T., Ushizaka, T., Uozumi, J., Asakura, T. (1997) Optical reproduction of sounds from old phonographic wax cylinders. Proc. SPIE **3190**, 304–313
3. Asakura, T., Uozumi, J., Iwai, T., Nakamura, T. (1998) Study on reproduction of sound from old wax phonograph cylinders using the laser. Proc. OWLS V
4. Nakamura, T., Asakura, T. (1998) Reproduction of sounds from an old Russian wax phonographic cylinder by various optical methods. Proc. OWLS V

Applications of Video Holographic Interferometry in Textile Conservation

P.S. Fowles*, A. Lord, Nguyen Huu Thanh†, and P.M.Boone†
The Conservation Centre, National Museums and Galleries on Merseyside, Whitechapel, Liverpool L1 6HZ, U.K. Tel. +44 (0)151 478 4904 Fax. +44 (0)151 478 4990
email: sculpture@nmgmc1.demon.co.uk
† University of Gent, Dept of Applied Mechanics, Sint-Peitersnieuwtraat 41, B-9000, Gent, Belgium
*Author to whom correspondence should be addressed.

Abstract. Video holographic interferometry is a powerful technique for the non-destructive testing of museum objects. In this paper, the possibility of applying this technique to the field of textile conservation is discussed. Results are presented from an initial study and indicate the direction in which further work might proceed.

1 Introduction

The Conservation Centre at the National Museums and Galleries on Merseyside, located in Liverpool, U.K, is a leading European centre for the conservation of works of art. The Centre deals with many aspects of art conservation, including objects constructed of organic and textile materials. The study of textiles has so far either concentrated on the properties of new threads and fabrics or on degraded textiles [1], but little work has been carried out on the interaction of new fibres with aged fibres which occurs in the process of conservation. Due to the relative fragility of textiles, much work is often needed to repair and strengthen works such as tapestries, fabrics, etc., against further deterioration. The techniques of such conservation are well established and often involve the stitching and mounting of artworks [2]. There is however little agreement as to the effect that using different types of thread and mount might have on a textile artwork. Many art objects are stored and displayed in areas in which the environment varies over time, often cyclically. The changes which can occur in variables such as temperature and humidity are known to effect the condition of textiles but no means has yet been found to quantify such changes. Currently, there is debate as to how closely the physical properties of the threads used in a repair should be matched to the original fabric. The choice of repair medium is likely to critically effect the response of the fabric following conservation. A method is required which will be able to provide information about an object as changes occur over time.
The application of Video Holographic Interferometry (VHI) to non-destructive testing (NDT) is well known and has already been applied to the study of cultural objects and details can be found elsewhere [3]. The aim of this paper is to detail an initial study into the suitability of the video holography interferometry (VHI) technique for use in the conservation of textiles. More particularly, whether VHI

can be applied to assess the properties of different stitching materials and techniques.

2 Experimental

Holography provides a method of recording an object which when used in conjunction with the technique of interferometry can provide information about the changes taking place over time. This is achieved by combining two holographic patterns to form an interferogram. The aim of VHI is to compare the light scattered by an object at one time with that at a later instant. If between the two recordings the object has undergone displacements due to changes in the local environment, fringes will appear on the reconstructed image of the interferogram. Such changes can occur naturally with changes in local environment but are more usually induced by altering temperature or humidity. Providing that the object studied is homogeneous then the fringes obtained will be regularly spaced representing an even change in displacement across the entire object. If any part of the object is inhomogeneous then this produces irregularities and discontinuities in the fringe pattern.
Applications of VHI have so far concentrated on relatively rigid items, such as ceramics and the technique has been found to be successful in determining cracks and showing areas of stress in a material. We are interested to see whether the technique of VHI can be used to evaluate different stitching threads and support fabrics commonly used in textile conservation. Stitching is one of the most fundamental methods used in textile conservation yet very little work has been done to understand the effects of stitching and its interaction with a textile. There are many variables: fibre type, stitch type, tension and method of application, size of needle, means of display, stitched to a board, hanging method (flat, vertical, at an angle), condition of object, etc.. In order to begin to assess the technique for textiles, a number of test samples were prepared using different method and materials to simulate the techniques of textile conservation. The samples were 100-150mm square fabrics of cotton or silk (naturally aged and new). The samples were either simply stitched using couching stitches, or in some cases supported on various backing, using silk, cotton or polyester threads. Where used, the support fabrics were silk, silk crepeline, nylon net and cotton. The samples were mounted in the VHI apparatus using a variety of methods, including fixing a single edge of the sample and allowing the sample to hang under its own weight and fixing the sample in place on a wooden board using adhesive.
The VHI system used was mounted on a damped optical bench in order to cut out external vibrations. The system is intended for use in the laboratory and is not mobile, although more portable systems do exist [4]. Such a portable system would be useful in conservation where it is frequently not possible to move art objects. The measurements are carried out in a room which may be blacked out to ensure there is no interference from external light sources. Heat could be applied to the sample using a number of techniques including a hot object and an IR lamp.

3 Results

A range of trials were carried out using the textile samples. It soon became apparent that studying samples of a complex and flexible material such as a textile is very different from objects previously studied by VHI. After a number of tests on the textile samples, we obtained results from which we were unable draw definite conclusions. Due to problems with mounting the sample and controlling the local environment it was not possible during this initial study to obtain reproducible results. Nevertheless, as a result of these initial tests, many useful lessons were learnt which can be applied to a further study of VHI for textiles. The series of VHI experiments indicated the following problems which need to be addressed:

- The textile fabrics and threads are very complex - The samples had in effect too many degrees of freedom to produce results that may be easily interpreted. Samples must initially be as simple as possible to see whether the technique of VHI can provide useful information.
- Sample mounting can be a problem - The flexibility of textiles means that mounting samples represents a considerable challenge and some form of mount which can be used to support and tension the samples is necessary. The question of mounting samples is one which is likely to prove most important to get right if VHI is to be a useful technique for such flexible materials.
- The common method for changing the environment by local heating are not very successful for textiles - Changing the environment by local heating proved difficult since textiles do not conduct heat well and are affected by the local convective airflow. Where used, the mounting boards for the samples also proved a problem because of their lack of heat conductivity. Ideally, better control of the local environment around the sample is required, probably in the form of an enclosed chamber. Changes in the temperature and humidity could then be controlled and the effects of local airflow eliminated.
- The resolution may not be high enough - A question which arose in the course of the experiments is what the resolution of the video camera / CCD combination must be to observe the small changes that might occur in some textiles. Currently, due to the other factors noted above, it is difficult to be confident as to whether or not there is a problem with current apparatus.
- Problems with colour - The colour and light absorption of the sample material can be a critical factor. Many fabrics are comprise threads of varied colours which form patterns or pictures. This can present a problem when using VHI, since the reflection of the monochromatic laser radiation which forms the hologram depends strongly on colour. In VHI, problems of this type are often overcome by applying a surface coating which reflects the laser light. It would be impractical to apply such a coating in the case of textiles since it would be virtually impossible to remove and therefore is not suitable as a conservation technique. The technique could therefore be colour dependent and require either a high power laser to ensure at least some reflection or the ability to use different laser sources.

4 Conclusions

The steep learning curve which resulted from the experiments enabled the initial problems of experimental parameters to be assessed. The two most pressing problems are those of effectively mounting specimens and then of controlling the local environment during experiments. The samples need initially to be very simple, this is most likely to take the form a single piece of textile with some simple stitching. For reproducibility, the first samples should probably be stitched by machine to try and ensure a uniformity in thread tension. These samples could then be mounted on a rigid frame. The local environment could probably be best controlled by means of a sealed, transparent container within which the temperature and humidity could be adjusted.
The experiments carried out on the textile samples provided very valuable insight into the problems which need to be assessed and overcome before the full potential of the VHI for textiles is known. Only when the initial problems have been solved will it be possible to begin the real evaluation of technique for textiles.

References

[1] S. Ellis, *A preliminary investigation of the tensile properties of yarns used for textile conservation*, Textile Conservation Newsletter Spring Supplement (1997)
[2] M. Brooks, D. Eastop, L. Hillyer, A. Lister, in *Lining and backing: the support of paintings, paper and textiles*, UKIC conference proceedings (UKIC, 1995)
[3] P.M. Boone and V.B. Markov, Examination of museum objects by means of video holography, Studies in Conservation, 40 (1995) 103-109
[4] Paoletti, D. and Schirripa Spagnolo, G., The potential of portable TV holography for examining frescos in situ, Studies in Conservation, 40 (1995) 127-132

Second-Harmonic Generation Imaging

R. Gauderon, P.B. Lukins and C.J.R. Sheppard

Department of Physical Optics, School of Physics, A28
University of Sydney, NSW 2006, Australia.
email:lukins@physics.usyd.edu.au

Abstract. Three-dimensional second-harmonic generation imaging using femtosecond pulses is demonstrated. Since second-harmonic generation results from coupling of the optical field to the second-order nonlinearity of the specimen, SHGI is a probe of local anisotropy and the distribution of molecular hyperpolarisability. The technique was implemented using a 3D scanning reflectance microscope together with a femtosecond Ti:sapphire laser. The efficiency of second-harmonic generation is enhanced by the use of short pulses (< 90 fs at 800 nm) since the detected average signal is inversely proportional to the laser pulse width. The method was characterised using crystalline nonlinear optical materials, 2D arrays and molecular crystals. We are currently applying the technique to biological specimens including isolated DNA, chromosomes and human skin.

1 Introduction

Second-harmonic generation (SHG) is an important phenomenon in nonlinear optics and has also been used in studies of materials [1,2] and biological systems [3,4]. Second-harmonic generation imaging (SHGI) has been achieved with CW illumination [5] and, more recently, two-dimensional SHG tomography of tissues using short pulses has been reported [6]. Here, we combine the techniques of true three-dimensional (3D) optical reflectance imaging and SHG produced using femtosecond laser pulses. In SHG, light of one frequency ω is converted by the nonlinear material into light at twice that frequency, 2ω. Using a simple plane-wave model, the harmonic power is [7]

$$P_{2\omega} \propto \frac{\omega^2}{A} L^2 d^2 P_\omega^2 \tag{1}$$

where A is the area of the focused spot, L is the interaction length, P_ω is the fundamental input power and d is the nonlinear optical coefficient.

Since the input pulse peak power is inversely proportional to the pulse width t, the SH peak power for a given average input power will be proportional to $1/\tau^2$ (Eq. 1) and therefore maximised by using ultrashort pulses. However, since we detect the integrated SH intensity, the signal will vary as $1/\tau$. In our experiments, a laser repetition rate of 82 MHz, a pulse width of $<$90 fs and an average power of 100 mW yields 14 kW pulse peak power which will generate sufficient SH radiation even for weakly nonlinear specimens. An unfortunate consequence is that $P_{2\omega}$ is also proportional to L^2 and so there is an effective reduction in the signal for high resolution SHGI.

SHG involves the interaction of light with the local nonlinear properties which are dependent on the molecular structure and hyperpolarisabilities. Therefore, the SH intensity, and hence the contrast, is a function of the molecular properties of the specimen and the orientation with respect to both the direction and the polarisation of the laser beam. Contrast in SHGI is guaranteed since the SH coefficients vary both between and within regions in a specimen and between different molecular systems which may occur within a given specimen. Normally, a phase-matching method is used to optimise the SH intensity. However, in 3D high resolution imaging, this method can only be exploited on a spatial scale smaller than the desired resolution which can be less than 1 μm.

2 Results

The experimental arrangement (Fig. 1) consists of a femtosecond Ti:sapphire laser (90 fs, 790 nm), a prism pair dispersion precompensation arrangement, and a stage scanning three-dimensional scanning reflectance microscope [8]. An adjustable group-velocity dispersion was used in order to compensate for the positive dispersion in the subsequent optics. The second-harmonic light at 395 nm generated by the specimen was separated from the fundamental beam, using dichroic and interference filters, and detected by a photodiode. 3D reconstructions were carried out using VoxelView. The observed square law dependence of SHG reduces the size of the effective point spread function by a factor of $\sqrt{2}$ giving a resolution of $\sim 0.61\lambda/(\sqrt{2}\,\mathrm{NA})$ or 0.43 μm which is close to the experimental value of 0.4 ± 0.1 μm.

A two-dimensional optical section taken for a sample comprising a LBO microcrystal suspension in agar (Fig. 2a) shows three microcrystals (A, B, C) of different shapes, sizes and orientations. There is an enhancement of contrast due to the short-range phase-matching and the orientation of crystal facets seen within the depth-of-field of the objective. The 3D SHG image (Fig. 2b) contains clear evidence of terraces on the crystal (A), isolated microcrystallites (B) and columnar stacking of microcrystals on the surface of the large LBO crystal (C). These columnar crystal structures are preferentially located at the edge steps on the terraced LBO surface.

Figure 3a shows a 2D SHG image of an aluminium grid deposited on a polished LBO crystal. The signal arising from the surface of the square aluminium regions is due to surface plasmon enhanced SHG, and its intensity depends on the electronic properties of the metal and the surface roughness. We interpret the variations in signal within these square regions as arising from variations in both surface roughness and the amount of oxide. This contrasts with the strong uniform SHG signal from the LBO substrate. The usefullness of this technique for depth profiling is again demonstrated by the 3D-SHG image (Fig. 3b). In this case, no signal is generated from the volume immediately below the aluminium regions because of the reflectivity of the metallic layer.

3 Conclusions

True 3D SHGI using femtosecond laser pulses was demonstrated and applied to a study of the 3D structure and second-order nonlinearity of LBO crystals and metal-coated grids. This reflectance mode SHGI configuration has great potential in the study of surfaces and interfaces as well as bulk materials. We have observed harmonic signals that arise from surface-enhanced, phase-matched and local SHG. In nonlinear microscopy using long (eg. infrared) wavelengths, the focal region is larger than is the case in confocal systems using visible light. In SHGI, optical sectioning is achieved by a soft-aperture effect arising from the spatial variation of the nonlinear interaction of the nonlinear properties of the specimen with the high power density at the focus region. Thus, the optical sectioning and enhanced resolution arise from the illuminating source, whereas in confocal, these properties arise from the use of a pinhole in the optical detection system. However, the spatial characteristics of the nonlinear processes in the focus region lead to overall resolutions similar to that of confocal.

We thank the Australian Research Council and the Swiss National Science Foundation.

References

1. Jentsch T., Jupner H. J., Ashworth S. H. and Elsaesser T. (1996) Second-order nonlinearities of polycrystalline molecular films studied on a 20 fs timescale. Opt. Lett. 21: 492-494
2. Vydra J. and Eich M. (1998) Mapping of the lateral polar orientational distribution in second-order nonlinear thin films by scanning second-harmonic microscopy. Appl. Phys. Lett. 72: 275-277
3. Bouevitch O., Lewis A., Pinevsky I., Wuskell J. P. and Loew L. M. (1993) Probing membrane potential with nonlinear optics. Biophys. J. 65: 672-679
4. Guo Y., Ho P. P., Tirksliunas A., Liu F. and Alfano R. R. (1996) Optical harmonic generation from animal tissues by the use of picosecond and femtosecond laser pulses. Appl. Opt. 35: 6810-6813
5. Gannaway J. and Sheppard C. J. R. (1978) Second-harmonic imaging in the scanning optical microscope Opt. and Quant. Elect. 10: 435-439
6. Guo Y., Ho P.P., Savage H., Harris D., Sacks P., Schantz S., Liu F., Zhadin N. and Alfano R.R. (1997) Second-harmonic tomography of tissues. Opt. Lett. 22: 1323-1325
7. Yariv A. (1967) Quantum electronics, Wiley, New York
8. Gauderon R., Lukins P.B. and Sheppard C.J.R. (1998) Three-dimensional second-harmonic generation imaging with femtosecond laser pulses. Opt. Lett. 23: 1209-1211

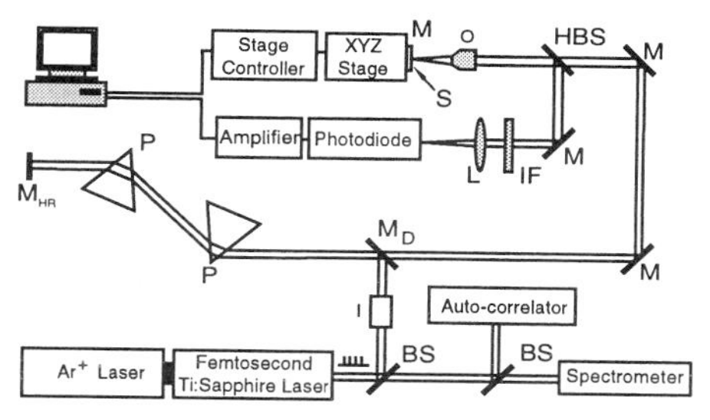

Fig. 1. Schematic diagram of the microscope. BS, beamsplitter; HBS, harmonic beamsplitter; I, isolator; IF, interference filter; L, lens; M, mirrors; O, NA = 0.8 objective; P, prism; S, specimen.

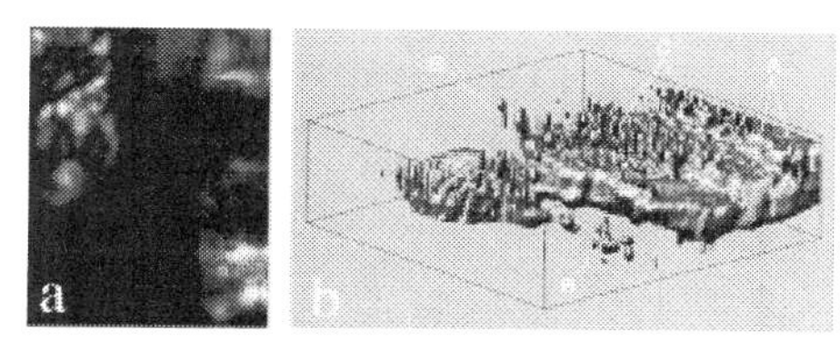

Fig. 2. SHG image (70 x 53 μm) of three LBO microcrystals embedded in agar (a) and a 3D SHG image (70 x 70 x 30 μm) of LBO crystal fragments (b) showing terraces on the crystal (A), isolated microcrystallites (B) and columnar stacking of microcrystals on the LBO surface (C).

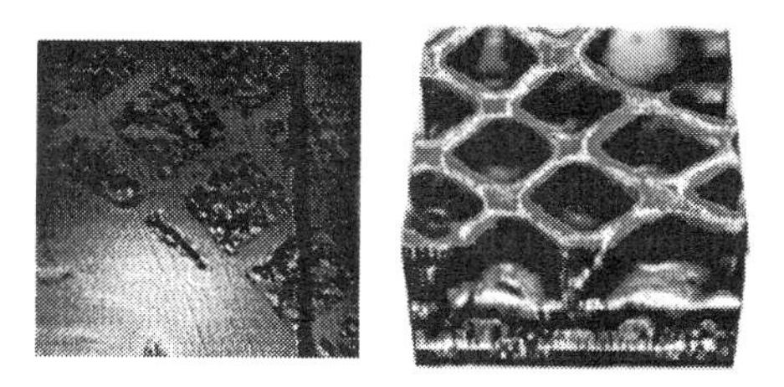

Fig. 3. 2D SHG image (70 x 70 μm) of the surface of an LBO crystal coated with a 240 nm thick aluminium grid (a) and a 3D SHG image (70 x 70 x 20 μm) of the same specimen.

Non-Steady-State Photocurrents in Indium-Oxide Thin-Film Holographic Recorders

S. Mailis[1], A. Ikiades, and N.A. Vainos
Foundation for Research and Technology-Hellas (FORTH), Institute of Electronic Structure and Laser (IESL), P.O. Box 1527, Heraklion 711 10, Greece E-mail: vainos@iesl.forth.gr
V.V. Kulikov and I.A. Sokolov
194021 A.F. Ioffe Physico-Technical Institute, Russian Academy of Sciences, Politekhnicheskaya 26, St.-Petersburg, Russia, E-mail: isok@math.ipme.ru
[1]Present adress: Optoelectronics Research Centre, University of Southampton, Southampton, SO17 1BJ, UK

Abstract: Non-steady-state photocurrents developing in non-stoichiometric indium-oxide thin films by illumination in the ultraviolet region of the spectrum (λ=325nm) have been observed. Simultaneous monitoring of the non-steady-state photo-electromotive force and holographic diffraction effects in these thin films, provides a powerful tools for the investigation of the photoinduced processes involved, aiming to the optimization of the material for information storage and processing applications. The observed effects may also lead significant advances towards the development of high-sensitivity compact interferometric sensors for the on-line monitoring of mechanical deformations and, of particular relevance here, structural changes induced in works-of-art and antiquities during laser cleaning or environmental cycling.

1 Introduction

Holographic recording is a powerful technique for measuring the fundamental parameters of photosensitive materials. In photorefractive crystals these methods permit the determination of the sign of the dominating photocarriers, their average transport lengths, the Debye screening length, the Maxwell relaxation time, the photogalvanic tensor components and other. This technique involves detection of light diffracted off the recorded hologram and additional independent measurements of, for example, the electro-optic tensor of the photorefractive crystal are required.

The above arguments cannot, however, be applied in the characterization of centrocymmetric, non-photorefractive, photoconductive materials and alternative effects, such as the non-steady-state photocurrent generation may be used [1]. In the case of photorefractives, a refractive index hologram, associated with a space-charge distribution, is recorded in a conventional manner. The photo-emf effect manifests itself as an alternating electric current J(t) observed through the short-circuited sample of the photoconductor, which is illuminated by a vibrating interference pattern. This pattern is formed between two waves one, of which, is phase modulated at frequency ω (Fig.1). The amplitude of the space charge is monitored by means of an electric current developing through the sample along the **K**-vector of a moving grating. This method does not involve light diffraction from the hologram and can therefore be used directly for centrosymmetric and even amorphous photoconductors.

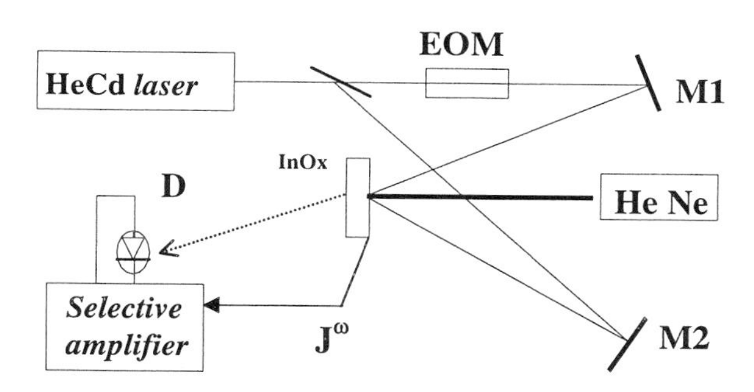

Fig.1. Schematic of the experimental setup used for holographic recording and non-steady-state photocurrent excitation (D- photodiode, M1, M2 -mirrors, EOM - phase modulator).

In the present contribution we apply this method to characterize indium-oxide thin films and present preliminary observations. This material exhibits interesting electrical and optical properties and a high potential for technological applications. Indium-oxide thin films grown by dc-magnetron or laser sputtering may have their conductivity modified upon illumination with ultraviolet radiation (photon energy>3.5eV), resulting in a variable electrical state of the material, from a resistive to a purely conductive. Conductivity states can also be cycled, with no fatigue, through an oxidizing exposure to ozone atmosphere and reduction by ultraviolet illumination. Holographic recording has also been recently demonstrated [2] aiming to information storage and processing applications. The development of thin-film materials by pulsed-laser deposition allows further tuning of the structural properties [3]. In fact, the behavior of the material, which may be classified in the mesoscopic phase, depends heavily on the size of micro- and nano-crystallites embedded in an amorphous matrix. Even though, the origin of the photoinduced absorption and refractive index changes is not yet fully understood, it is considered that they are due to reversible photoinduced reactions and charge transfer in the crystallite boundaries, which are most probably relating to reconfiguration of oxygen-bonds. Such arguments are also corroborated by recent results of holographic performance obtained at various environments [4].

2 Experiments and Discussion

A typical holographic setup, shown in Fig.1, was used for the realization of holographic recording and non-steady-state photocurrent measurements. Simultaneous measurements of the non-steady-state photocurrents and holographic recording in indium-oxide thin films at the He-Cd laser wavelength of 325 nm have been performed. Two HeCd laser beams formed a holographic grating in the thin film. One of the beams was phase modulated at frequency ω. A Helium-Neon laser beam was monitoring the holographic diffraction efficiency, by detecting the two scattered beams of equal intensity, corresponding to the ±1 diffraction orders of the recorded sinusoidal grating. The intensity of the beams was continuously monitored.

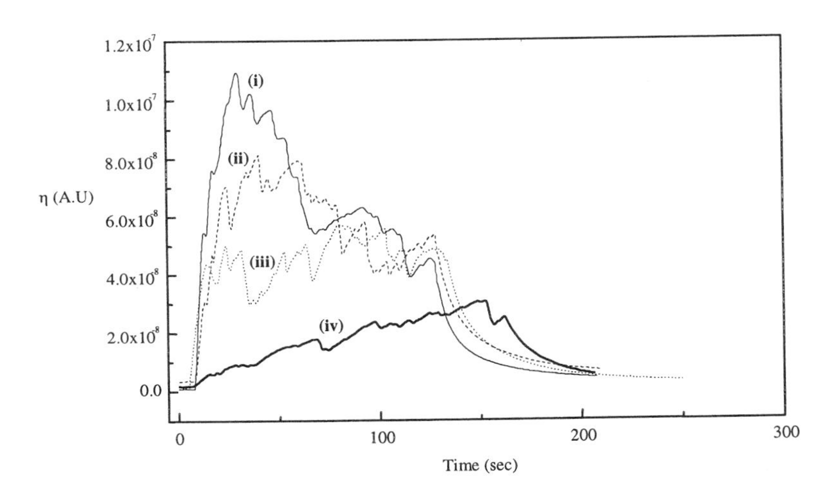

Fig. 2. Typical hologram development and decay at various laser intensities. Recording wavelength is 325nm. Curves are corresponding to (i): 58%, (ii): 64%, (iii): 68%, (iv): 83% reduction of total intensity.

The typical hologram recording and decay for various total intensities at 325nm is depicted in Fig. 2. Starting intensity is in the range of $70 mW/cm^2$ and the modulation index is about 70%. The diffraction efficiency of the recorded holograms is of the order of 10^{-4}. It is noted here, that conductivity changes are also observed simultaneously and the fast rise of the diffraction efficiency is highly correlated with those.

The non-steady-state photocurrent was also detected by applying lock-in techniques. Several InO_x thin films grown under different conditions have been investigated. A pronounced photocurrent flowing through them is always observed.

Figure 3 depicts the typical temporal behavior of the signal measured for different spatial frequencies of the interference pattern. As it is observed in the figure, at least two significant regions of photocurrent dynamics are present. A fast rise to the maximum value is first observed, followed by a slow decay, with decay time constants for reaching steady-state amplitude of the order of 50-100 seconds. Such effects may be correlated with respective conductivity changes.

In further experiments, non-steady-state photocurrent measurements were performed in the presence of the HeNe laser beam. The power level of the red light was about 5mW. A reduction of the photocurrent signal by a factor of 0.3 was observed. Detailed experiments are under way for understanding the origin of this effect. At a first sight, the existence of intermediate relatively shallow levels, which are probably ionized in the red, may be assumed.

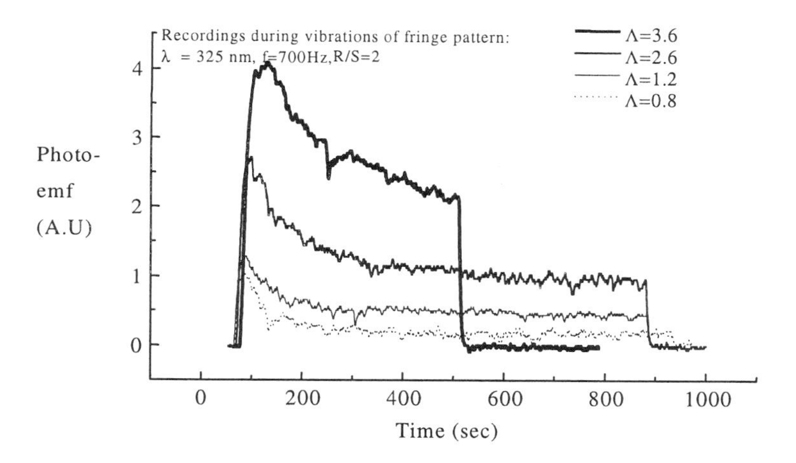

Fig. 3. Photo-emf current observed at various holographic fringe spacings, Λ(μm).

3 Conclusions

The observation of non-steady-state photocurrent signals in indium-oxide thin films may aid the understanding of holographic recording in this and the like materials. These materials, which may be classified in the mesoscopic phase, would be proved useful in a variety of optics and optoelectronics applications including holographic storage and chemical sensing. Furthermore, their nonlinear optical properties are yet to be explored and may lead to the exploitation of their optical switching behavior.

In this work, preliminay results are presented indicating the large potential of the methods. The characteristic decay time constants of photocurrent and their relation to the thin film photoconductivity values and respective hologram diffraction efficiency is under investigation. The influence of red light illumination on the photocurrent signal indicates the existence of a series of intermediate relatively shallow levels not observed before. The potential for developing low-cost, high-sensitivity interferometric vibration sensors based on this material and respective effects should be noted. These aspects should also be examined in conjunction to the potential on-growth tailoring of the properties of the material, which enables it to comply with specific application requirements. The latter would be imposed in the development of information storage devices or the construction of low-cost compact optical sensors appropriate for the structural monitoring of sensitive cultural heritage.

4 Acknowledgements

This study was supported by the EU in the frame Ultraviolet Laser Facility at FORTH and the NATO Science Programme, through Linkage Grant HTECH.LG 970314. The collaboration with C. Moschovis, G. Kiriakidis and H. Fritzsche is gratefully acknowledged.

References

1. I.A. Sokolov and S. I. Stepanov, J. Opt. Soc. Am. **B 10,** 1483 (1993)
2. S. Mailis, L. Boutsikaris, N.A. Vainos, C. Xirouhaki, G. Vasiliou, N. Garawal, G. Kyriakidis and H. Fritzsche., Appl. Phys. Lett. **69**, 2459 (1996).
3. C. Grivas, D.S. Gill, S. Mailis, L. Boutsikaris and N. A. Vainos, Appl. Phys. **A 66,** 201(1998)
4. K. Moschovis, E. Gagaoudakis, E. Chatzitheodoridis, G Kiriakidis, S. Mailis, E. Tzamali,
5. N.A. Vainos and H Fritzsche, Appl. Phys. **A 66**, (6), 651-4 (1998)

Bending Loss in Mid-Infrared Waveguides and Fibers

A. A. Serafetinides[a], K. R. Rickwood[a], E. T. Fabrikesi[a], G. Chourdakis[a], N.Anastassopoulou[a], Y. Matsuura[b], Yi-Wei Shi[b], M. Miyagi[b] and N. Croitoru[c]
[a]Department of Physics, National Technical Univ. of Athens, 15 780, Athens, Greece
[b]Department of Electrical Communications, Graduate School of Engineering, Tohoku University, Sendai 980-8579, Japan
[c]Department of Electrical Engineering - Physical Electronics, Faculty of Engineering, Tel Aviv University, Tel Aviv 69978, Israel

Abstract. A pulsed Er:YAG laser and a preionised pulsed HF laser were used for the evaluation of the transmission properties of fluoride glass fibers as well as a cyclic olefin polymer-coated silver (COP/Ag) hollow glass and a silica hollow glass waveguide. Both lasers provided an output of 100 mJ/pulse, with a repetition rate of 1Hz at 3μm wavelength. The pulse duration of the two lasers was 200 ns for the HF and 80μs for the Er:YAG. The evaluation of the fiber and waveguide properties include both straight and bend loss characteristics.

Keywords: waveguides, optical fibers, HF laser, Er:YAG laser, mid-infrared lasers, laser radiation transmission.

1. Introduction

In the last few years there has been an increasing interest in lasers emitting at 3.0 μm, as this wavelength is highly absorbed by water [1]. Therefore, the Er:YAG laser and the HF laser, both emitting in the mid-infrared, are ideal for medical applications, such as hard and soft tissue treatment [2].

To enable application of these mid - infrared lasers in medical applications, flexible fibers [3] or waveguides [4], which have low attenuation, can deliver high power or high energy and can stand short duration laser pulses, are necessary. The research on the development of delivery systems for the 3.0 μm wavelength has been progressed a lot in the last few years with the production of sapphire fibers and dielectric-coated hollow waveguides.

One issue of great importance is the transmission of the straight fibers or waveguides and the loss induced when these are bent. In this work the transmission bend losses of fluoride glass fibers as well as dielectric coated and silica glass waveguides are examined. These results can be useful for the development of design information for a 3μm wavelength laser delivery systems used in medical applications.

2. Materials and Methods

The experimental arrangement used to obtain practical values of transmission and bending losses is shown in previous work [5]. The lasers used were an HF with a pulse energy of 100 mJ and pulse duration of 200 μm and an Er:YAG with a pulse energy of 100 mJ and pulse duration of 80 μm laser both emitting in the mid-infrared. The two lasers were designed and manufactured in the laboratory of NTUA ,Physics Dep., Laser Group.

The fibers used were Fluoride glass multimode step index fibers with 600, 450 and 350 μm core diameter, 1.6 m length and 0.2 N.A. The waveguides were of two types: i) Cyclic Olefin Polymer-coated Silver (COP/Ag) hollow glass fabricated in Japan and ii) Silica hollow glass waveguide fabricated in Israel. For the evaluation of the bend loss relative to bend radius the fibers (waveguides) were bent by curving them along disks with different radii starting from 30 cm down to 5cm.

3. Results

The bend loss was measured for the fluoride glass fibers. In Fig.1 bend loss for the three fibers of different core diameter as a function of the bend radius for the HF laser is shown. In Fig. 2 the same results are shown for the Er:YAG laser.

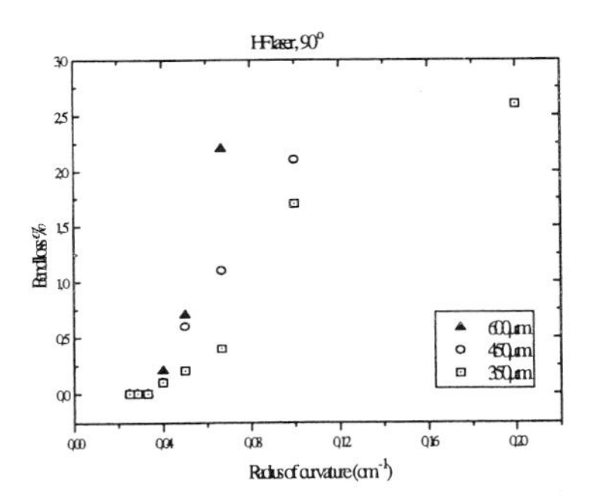

Fig 1: Bend loss of HF laser as a function of bend radius of curvature at 90° for the fluoride glass fibers with 600, 450 and 350 μm core diameters.

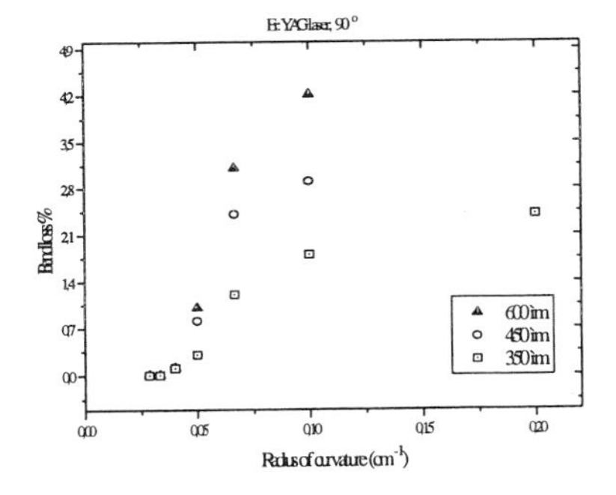

Fig 2: Bend loss of Er:YAG laser as a function of bend radius of curvature at 90° for the fluoride glass fibers with 600, 450 and 350 μm core diameters.

The problem of bending losses from optical fibers has been extensively studied by a number of authors. A model proposed by Schevchenko [6], shows that the power loss attenuation coefficient of a guided mode propagating at an angle θ to the axis for a step index fiber is given by

$$a = 2n_1 k\left(\theta_c^2 - \theta^2\right)\exp\left[-\frac{2}{3}n_1 kR\left(\theta_c^2 - \theta^2 - \frac{2r}{R}\right)\right]^{\frac{2}{3}} \quad (1)$$

where n_1 is the refractive index of the fiber core, $k = \frac{2\pi}{\lambda}$,the propagation constant, θ_c is the critical angle of the fiber, r is the radius of the core and R is the radius of the bend. For higher-order modes of the the fiber which travel near critical angle θ_c, it is evident that for increased fiber core radius and decreased bend radius, the power loss attenuation coefficient increases resulting in higher bend loss.

Figures 1 and 2 clearly show that as the radius of curvature is increased above a certain limit the bend loss is also increased. In Fig.3 bend loss as a function of core diameter for both HF and Er:YAG laser is presented. These data correspond to a bend radius of R=10 cm. An exponential best fit performed for the case of the HF laser for a bend radius of 10 cm shows that the bend loss increases as the fiber core diameter increases following the function B.L. = 0.14 + 0.3 exp(r /124).

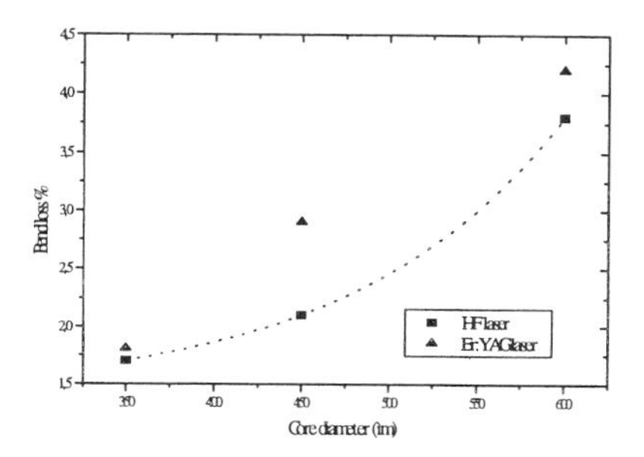

Figure 3 : Bend loss as a function of core diameter of both HF and Er:YAG laser for a fluoride glass fiber at R=10 cm bend radius.

Bend loss as a function of the bending radius of curvature was measured for the waveguides. The waveguides were bent at 90° and at 180°, with a bending radius varying from 30 cm to 10 cm. Results are shown in the following figure 4.

As seen from the graphs, the bend loss increases as the bend radius of curvature increases. This result is in agreement with the theory that describes losses in hollow waveguides [7], according to which the attenuation coefficient a is proportionally reverse to the waveguide bending radius R, i.e.

$$a \propto 1 / R \qquad (2)$$

Bending at 90° angle results in higher transmission values than bending at 180°. This result is expected, as at 180° the propagating distance around a bend is two times the distance for the 90° bending angle.

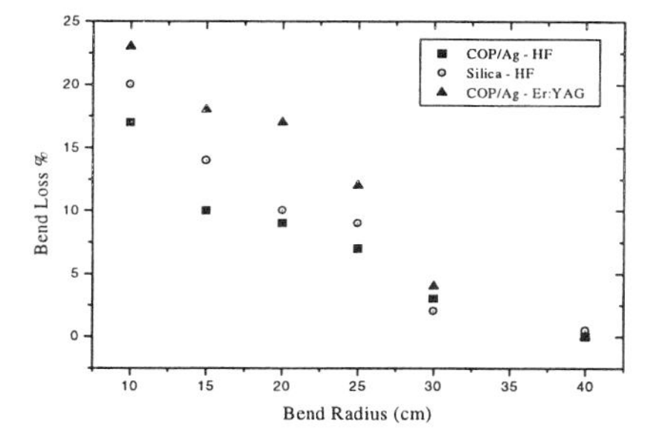

Figure 4 : Bend loss of HF and Er:YAG laser as a function of bend radius for the COP/Ag and the Silica hollow glass waveguide

4. Discussion

The transmission characteristics of fluoride glass fibers and two different types of waveguides have been investigated using an HF and an Er:YAG laser. The best transmission achieved was 94% for the fluoride fibers and 97% for the COP/Ag hollow waveguide. These values are in accordance with the specifications provided by the suppliers of the fibers and waveguides. Higher values of transmission were obtained for fibers with larger core diameter. The transmission of the HF laser is higher compared to Er:YAG. Since the two lasers have similar operation characteristics (wavelength, pulse energy and repetition rate) that difference can only be attributed to the different pulse duration and possibly the different beam profile. Bend loss depends on bend radius of curvature and fiber parameters such as fiber core radius. As the bend radius of curvature increases the loss increases. As the core radius of the fibers increase the loss increases also following an exponential growth. The bend loss of the waveguides (up to 20%) is significantly higher compared with the bend loss of the fibers (up to 3.5%). These results suggests that when bend loss has to be kept in a minimum bend have to be restricted to the largest possible radii. Additionally a fiber with smallest core diameter consistent with power density has to be employed.

References

1. J.S. Nelson, A. Orenstein, L.-H. Liaw and M.W. Berns, "Mid-infrared Er:YAG laser ablation", *Lasers Surg. Med.*, Vol. 9, pp. 362-374, 1989.

2. H.A. Widgor, J.I. Walsh, J.D.B. Featherstone, S.R. Visuri, D. Fried and J.L. Waldrogel, "Lasers in dentistry", *Lasers Surg. Med.*, Vol.16, pp. 103-133, 1995.

3. G.M. Clarke, O. Chadwick, R.K. Nubling and J.A. Harrington, "Sapphire fibers for the three micron delivery systems", *Proc. Soc. Photo-Opt. Instrum. Eng., S.P.I.E.*, Vol. 2396, pp. 54-59, 1995.

4. A. Hongo, M. Miyagi, Y. Kato, M. Suzumura, S. Kubota, Y. Wang and T. Shimomura, "Fabrication of dielectric coated silver hollow glass waveguides for the infrared by liquid-flow coating method", *Proc. Bio-medical Opt. Europe 95, S.P.I.E.*, Vol. 2677, pp. 55-63, 1995

5. A.Serafetididis, E. Fabrikesi, G. Chourdakis, N. Anastassopoulou, "Pulsed HF nad Er:YAG laser radition transmission through sapphire and fluoride glass fibers", Proceedings BiOS '98, Stockholm. To be published.

6.V.V. Shevchenco, *Izv. Vuz.Radiofiz.* 14,768, (1971).

7. T. Abel, J. Hirsch and J. A. Harrington, "Hollow glass waveguide for broadband infrared transmission", *Optics Letters*. Vol. 19, No 14, July 15,1994.

Twin-image Elimination in Digital In-line Holography

Songcan Lai*, Björn Kemper and Gert von Bally
Laboratory of Biophysics, Institute of Experimental Audiology, University of Münster
Robert-Koch-Str. 45, D-48129 Münster, F. R. Germany
E-mail: biophys@gabor.uni-muenster.de
* On leave from Zhengzhou University of Technology, P. R. China

Abstract. A new method for twin-image elimination based on the principle of off-axis reconstruction of in-line holograms is presented in this paper. Simulated and experimental results show that undisturbed reconstructed real images can be obtained by this method.

1 Introduction

The main disadvantage of in-line holography is the well-known inherent twin-image problem. Efforts have been made to remove the disturbance of twin-images[1]-[4]. But there are still drawbacks and limitations in the previous methods. In this paper, a new technique for twin-image elimination in digital in-line holography is proposed. This method is based on the principle of off-axis reconstruction of in-line holograms which is a new approach in the field of optical holography. Principle analysis and computer simulation, as well as experimental results showed that it is a simple but efficient method to eliminate the twin-images of digital in-line holograms. It is expected that this method is promising for fluid-field measurement and soft X-ray holography that are applicable to medical investigations.

2 Principle

An in-line hologram is formed by the interference of a plane wave and the diffracted wave of an object which is illuminated by the plane wave. The recording geometry of in-line holograms, as shown in Fig. 1, determines that both the size of the object and the recorded object wave cannot exceed the spatial boundary of the reference wave, which means the reference wave and the object wave propagate almost in the same direction. This is the origin of the term "in-line holography".

However, if we carefully analyze the recording geometry of in-line holograms, as shown in Fig. 2, we will find that the in-line hologram is not absolutely in-line between reference wave and object wave. Let us take part ***AB*** of the hologram plane as a sub-hologram and consider the reconstruction of point a and b of the object. The sub-hologram is still an in-line hologram for point a but at the same time, it can be regarded as an off-axis hologram for point b. Therefore, if we only illuminate the sub-hologram with a narrow reference beam, the reconstructed image of point a will be a real image a′ disturbed by the twin-image beam, while the reconstructed image of point b will be an undisturbed real image b′. The direction of the twin-image beam which diverges as from point b is different from that of the real image beam which converges to point b′. Thus, the real image is separated from the twin-image beam. For any object point, we can always find an area as an off-axis sub-hologram of the point and reconstruct it without twin-image disturbance. Therefore, an undisturbed real image of the whole object can be obtained by scanning and reconstructing the whole holograms. This idea is referred to as off-axis reconstruction of in-line holograms. In practice, we can reconstruct the whole hologram area by area and obtain the corresponding reconstructed images. Each image is composed partly of an undisturbed real image and partly of a disturbed real image. Selecting the undisturbed part of each image and piecing these parts together will result in a complete real image of the original object without any twin-image disturbance.

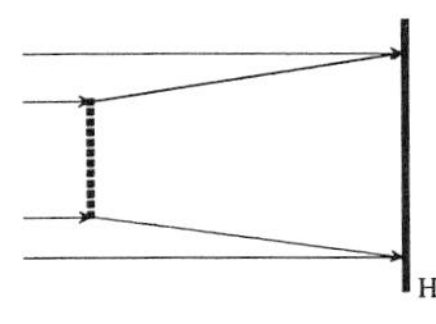

Fig. 1 Formation of in-line holograms (for details see text).

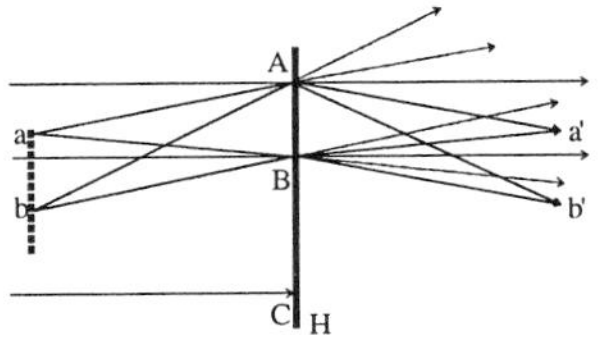

Fig. 2 Principle of off-axis reconstruction of in-line holograms (for details see text).

3 Computer simulation results

Fig. 3-(a) is a simulated object of a dot array. Fig. 3-(b) is the conventional reconstructed image with twin-image. By using the method presented in this paper, we obtained the undisturbed real image of the central part and outer parts of the object that are shown in Fig. 3-(c), (d), (e), (f) and (g) respectively. Each real image was reconstructed by blocking the corresponding area and illuminated the remained area of the hologram. Fig. 3-(h) indicates the reconstructed image of the whole dot array by superimposing Fig. 3-(c), (d), (e), (f) and (g). We can see that the reconstructed images are real images without the disturbance of twin-images.

4 Experimental results

An in-line hologram of a pair of hairs with a diameter of 60 μm was recorded on a CCD sensor array. Fig. 4-(a) shows the digitally reconstructed image with twin-image. It is not possible to determine which line is the object. Fig. 4-(b) is the reconstructed image by blocking the central strip of the hologram. The hairs are clearly seen in the middle area.

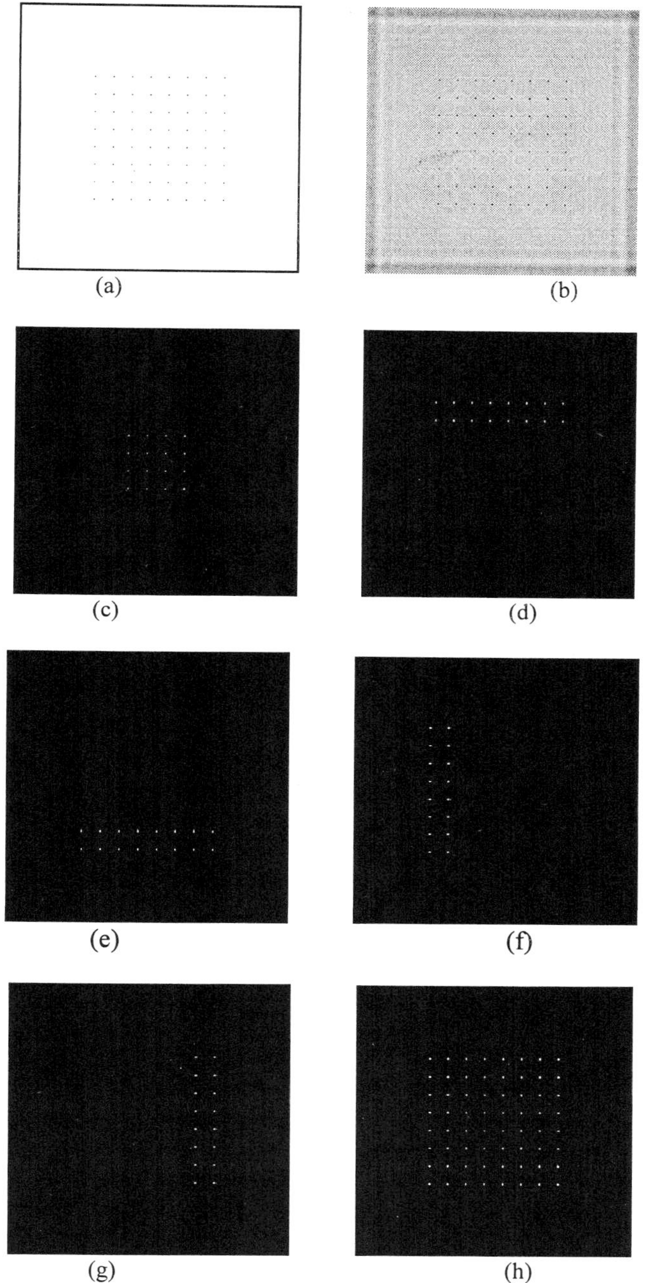

Fig. 3 Computer simulation: (a) simulated dot array; (b) reconstructed image with twin-image of the dot array; (c), (d), (e), (f), and (g) reconstructed image without twin-image of central, upper, lower, left and right part respectively; and (h) superimposed reconstructed image.

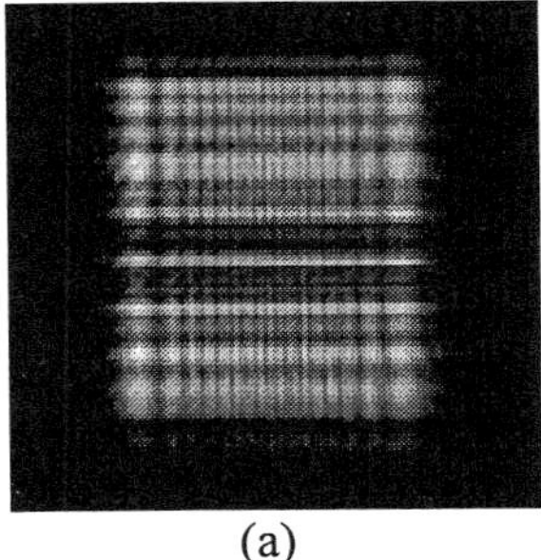
(a)

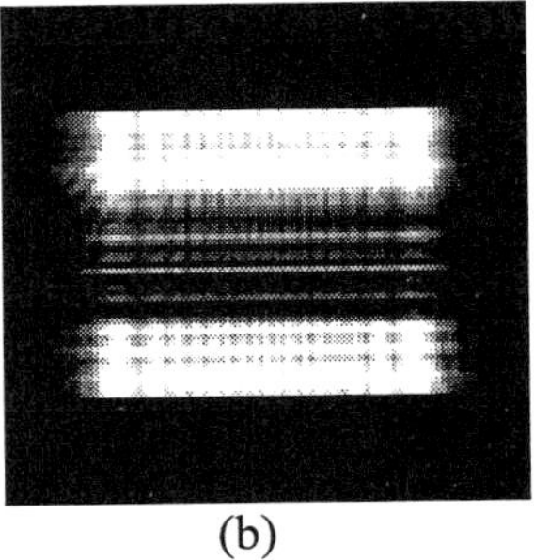
(b)

Fig. 4 Experimental results: (a) reconstructed image of hairs with twin-image; (b) reconstructed image of hairs without twin-image.

5 Conclusions

A new method to eliminate the twin-image in in-line holography is presented. This method is based on the principle of off-axis reconstruction of in-line holograms. It relies on the holographic reconstruction itself, rather than other physical or numerical ideas such as phase retrieval and linear filtering. Therefore, this simple but efficient method has promising applications to traditional holography used in particle-field analysis, fluid-field measurement and aerosol measurement as well as to digital holography with CCD detectors, X-ray holography and electron holography in which in-line holographic recording and reconstruction are necessary. A limitation of this method is that because only part of the hologram is used to reconstruct, the resolution of the reconstructed real image is degraded compared to other twin-image elimination methods in which holograms are processed in total frame.

Acknowledgments

The authors wish to express their thanks for financial support from the Alexander von Humboldt Foundation of the F. R. of Germany.

References

1 L. Bragg and G. L. Rogers, Nature, **167**, 190 (1951).
2 T. Xiao, H. Xu, Y. Zhang, J. Chen, and Z. Xu, J. Modern Optics, **45**, 343 (1998).
3 A. Nugent, Opt. Commun., **78**, 293-299 (1991).
4 Liu and P. D. Scott, J. Opt. Soc. Am. A, **4**, 159 (1987).

Examination of Jade Conditions by Laser Technology

Maung H. Khin1, Yasufumi Emori1, Hitoshi Ohzu 2
1 Advanced Research Institute for Science and Engineering, Waseda University
2 Dept. of Applied Physics, Waseda University
3-4-1, Okubo, Shinjuku-ku, Tokyo 169-0072, JAPAN
e-mail: khin@mn.waseda.ac.jp

Abstract. A new examination apparatus, which consists of the two kinds of laser illumination system and the detecting system using by high-grade CCD camera, was developed for the examination of Jade stone conditions. It can be obtained that the images of reflecting and scattering of various jade stones which have different light distribution. It can be examined that the degree of clarity, the different contrast of images by noise in jade stone.

1 Introduction

1.1 Nature of Jade

The term "Jade" ($Na(Al,Fe^{3+})Si_2O_6$) often refers to two different jade categories: nephrite and jadeite. In fact, there was a kind of confusion to the term "jade" before Han dynasty. According to the famous dictionary written in Han explained that jade was "a kind of beautiful stones". The "jades" before Han actually contains agate, turquoise, crystal, nephrite and other local stones as long as it reveals beautiful nature. Jadeite is called "hard jade" and nephrite refers to "soft jade" in modern China by its level of hardness. Furthermore, jadeite appeared around mid-seventeenth century and nephrite already had been used in early Neolithic period in China. The most important jadeite deposits are in upper Myanmar (Burma).

The nature of jade stones contains two important aspects: quality and character.

1.2 Jade Quality

The standard factors for jade quality include visual sense and tactile sense. The visual sense of a good jade stone possesses the characteristics of gentleness, luster and cleanness. The tactile sense of a good jade stone possesses the characteristics of fine and smooth, soft and mild by touching, and is felt extremely heavy at hand.

1.3 Jade Character

On the other hand, the standard factors for jade character include the infrastructure and the color of the jade stone. The harder and solider the jades, the better the infrastructure. The colors of the jade stones have three big categories: white jades, yellow jades and green jades. It should be noticed that, generally speaking, yellow jades do not possess the pure yellow color but rather it has the color of greenish yellow.

Fig. 1 Jade production area in Myanmar (Burma)

Fig.2 Rough jade

It is very difficult to know the good conditions of inside from appearance of jade rough stone. We report herein the development of a new examination apparatus for jade conditions by laser technology. The purpose affects the preservation and promotion of the local environment and the natural resources. It is useful for the estimation of the clarity and cleanness from appearance of various jade stone.

2 Experiments and Samples

The new examination apparatus is composed of two components; light source unit which consists of the two kinds of laser illumination system, and image detecting

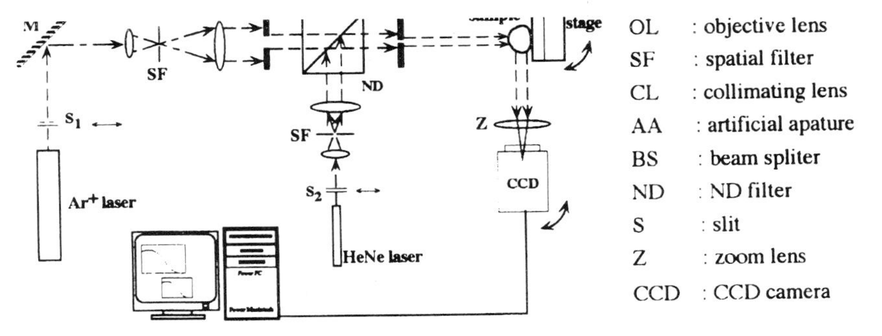

Fig.3 Experimental apparatus

unit using by high-grade CCD camera as shown in Fig. 3. The light source unit has two types of laser; green color Ar^+ laser (514.8nm) and red color He-Ne laser (632.8nm), which illuminate to the sample by alternately. The image captured by the high-grade CCD camera. The detected images are converted to an Apple Power Macintosh computer on which the image processing and analysis 2 are performed using the IP Lab Spectrum Program (written by Signal Analytic Corporation, USA). .

The samples, three kinds of different color of jadeite, which are founded at the Hpacant area in Myanmar (Burma) are selected. The crystallization structures of those are conformed by x-ray analyzer.

3 Results and Discussion

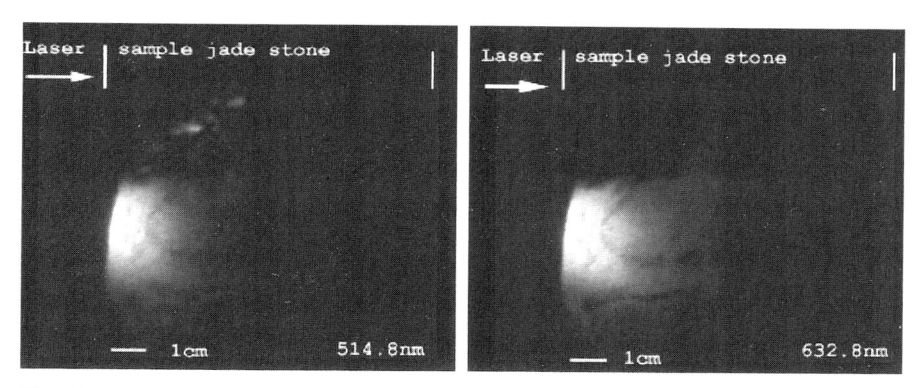

Fig. 4 Depth of the light transmission by two different wavelengths (a) 514.8nm (b) 632.8nm through the jadeite stone sample

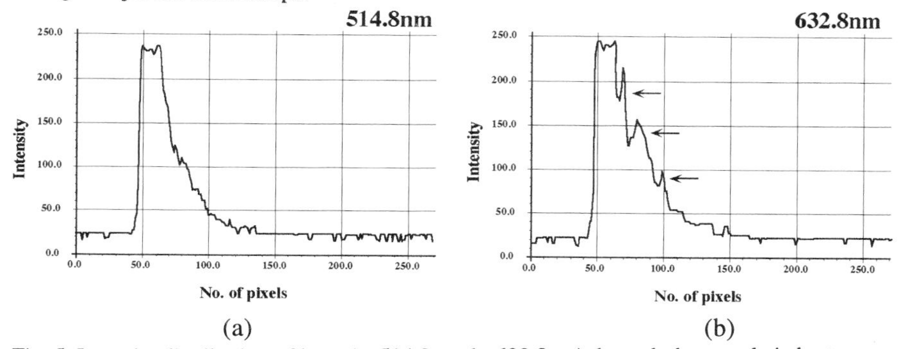

Fig. 5. Intensity distribution of laser (a: 514.8nm, b: 632.8nm) through the sample jade stone

Figure 4 shows the depth of the transmission light by two kinds of different wavelength (a; 514.8nm, b; 632.8nm) through the same sample jadeite stone. The arrow shows the incident direction of laser and, the scale bar, bottom, shows the size of sample. The stimulating lasers are incident perpendicularly on the window of rough stone and cross-sectional of the sample is illuminated by the transmitted laser, and lights are induced inside of sample jadeite stone. The results understand that the degree of transmission, in other words, the degree of absorption is different by the various wavelength when the same laser intensity. In this case, it is considered that the He-Ne laser is convenient to use for the examination of the clarity of jadeite sample.

The intensity distributions between Ar+ laser and He-Ne laser of the sample jadeite stone are shown in figure 5. Arrows in the figure show the difference of contrast by noise, that is, crack, the nature of matter and so on. These results show not only the clearness but also the difference of contrast, which means cleanness of jadeite sample.

In conclusion, the wavelength near the infrared region is convenient to examine the conditions of good Jadeite such as cleanness and clearness by laser.

4 Acknowledgements

The author would like to thank to the Government of the Union of Myanmar for helpful comments and supports, and to Takashi Mitsuishi, Joytec Co., Ltd., for helpful advice on sample jadeite stones.

References

1. Kazuya Chihara (1987) Science of Jadeite, Chihara Memorial Foundation, Japan.
2. Ohzu, H., Khin, M.H., et al.(1997). Residual stress in PRK operated cornea and evaluation of the retinal image. Basic and Clinical Applications of Vision Science, 1997 Kluwer Pub., The Netherlands, 115-118.

Laser and Optics in Art Conservation

Application of UV-Lasers in Historical Glass Cleaning

Klaus Dickmann[1], Jens Hildenhagen[1], Farideh Fekrsanati[1], Hannelore Römich[2], Carola Troll[2], and Ursula Drewello[3]

[1] Lasercenter FH Münster (LFM), Dpt. Physical Technique
D-48565 Steinfurt, E-mail: laserlab@fh-muenster.de
[2] Fraunhofer-Institut für Silicatforschung (ISC) Würzburg
D-97877 Wertheim, Bronnbach Branch, E-mail: roemich@isc.fhg.de
[3] University of Erlangen-Nürnberg, Inst. of Materials Science
D-91058 Erlangen, E-mail: drewello@aol.com

Abstract. Fundamental studies were carried out on laser removal of various deposits from glass. An excimer laser $\lambda = 248$ nm was used to irradiate glass samples on which cleaning problems of stained glass windows were simulated. Ablation thresholds were estimated for various deposits to reveal the potential of excimer lasers in glass conservation.

1 Introduction

Historical stained glass windows are suffering strong deterioration due to a complex corrosion process. Beside various gaseous pollutants and acid rain also deposits of soot, dust and even microorganisms attack the glass [1–4]. As a result a layer of corrosion products and dirt covers the painted glass surface, diminishing its historical and aesthetic value. These surface deposits reduce the transparency of the glass and may retain water, which may cause further corrosion. Removal of these encrustations is required to slow down the degradation process of the glass, to regain its transparency and prepare the surface for further conservation measures. State of the art cleaning is based on various mechanical or chemical methods. The degree of cleaning is critical to control. Especially if encrustations are firm and dense, damage on the glass and the paint layers can not be avoided. UV-Laser radiation as a *non-contact tool* is a promising alternative for cleaning of historical glass [5–8].

2 Experimental strategy

2.1 Laser equipment and optical set-up

As a UV-Laser source a KrF-Excimer-Laser ($\lambda = 248$ nm) was used (Type: LPX 315i) with maximum average output power of 100 W, maximum repetition rate of 150 Hz and a pulse duration of appr. 40 ns. A specific optical set-up was designed to form the rectangular output beam with strong inhomogenous intensity distribution into a square beam profile with nearly uniform intensity (see Fig. 1). The projection technique was used to irradiate the sample with a beam geometry given by the projection mask.

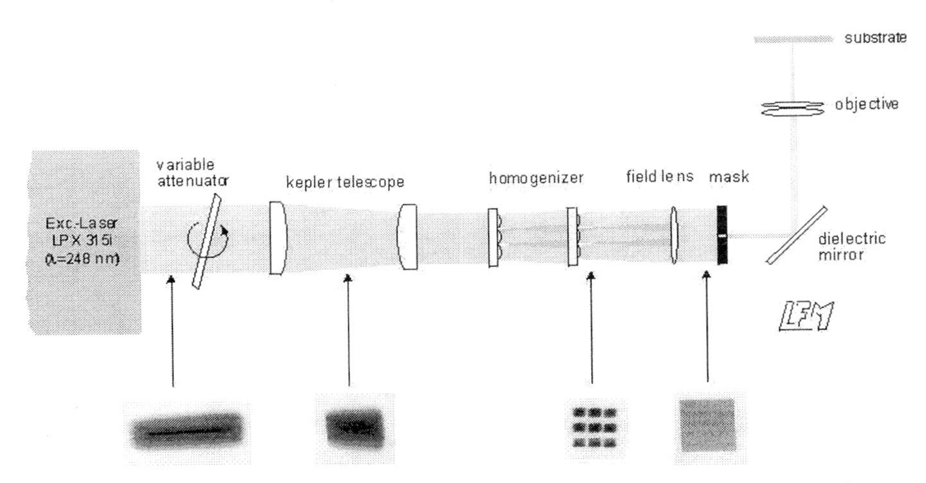

Fig. 1. Schematic arrangement of the optical set-up and intensity distribution at specific positions along the optical axis

2.2 Corrosion layer on glass and demands for laser cleaning

The cross-section of a corroded historical glass shows a complex composition of various types of materials, as depicted in Fig. 2. Not only environmental deposits as dirt, microorganisms and corrosion products cover the glass/paint, but also organic conservation materials from former restoration campaigns. In order to demonstrate the structure of a typical corroded glass surface a cross-section of an original glass fragment is given in Fig. 3. The *gel layer* is a result of ion exchange reactions, due to an acidic attack of the environment on glass.

Especially medieval glasses, characterised by a low content in silica but high contents of potassium and calcium may build up gel layers with a thickness to 100 μm. This layer has to be replaced by any cleaning method since it acts as a natural barrier to protect the bulk glass underneath [3,4]. Potassium and calcium ions, leached out from the glass can react with gaseous pollutants to form carbonates or sulfates, buiding up the crust of corrosion products, which should be removed [2,9].

In order to avoid experiments on valuable originals, basic studies on laser cleaning of corroded glasses were carried out on potassium-rich model glasses [11]. The preparation process of the glass samples was optimised for specific characterisation of laser irradiation effects. By accelerated weathering in a climate chamber corrosion processes leading to the formation of gel layer and corrosion crusts, can be induced on the glass samples. Besides, coating materials can be applied or the model glass samples can be subjected to the attack of microorganisms. By systematic sample preparation the ablation rate of corrosion products, coatings and biolayers can be studied seperately, to optimise the laser parameters for cleaning more complex surface layers on originals.

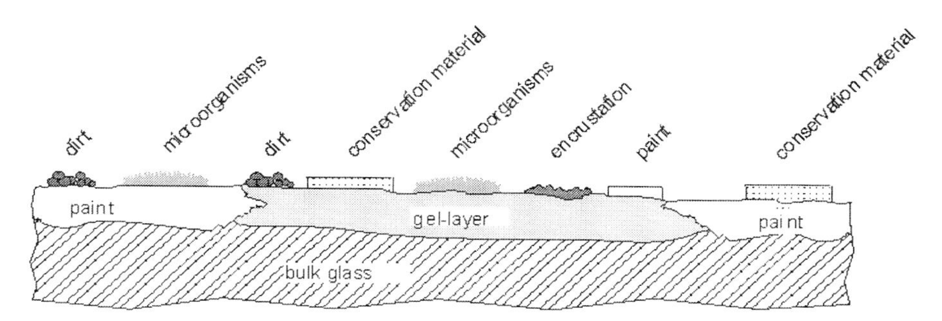

Fig. 2. Overview on cleaning problems on historical glass

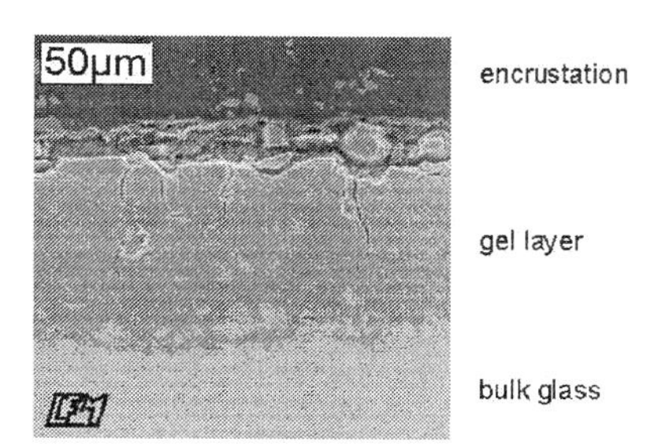

Fig. 3. Cross-section of an original glass fragment (Cathedral of Cologne, 13th century)

3 Results

The removal process of all materials mentioned above was examined by measuring the ablation rate (depth per pulse) versus fluence. Therefor a *parameter matrix* was produced onto the individual samples by variing the fluence H and repetition rate f_p, respectively (Fig. 4). The ablation depth was measured with a laser profilometer. An exemplary ablation graph is shown in Fig. 5.

The penetration of laser fluence $H(x)$ into the material is given by Beer's

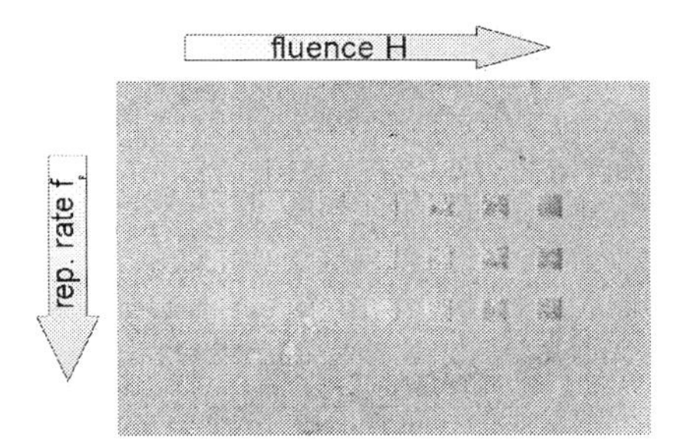

Fig. 4. Top view onto carbonate sulfate encrustation sample showing the *parameter matrix*

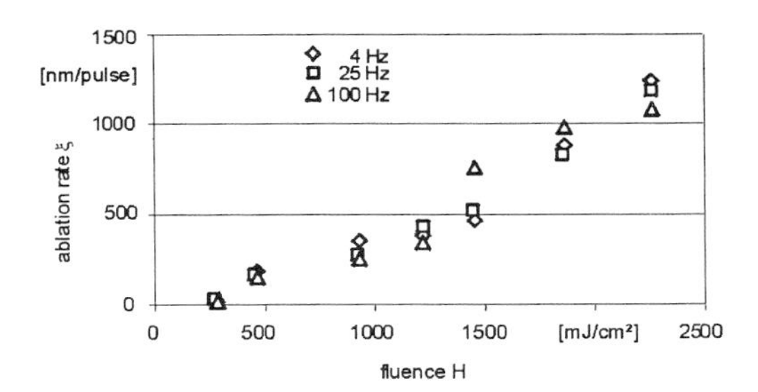

Fig. 5. Ablation rate of encrustation (sulfate Type 116)

law.

$$H(x) = H_0 \, e^{-\alpha \cdot x} \tag{1}$$

(H_0 is the fluence on the surface, α is the absorption coefficient, x is the coordinate into the material)
The ablation depth ξ is characterized by a certain fluence H_{th}, which is a threshold for interrupting the ablation process (see also Fig. 3). Thus from Eq. (1) we obtain:

$$\xi = \alpha^{-1} \ln \frac{H_0}{H_{th}} \tag{2}$$

In order to protect underlying materials during laser cleaning, the knowledge of H_{th} is of great importance for all materials mentioned above.
Our studies have shown that encrustations of carbon and conservation materials e.g. wax, organic and inorganic-organic polymers have $H_{th} \leq 1 \text{ J/cm}^2$.

This in general is lower than the threshold fluence of our bulk glasses and gel-layers with 1 J/cm^2 < H_{th} < 4 J/cm^2. However it should be emphazised that the paint layer starts ablating already at low fluence $H_{th} < 0,4J/cm^2$ which may cause a problem if the painted side of the window has to be cleaned. Microorganisms were taken from stained glass windows of various German Churches (e.g. *Altenburg, Bad Driburg*) and have been cultivated on special substrates for laser cleaning investigations. The activity of microorganisms could be stopped or in ideal cases fungis and bacteria could be removed entirely at certain thresholds. Fig. 6 depicts the result at H=1,3 J/cm^2.

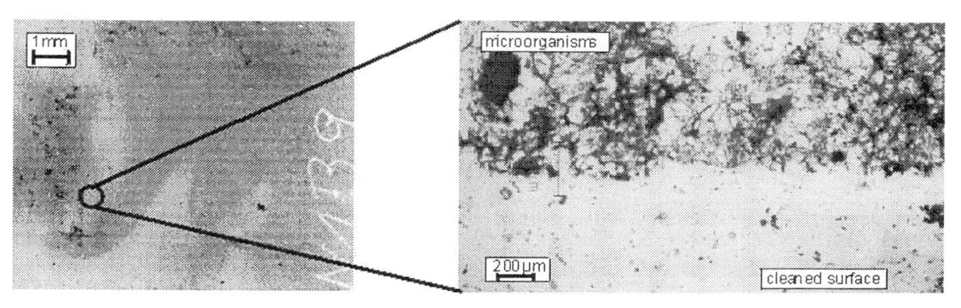

Fig. 6. Top view onto microorganism layer from the *Church of Bad Driburg*/Germany. Laser removal is achieved at $H_{th} > 1,3J/cm^2$

4 Conclusion

Within our studies we have worked out a catalogue with ablation thresholds for individual deposits, typical for ancient glass. As a result is has turned out, that there are several materials (e.g. conservation materials) which can be removed uncritical with excimer laser radiation. This is due to a lower ablation threshold in contrast to the underlying material. In the opposite case an on-line monitoring system or sensor based closed loop control will be necessary. In first attempts laser induced breakdown spectroscopy (LIBS) has shown promising results for been used therefor [10]. Future research will concentrate on a transfer of our results to original glass fragments.

5 Acknowledgement

The financial support of this interdisciplinary research project provided by the *Deutsche Bundesstiftung Umwelt* (DBU), Grant No. Az: 11 472 is gratefully acknowledged. We are also grateful to Mrs. Ulrike Brinkmann from the *Kölner Dombauhütte* and Mr. Peters and Mr. Sanders (*Glasmalerei Peters/Paderborn*) for the dicussions and consulting during our experiments as well as for the delivery of original glass fragments.

References

1. Drewello R., Weissmann R. (1997) Microbially influenced corrosion on glass. Appl Microbiol Biotechnol 47:337-346
2. Römich H. (in print)Historic glass and its interaction with the environment. James & James Science Publishers Ltd.
3. Müller W., Torge M., Adam K. (1994) Ratio of CaO / K_2O as evidence of a special rhenish type of medieval stained glass. Glastechn Ber Glassm Sci Technol **67**, 2, 45-48
4. Müller W., Torge M., Adam K. (1995) Primary stabiliyation factor of the corrosion of historical glasses: the gel layer. Glastechn Ber Glassm Sci Technol **68**, 9, 285-291
5. Leissner J., Drewello R., Weissmann R. (1998) Glasmalereien mit dem Excimerlaser behandeln? Restauro 5, 324-329
6. Leissner J., Fuchs D.R., Barkhausen W., Wissenbach K. (1995) Examination of Excimer-Laser treatments as a cleaning method for historical stained glass windows. Glastechn Ber Glassm Sci Technol **68**, 332-339
7. Olaineck C., Dickmann K. (1997) Reinigen historischer Kirchenfenster mit dem Excimerlaser. Restauro 2, 108-111
8. Olaineck C., Dickmann K. et al. (1997) Restoring of cultural values from glass with excimer laser. Restauratorenblätter Sonderband LACONA I, 89-94
9. Römich H., Fuchs D.R. (1995) Historsiche Glasmalereien: Risiken durch Reinigungsmassnahmen. Jahresberichte Steinzerfall-Steinkonservierung, Verlag Ernst & Sohn, 199-205
10. Maravelaki P.V., Zafiropulos V. et al. (1997) Laser induced-breakdown spectroscopy as a diagnostic technique for the laser cleaning of marble. Spectrocim. Acta Part B **52**, 41-53
11. Römich H. (1998) Simulation of corrosion phenomena of glass objects on model glasses. Proceedings of the 18th Int. Congress on Glass

Humanoid Machine Vision Systems and Underwater Archeology

Pal Greguss
Department of Manufacture Engineering, Technical University Budapest
H-1521 Budapest, Hungary
e-mail:greguss@manuf.bme.hu

Abstract. The Humanoid Machine Vision System (HMVS) is a further development of the Panoramic Annular Lens (PAL-optic) reported already at previous OWLS conferences. It provides - in contrast to PAL-optic - not only a centric minded image but also a central foveal view of a selected section of the periphery. This imaging situation is similar to but not identical with the function of human vision: a portion of the peripheral view can be brought "in focus" for better inspection and further processing. Its possible use in underwater research is described.

1 Introduction

In astronautical vocabulary, the word "microspace" refers to small satellites and associated user equipment [1] Microspace systems are usually not only physically relatively *small*, but also financially rather inexpensive, because - in spite of the fact that they include early use of advanced technolgy - most of their components are off-shell products. A typical example of such a microspace system is the optical attitude determination system, PALADS, used in the SEDSAT program [2], to describe the spacecraft attitude and position via an optic, called Panoramic Annular Lens (PAL) [3], on which we already reported on previous OWLS Conferences.

A recent version of PAL-optic provides not only a 360° panoramic view of the space around the system, but also a centric (foveal) image is simultaneously created in the central empty spot of the annulus. Thus, the system's behavior may be similar to that of the human vision, a portion of the peripheral view - which is 360° in contrast to the human's 40-50° - can be brought "in focus" for more detailed investigations. This is achieved by exploiting one of the unique properties of PAL-optics, namely, that its center region around the optical axis does not take part in forming the panoramic annular image: it serves only to ensure undisturbed passing through of the image forming rays. Thus, by removing the reflecting coating from the mentioned region of the PAL-optic and putting in front an imaging lens in such a manner that its image plane coincides with the annular image plane *inside* the PAL-optic, the resulting image in the annulus will complete the viewing angle around the optical axis. If now a zoom lens is used, the system's behavior may be similar to that of the human vision, namely, the target of interest can be visualized in the central empty spot of the annulus, which then can act as a fovea. Therefore, this system is called Humanoid Machine Vision System

(HMVS). [4] Combining this lens with a mirror having a variable inclination to the optical axis, any portion of the peripheral field of view can be brought into the fovea.

2 Aquapal

AQUAPAL is an underwater viewing system that we have proposed and is under development in our laboratory. It has not only optical ranging capability but may also display image data from the sea surface and seabed simultaneously, which is achieved through the use of HMVS.

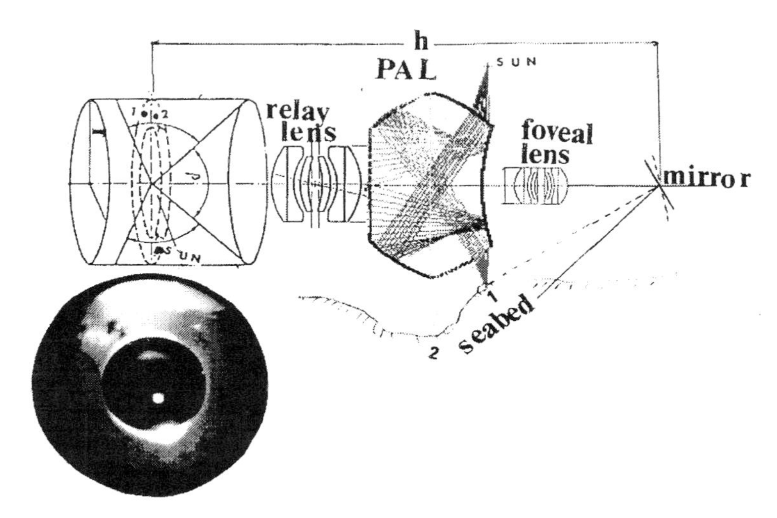

Fig. 1. Operating principle of AQUAPAL using HMVS.

On hand of the sketch of Fig. 1 - which is far from being scaled - we explain how the HMVS operates in the microspace system of AQUAPAL. Let us assume that in this example the optical axis of AQUAPAL is more or less parallel to the sea surface, which means that in the peripheral annular image resulting from the fact that the PAL optic considers its field of vision to be cylindrical and is projected onto the target of the CCD camera, will display data from the *sea surface* and the *seabed* simultaneously. Placing a mirror having an inclination β to the optical axis in front of the foveal lens, a selected section of the peripheral view can be displayed in the fovea, in our example, the seabed part marked 1 and 2.

The distance R of those image points - in our example seabed points 1 and 2 - which are displayed simultaneously both in the peripheral and the foveal images, from the optical axis of AQUAPAL can easily be calculated by using well known trigonometric relationships since the inclination angle β of the mirror and the distance h of the mirror from the image plane is known, and the width s of the annular image is a function of the viewing angle of the PAL-optic. It can be described as a sum of the viewing angles α_1 below and α_2 above the "*horizon of the optic*" (HO), which is the plane where the curved refracting and reflecting surfaces of the PAL-optic intersect. Thus, when projecting this annular image formed *inside* the optic is projected onto the CCD target via a relay lens, each pixel along the radii of the projected annular image will represent the direction of an object point in space above and below the HO. Thus, knowing the number of the pixels along the width s of the annulus, the peripheral viewing angle, and the sum of α_1 and α_2, the position of the PAL-optic's HO displayed in the annular image as a circle can easily be calculated. Thus, e.g., if $s = 960$, $\alpha_1 + \alpha_2 = 72°$, the circle representing the HO will be 640 pixels apart from the interior rim of the annular image, and 320 pixels apart from the outer rim of the annular image, and each pixel will represent a viewing angle of 0.075°. Thus the condition for panoramic range imaging is established. Only one has to remember that the annular image is not a cross section, it is a peripheral view representing the two-dimensional skeleton of the three-dimensional space. We strongly believe that this feature of AQUAPAL could be exploited in underwater archeology when searching for underwater ruins and locating their exact 3-D position.

2.1 System design

The preliminary breadboard realization of AQUAPAL can be considered as a tethered underwater inspection and mapping system. It is made of a hard anodized heat-reflecting bright-finish aluminum alloy cylinder, having a diameter of 100 mm, and a length of 210 mm. Using appropriately designed filler rings, the sealing is such that the structure flooding is not possible under proper care.

The PAL-optic-foveal optic ensemble of the HMVS is mounted on that side of the transparent covering surface of the cylinder which is in contact with the water, while the relay lens on the other side, and it fits with a C-mount to a CCD camera. At present, we use an inexpensive CCD camera, nevertheless, we have obtained rather good resolution. In the fluid-tight housing there is room enough for rechargeable batteries. As an option, a floating video transmitter can also be attached to the housing, making superfluous the use of a tether.

2.2 AQUAPAL as optical remote target signature analyzer

The characteristics of HMVS, namely, that the foveal image is displayed in the same see-through-window (STW) imaging strategy as we are accustomed to in spectral analysis and,at the same time, the display of the peripheral field of view uses centric minded imaging (CMI) strongly suggest that such a suystem may be a good candidate for target analysis.

Panoramic spectroscopy. More than 30 years ago, G.W.Stroke and A.T. Funkhouser [5] suggested to abandon the use of a moving mirror in Fourier spectroscopy, and to use, instead, two steady mirrors with a light inclination so that a hologram of the spectral intensity distribution can be formed, allowing so spectroscopy without computation.

Instead of using two steady mirrors, a glass block that resembles for the first sight to a conventional beam splitter cube is used, however, its far side facing the incoming collimated beam is not parallel to the front surface of the cube but it is oblique, a hologram of the spectral intensity distribution inside this block will be formed.

Since this situation is somewhat similar to PAL-optic imaging, in both cases, namely, the "images" are created inside the optical block, combining the HMVS with this holographic Fourier spectrometer, the spectral intensity distribution of the foveal image can be recorded using a Fourier lens, provided that the following relation between D, the diameter of the Fourier optic, H, the size of the interference pattern, the optical distance of this interference pattern from the optic holds:

$$D = H + 2L\ \mathrm{tg}(\Theta/2),$$

where Θ is the angle between the two oblique reflecting surfaces.

Two-color target analysis. Starting from the recognition that two objects showing the same total radiative energy emitted in a broad band may emit different amounts of energy in another narrower band, however, seen within this broad band, a HMVS can be designed which may recognize targets via so-called *two-color* target signature analysis method, which is the most advanced way to make target signatures. At present, two-color target signature analysis is achieved by using two cameras, the HMVS, however, needs only one single camera. It uses a ring-shaped broadband filter which is linked to that part of HMVS that is responsible for the 360° peripheral panoramic field of view, and a narrower band, however, which has its frequency band within that of the broadband filter and is linked to the foveal image. Thus, it is possible to distinguish between the two objects. The data resulting from the foveal image can also be regarded as a target signature.

Panoramic polarimetry. Although the direct sunlight is not polarized, a substantial amount of light reflected from various surfaces may be polarized in one way or another. This means that, unlike the sky where the E-vector may be vertical at sunrise and sunset, the greatest angle the E-vector can tilt from the horizontal is the angle from the vertical marked by the reflection characteristics of surfaces such as soil, ore, minerals, etc. [6] Thus, the polarization pattern of the reflected surface can be considered as target signature. This fact was used, among others, by G. Horvath [7] to show why Kuvait oil lakes acted as insect traps. We have shown that, if a linear polarizer is placed between the PAL-optic of the HMVS and the relay lens, by rotating the polarizer the reflected E-vector distribution can easily be mapped and used as target signature.

Underwater Sun compass. Knowing that when the sun elevation above the horizon increases, the maximum E-vector tilt would be decreased to zero if the Sun should reach the Zenith exactly (or if there is a heavy complete overcast), Sun compasses have been designed. Except at such moments, the Sun's compass direction can readily be found by locating the directin of tilt toward the Sun bearing or azimuth. If the observer maintains a stable spatial orientatin relative to gravity-horizon, and is capable of analyzing E-vector direction, a Sun compass is available, even though the glitter pattern or its depth prevents it from seeing the Sun directly. Thus, the previously described HMVS equipped with a polarizer may be of help in underwater navigation.

3 Acknowledgements

The author wishes to thank Antal K. Bejczy of Jet Propulsion Laboratory of NASA for the fruitful discussions, and to Marius Patko of the Department of Manufacture Engineering of the Technical University Budapest for his help in the experimental work.

References

1. David W. Thompson, Microspace systems for military missions, The Space Times, Vol. 31(No. 4):12-14 (1992).
2. http://146.229.5.181
3. Hung. Pat. 192125; USA Pat. 4566763; Japan Pat. 1962784; German Pat. 3402847.
4. Patent Pending.
5. G. W. Stroke, A. T. Funkhouser, Fourier-transform spectroscopy using holographic imaging without computing and with stationary interferometers, Phys. Lett. Vol. 16:272 (1965).
6. Egan, W.G., *Photometry and polarization in remote sensing*. (Elsevier, New York NY. 1985)
7. Horvath, G., Zeil, J., Kuvait oil lakes as insect traps, Nature **Vol. 379**, 303-304 (1996)

Near-Ultraviolet Pulsed Laser Interaction with Contaminants and Pigments on Parchment: Spectroscopic Diagnostics for Laser Cleaning Safety

Wolfgang Kautek[1*], Simone Pentzien[1], Pascale Rudolph[1], Jörg Krüger[1]
Claus Maywald-Pitellos[2], Helmut Bansa[2], Heinz Grösswang[3], Eberhard König[4]

[1] Laboratory for Thin Film Technology, Federal Institute for Materials Research and Testing, Unter den Eichen 87, D-12205 Berlin, Germany

[2] Institut für Buch- und Handschriftenrestaurierung, Bayerische Staatsbibliothek, Ludwigstraße 16, D-80539 München, Germany

[3] PHYMA GmbH, Ferdinand-Waldmüller-Gasse 6, A-2531 Gaaden, Austria

[4] Kunsthistorisches Institut, Freie Universität Berlin, Morgensternstrasse 2-3, D-12207 Berlin, Germany

* Correspondence: Wolfgang.Kautek@bam.de; Phone/Fax (+49-30) 8104-1822

Abstract. Laser cleaning is a contactless and dry process. The absence of chemical agents, spectroscopic selectivity, micro-precision, computer-aided handling, and the combination with *in-situ* diagnostic techniques makes it attractive not only for industrial but also for parchment and paper restoration applications. Potentials and limitations of the near-UV pulsed laser cleaning of historical parchment manuscripts (λ = 308 nm, τ = 17 ns) have been investigated by experiments at contamination/pigment/parchment model systems. Major attention is being payed to spectroscopic diagnostics which allow quantification of laser-induced degradation reactions, and thus support the cleaning safety. First results by diffuse reflection absorption FT-IR microscopy, spectrophotometry, colour metrics, laser-induced fluorescence and plasma spectroscopy are presented.

1 Introduction

Laser cleaning of stone artifacts, wall paintings, and facades has reached a high technological level. Decades of experience and commercial laser systems are already available for innovative restorers [1,2]. The status for varnish laser removal of paintings is by far not as advanced [1,3,4], but may be commercially available soon [5].

The contactless cleaning of fibrous biogenetic surfaces like e.g. parchments and paper by laser radiation, on the other hand, has been approached only in the very recent years [6,7,8]. There are few studies on laser cleaning of paper, i.e. cellulose fibre materials. However, degradation of the paper structure was observed when the ablation of modern ink was attempted with the fundamental wavelength (1064 nm) of a Nd:YAG laser [9]. Removal of fungus-induced stains [10,11] and glue spots [6,7,8,12] on paper with excimer lasers was reported.

Parchment is made from skin. Therefore, laser cleaning of parchment can be considered in some analogy to dermatological laser applications which are regarded as a multi billion dollar market [13]. There, selective photothermolysis of pigmented subsurface structures, such as melanin particles, enlarged blood vessels, and tattoo ink particles, or char-free vaporization of skin, takes place. Particles photocoagulate or photodisrupt. Then the skin's natural physiological mechanisms break down and remove the laser-altered remnants. Further, laser hair removal relies on selective photothermolysis interaction with hair follicles. In parchment cleaning, however, natural resorption of laser-altered remnants is absent, and contaminants have to be completetely removed by the laser action.

Traditional conservation techniques for parchment and paper include mechanical scratching, the use of erasers or solvents [14]. Erasers (e.g. polyvinyl chloride) can loosen dirt from the surface but cause partial mechanical surface disruption. The pressure applied may lead to closer fibre packing and/or different degrees of hydration producing irreversible unacceptable changes to the appearance of the parchment. The application of alcohols or water can lead to hydrolysis of the collagen bonds of parchment resulting in protein degradation or even gelatine formation. Such unacceptable restoration attempts can be encountered repeatedly, and have to be taken as a quality reference for laser cleaning experiments on original parchment samples [8,12]. Laser cleaning of parchment with a near-ultraviolet excimer nanosecond pulse laser (308 nm) may have the potential to increase the quality of conservation work because toxic cleaning fluids with the potential of long-term degradation can be avoided. Moreover, a laser beam represents a contactless scalpel and eraser of unreached lateral precision.

2 Mechanism and Cleaning Safety

The most important parameter for destruction-less laser cleaning is the ablation (and destruction) threshold laser fluence F_{th} (pulse energy per spot area) of the respective materials, either contamination or parchment [6,7,8]:

$$F_{th} \approx \rho \times (1-R)^{-1} \times \alpha^{-1} \times h_{sub} \tag{2.1}$$

where the density is ρ, the reflectivity R, the absorption coefficient α, and the specific sublimation enthalpy h_{sub}. With the assumption that the ablation depth is equal to the penetration depth of light, α^{-1}, α and F_{th} of the model contamination graphite and of parchment were esperimentally determined (Table 1). This shows that the laser cleaning working condition is [6,7,8][6,7,8]

$$F_{p,th} / F_{c,th} = \alpha_c \, h_{sub,p} / (\alpha_p \, h_{sub,c}) >> 1 \tag{2.2}$$

(suffix p for parchment, c for contamination), or is simply the fluence F when

$$F_{p,th} > F > F_{d,th} \tag{2.3}$$

is fulfilled.

	Parchment [a]	**Graphite** (soot)
α [cm^{-1}]	≤ 400	$2\cdot\times 10^5$
α^{-1} [μm]	≥ 25 μm	0.05
F_{th} [J cm^{-2}]	1.8	0.4

Table 1: Absorption coefficients and ablation thresholds at 308 nm (experimental data [7,8]). [a] Goat: Carl Wildbrett, Waldstrasse 20, D-86399 Bobingen, Germany)

3 Experimental

Parchment is commonly manufactured from the dermis layer of animal skin (calf, sheep, goat) after strong alkaline removal of the epidermis and the subcutaneous tissue layers. Two-dimensional reorientation of the collagen fibers take place in a stretching and drying process controlled by $CaSO_4$, CaO, and $CaCO_3$ additives. Often a plastizising rehydration end-treatment is applied using egg white and linseed oil. The model film systems consisted of modern calf or goat parchment samples (Carl Wildbrett, Waldstrasse 20, D-86399 Bobingen, Germany) coated with synthetic contaminations like e.g. commercial watersoluble "black" from Herlitz® or a blend of yellow ochre pigment, soot, and gummi arabicum binder. Pigments were tested either as wax crayons ("chromium yellow" from Jaxon®), or as various self-made binder mixtures of pigments (Kremer®) common in medieval book illuminations.

An excimer laser (Lambda Physik, model EMG 150) filled with XeCl gas emitting at 308 nm was used. The pulse duration was 17 ns with a repetition rate of 1 - 2 Hz. The beam was focused by a quartz cylinder lense (156 mm focal length) to a spot dimension of 0.17 mm x 5.50 mm (area 0.01 cm^2) on the parchment substrates. These were mounted and scanned on a computer controlled x-y-z stage (LOT). Fluences ranged between 0.2 - 1.4 J cm^{-2}. For optical laser plasma emission spectroscopy (LIPS) and laser-induced fluorescence spectroscopy (LIF), the plasma and/or the irradiated area was imaged through a 50 mm focal quartz lens onto a quartz fiber bundle which was coupled with the entrance slit of a 320 mm Czerny-Turner monochromator (300 lines/mm grating blazed at 550 nm, SA Instruments, Model HR-320). A two-dimensional multi-channel-plate-intensified CCD detector was connected to the exit slit. An optical spectroscopic multi-channel analyser system (OSMA, Spectroscopy Instruments, Inc.) served to collect emission spectra and to store data. Diffuse reflection absorption FT-IR microscopy has been done with a Nicolet® Type 800 spectrometer with an Olympus® microscope module.

4 Degradation Diagnostics

Modern analytical chemistry offers valuable tools to investigate laser-induced degradation, i.e. morphological and chemical changes, of biogenetic fibre materials. Besides imaging techniques like optical microscopy and scanning electron microscopy, chemical analyses on the basis of diffuse reflection absorption FT-IR microscopy, UV-VIS spectrophotometry together with colour metrics, *in-situ* laser-induced plasma spectroscopy (LIPS) and *in-situ* laser-induced fluorescence (LIF) have been employed in the present study. Other techniques, like scanning force microscopy (SFM), small spot ESCA, shrinkage temperature testing, etc. are currently being applied in this laboratory.

4.1 Degradation Processes

Common degradation of cellulose is caused by autoxidation due to the formation of radicals by air oxygen [15,16,17,18]. A first assessment on the near-UV pulsed laser-induced paper degradation is under way [19]. Oxidative breakdown processes of parchment base on heat and light [20,21]. Parchment starts degrading at tripeptides in clusters of charged amino acids following the pattern: (1) loss of mainly the basic amino acids Lysine, Arginine, Hydroxylysine, and the imino acids Proline and Hydroxyproline, (2) gain of acidic amino acids, (3) formation of small amounts of breakdown products. That means that acidic breakdown causes hydrolysis of the peptide bonds in the peptide chains, and amino end groups are generated. Autoxidation of parchment occurs only in the presence of light [22]. It leads to the formation of polar groups, particularly carboxylic acid groups and to conjugated double bonds. Yellowing of parchment, on the other hand, is a secondary non-oxidative thermal process which takes place among autoxidation products, i.e. aldol condensation among carbonyl species and dehydration reactions. Fluorescence strongly increases. Interestingly, these yellow stains can be bleached with visible light.

4.2 Diffuse Reflection Absorption FT-IR Microscopy

IR bands of hydrogen-bonded OH (3445 cm^{-1}), carboxyl OH (3300-2500 cm^{-1}), unsaturated carbonyl (1710 cm^{-1}) are indicators for the above degradation changes of parchment [23][1]. The CH stretch vibration was taken as an internal reference (2950 cm^{-1}). The oxidation product, mainly carboxylate groups (1567+1384 cm^{-1}), were superimposed on the CH_3 bending signal. Yellowing could be detected at the long wavelength side of C=O (1706 cm^{-1}) due to unsaturated carbonyl functions.

4.3 Spectrophotometry and Colour Metrics

Pigments used in medieval manuscripts were both organic and inorganic (Table 2 and 3) [24,25]. Spectrophotometric and colour metric results in Table 2 represent a quantification of their stability towards 308 nm pulses laser light below the threshold fluence of parchment (<1.5 Jcm^{-2}). Multiple sensor segment spectrophotometric measurements resulted in spectral reflectance graphs which numerically were converted into numerical color data in the $L^*a^*b^*$ colour space.

Pigment	L^*_0	ΔL^*	a^*_0	Δa^*_0	b^*_0	Δb^*	ΔR	**Eye**
Indigo	26.68	+1.54	0.31	-1.89	-12.02	+0.79	0	Ø
Lapis	43.50	+4.12	-7.31	-1.34	-21.95	+0.81	0	Ø
Saftgrün	39.84	+0.62	-13.98	-0.38	21.37	+5.54	-<470<0	Ø
Gold Ochre α-Fe_2O_3 (Hämatit) & α-FeOOH (Goethit)	57.03	-4.16	12.82	-1.69	33.43	-13.39	+<510<-	↑
Azurite $2CuCO_3$ $Cu(OH)_2$	42.95	+1.02	-6.65	+1.54	-3.96	+14.68	-<520<+	↑
Grünspan	63.56	-5.59	-21.37	+4.12	6.61	+6.97	-<650<0	↑
Dark Ochre	43.86	-3.37	17.77	-4.47	23.57	-9.50	+<520<-	→
Malachite $CuCO_3$ $Cu(OH)_2$	47.81	-11.68	-7.46	+4.27	11.11	-0.24	+<380<-	→
Minium Pb_3O_4	59.16	+2.03	46.08	-3.63	47.41	-10.16	+<550<0	→
Parchment & "Synth. Contam"	83.62	+35.90	-0.84	-4.31	1.53	-13.81	++	↑↑

Table 2: Laser-induced colour changes of pigments on parchment; $F < 1.5$ J cm^{-2}. ΔR: UV-VIS reflectivity change; „Eye": aspect by naked eye; ↑↑ strong, ↑ medium, → low; Ø at detection limit

The $L^*a^*b^*$ colour space [26][2] is presently one of the most popular colour spaces in virtually all fields. L* indicates lightness, a^* and b^* are the chromaticity coordinates ($+a^*$ ... red, $-a^*$ ... green, $+b^*$ ... yellow, $-b^*$ ... blue). Laser-induced spectral changes are given as changes of L^*, a^*, and b^*: ΔL^*, Δa^*, and Δb^*. Δa^* and Δb^* assume <|2| when the naked eye ('Eye') cannot sense the difference. One can see that the human eye is much less sensitive for lightness changes than for colour changes. The spectrum region where substantial reflectivity changes occured are indicated as ΔR. Indigo and Lapis are practically tolerant against laser irradiation whereas all other pigments show spectral and colour changes - iron and copper compounts the most, and lead oxide the least. The laser removal of synthetic contamination on parchment exhibits of course a substantial colour and lightness change.

4.4 Laser-Induced Fluorescence (LIF) and Plasma Spectroscopy (LIPS)

Physical diagnostic tools which are independent of the restorer's naked eye control have recently been searched for. It is required that they allow to quantitatively detect the destructive interaction of the laser beam with the substrate. In this context, LIF could be applied to the non-destructive analysis below the ablation (destruction) threshold of molecular materials such as paint varnishes consisting of e.g. mastic, dammar resin, and linseed oil [27,28]. In LIPS, on the other hand, the material like e.g. a varnish on a painting or carbonaceous material on parchment (as a model substance for dirt on a manuscript) [29]is transformed into a gaseous plasma which emits radiation spectroscopically typical for the vaporized material. Since LIPS relies on the destruction of a substrate, it seems to provide limited applicability in the *in-situ* monitoring of non-destructive laser cleaning of parchment and paper.

In-situ emission spectroscopic experiments during the ablation process on parchment model systems such as multilayer composites of carbon containing black material, pigments and parchment are presented in Table 3 [7,8,23]. A substantial list of mainly inorganic pigments shows strong emission intensity and provide good spectroscopic separation from the LIF emission of the parchment ($\Delta\lambda$). This pigment category also shows moderate colour changes [23]. Those pigments exhibiting strong discoloration (see Table 2), however, show moderate til bad suitability for LIPS registration on parchment. Carbonaceous contamination results in a weak emission, too. LIPS/LIF is less sensitive for the less stable pigments. That shows that this diagnostic tool is a relatively unfavourable candidate for *in-situ* diagnostics of laser cleaning of pigmented parchment and paper.

Pigment	**Intensity**: LIF+LIPS	$\Delta\lambda$: Pigment/Parchment
Cobalt Green, $CoO \cdot Cr_2O_3$	↑↑	↑↑
ChromeYellow, $PbCrO_4$ (PbO)	↑↑	↑ (547 & 407 nm)
Cinnabar, Vermillion, HgS	↑↑	↑
Saftgrün/Kreuzdorn	↑↑	↑ (515 & 527 nm)
Schüttgelb	→	↑ (~550 & 400 nm)
Ultramarine, ~$Na_8(AlSiO_4)_6S$	↑↑	→
Cobalt Blue, $CoO \cdot Al_2O_3$	↑↑	→
Carmine	↑	→

Indigo	↑↑	Ø
Gold Ochre α-Fe_2O_3 (Hämatit) & α-FeOOH (Goethit)	↑	→
Malachite, $CuCO_3$ $Cu(OH)_2$	→	↑↑ (~ 500 nm)
Azurite, $2CuCO_3$ $Cu(OH)_2$	↑↑	Ø
Minium Pb_3O_4	→	→
Herlitz-Schwarz	↑↑	Ø
"Synth. Contamin.", α-FeOOH & Soot	→	Ø

Table 3: LIF/LIPS Intensities; Wavelength Difference between Pigment and Parchment Emission. ↑↑ strong, ↑ medium, → low, Ø at detection limit

References

1 W. Kautek and E. König (Eds.), *Lasers in the Conservation of Artworks I*, Restauratorenblätter, Special Issue (Verlag Mayer & Comp., Wien, 1997).
2 *Lasertechnik in der Restaurierung*, Restauro 104 (Special Thematic Issue 6/1998).
3 S. Georgiou, V. Zafiropulos, D. Anglos, C. Balas, V. Tornari, and C. Fotakis, Appl. Surf. Sci. 127-129, 738 (1998).
4 A.E. Hill, A. Athanassiou, T. Fourier, J. Andersen, and C. Whitehead, this volume.
5 J.H. Scholten, J.M. Teule, V. Zafiropulos, and R.M.A. Heerens, this volume.
6 W. Kautek, S. Pentzien, J. Krüger, and E. König, in Lasers in the Conservation of Artworks I, Restauratorenblätter (Special Issue), (Eds.) W. Kautek and E. König (Verlag Mayer & Comp., Wien, 1997) p. 69.
7 W. Kautek, S. Pentzien, P. Rudolph, J. Krüger, and E. König, Appl. Surf. Sci. 127-129 (1998) 746.
8 P. Rudolph, S. Pentzien, J. Krüger, W. Kautek, and E. König, in Lasertechnik in der Restaurierung, Restauro 104 (Special Thematic Issue 6/1998) p. 396.
9 J. Caverill, I. Latimer, B. Singer, The Conservator 20 (1996) 65.
10 H.M. Szczepanowska and W.R. Moomaw, J. Am. Inst. Conservation 33 (1994) 25.
11 T.R. Friberg, V. Zafiropulos, Y. Petrakis, and C. Fotakis, in Lasers in the Conservation of Artworks I, Restauratorenblätter (Special Issue), (Eds.) W. Kautek and E. König (Mayer & Comp., Wien, 1997) p. 69.
12 W. Kautek, S. Pentzien, P. Rudolph, J. Krüger, and E. König, in Lasers in the Conservation of Artworks II, Restauratorenblätter (Special Issue), (Eds.) W. Kautek and E. König (Verlag Mayer & Comp., Wien, 1999), in press.
13 J.G. Manni, Biophotonics International, May/June (1998) 40.
14 R. Reed, Ancient skins, parchments and leathers (Seminar Press, London and New York, 1972).
15 G.J. Leary, J. Pulp Pap. Sci. 20 (1994) 154.
16 C. Heitner, in Photochemistry of Lignocellulosic Materials, Advances in Chemistry Series 531 (Pointe Claire, Canada, American Chemical Society, 1993) p.192.
17 I. Forsskahl, Trends in Photochem. Photobiol. 3 (1994) 503.
18 J. Kolar, Restaurator 18 (1997), 163.
19 J. Kolar, M. Strlic, D. Müller-Hess, A. Gruber, K. Troschke, S. Pentzien, M. Röllig, E. König, and W. Kautek, in Lasers in the Conservation of Artworks III, Restauratorenblätter

(Special Issue), (Eds.) W. Kautek and E. König (Verlag Mayer & Comp., Wien, 2000), in preparation.
20 F. O'Flaherty, W.T. Roddy, and R. Lollar, The Chemistry and Technology of Leather, Vol. 4 (Reinhold Publishing Corporation, New York, 1965).
21 R. Larsen, M. Varie, and K. Nielsen, in STEP Leather Project, (eds.) R. Larsen, M. Vest, U.B. Kejser, Protection and conservation of European Cultural Heritage, Res. Report No. 1 (Directorate - General for Science, Research and Development, European Commission, 1994) p. 151.
22 E. René de la Rie, Studies in Conservation 33 (1988) 53.
23 W. Kautek, S. Pentzien, P. Rudolph, M. Röllig, J. Krüger, C. Maywald-Pitellos, M. Bansa, H. Grösswang, and E. König, in Lasers in the Conservation of Artworks III, Restauratorenblätter (Special Issue), (Eds.) W. Kautek and E. König (Verlag Mayer & Comp., Wien, 2000), in preparation.
24 M.V. Orna, J. Chem. Education 55 (1978) 478.
25 H.G. Friedstein, J. Chem. Education 58 (1981) 291.
26 Color space defined by the Commission Internationale de l'Eclairage (CIE) 1976 (also referred to as CIELAB)
27 T. Miyoshi, Jpn. J. Appl. Phys. 24 (1085) 371.
28 D. Anglos, M. Solomidou, I. Zergioti, V. Zafiropulos, T.G. Papazoglou, and C. Fotakis, Appl. Spectrosc. 50 (1996) 1331.
29 O. Derkach, S. Pentzien, and W. Kautek, SPIE Proceedings Vol. 2991, *Laser Applications in Microelectronic and Optoelectronic Manufacturing II*, (1997) 48.

---

1 W. Kautek, S. Pentzien, P. Rudolph, M. Röllig, J. Krüger, C. Maywald-Pitellos, M. Bansa, H. Grösswang, and E. König, in Lasers in the Conservation of Artworks III, Restauratorenblätter (Special Issue), (Eds.) W. Kautek and E. König (Verlag Mayer & Comp., Wien, 2000), in preparation.
2 Color space defined by the Commission Internationale de l'Eclairage (CIE) 1976 (also referred to as CIELAB)

A Preliminary Study into the Suitability of Femtosecond Lasers For the Removal of Adhesive from Canvas Paintings

J. Shepard
Conservation Dept., Courtauld Institute of Art, Somerset House, London, WC2R ORN.
C.R.T. Young*
Conservation Dept., Tate Gallery, Millbank, London, SW1P 4RG. (*Author to whom correspondence should be addressed)
D. Parsons-Karavassilis
Physics Dept, Imperial College, Prince Consort Rd., London, SW7 2AZ.
K. Dowling
Physics Dept, Imperial College, Prince Consort Rd., London, SW7 2AZ.

Abstract: The numerous attempts to use lasers in different fields of conservation reflect awareness of their potential as non-contact tools which may be tailored to specific cleaning problems. Ultrashort laser pulses, in the order of pico- and femtoseconds have been used for precision ablation of biological tissue and have been shown to eliminate collateral damage normally caused by thermal diffusion of the laser pulse. Thermal damage and photochemical changes have both been reported as by-products of laser cleaning experiments with Excimer lasers and thus femtosecond lasers potentially offer significant benefits in this application. In the work reported here a specific problem within painting conservation has been addressed to assess the suitability of femtosecond lasers; namely the removal of glue-paste adhesive from the reverse side of a painting on canvas. This study is based on a collaborative project between the Conservation Departments of the Courtauld Institute of Art, the Tate Gallery, and the Femtosecond Optics Group at Imperial College of Science, Technology and Medicine. In this study 10ps, 830nm pulses derived from a Ti:Sapphire laser were used to ablate marks with a diameter of 5□m and a depth of 10□m in a glue-paste layer on canvas. The surface characteristics of the ablation marks on glue-paste, canvas and priming were investigated using microscopy. Ablation occurred irrespective of material type and no thermal damage was observed.

1. Introduction

The materials and methods used in the preparation of paintings on canvas supports have varied considerably from one era to another. However, these complex structures can briefly be summarised as consisting of woven fabric, most commonly linen, which is stretched taut and given a coating of animal glue size. This prepares the canvas for priming and subsequent paint layers consisting of pigments ground in a medium, usually egg or oil, and finally varnish layers.

Degradation of the canvas may occur with time, reducing its strength and its ability to function as a support for the paint layers. One solution to this problem is to line the painting, providing reinforcement by attaching a new canvas to the reverse of the original canvas with an adhesive. During the nineteenth century use of glue-paste lining adhesive, based on animal glue mixed with flour, was widespread. Typically, after 50-100 years a glue-paste lining itself ceases to provide appropriate strength or stiffness and its removal becomes necessary. Scalpels are used to cut away the lining canvas and scrape the old glue-paste from the surface and weave interstices of the original, sometimes with the addition of moisture to soften hardened glue. No alternative has been proposed to this method, which subjects a painting to unacceptable risk of damage.

2. Lasers in Conservation Applications

Research into the laser conservation of paintings has focused on use of Excimer lasers operating at nanosecond pulse widths to clean paint films [1-6], and more recently, on integrated laser-based monitoring techniques [7-9]. Cleaning is theoretically self-limiting, based on the principle that penetration of laser radiation to the paint film is minimised by the strong UV absorptivity of dirt and varnish to be removed and comparative UV reflectance of the underlying layers [10-12]. The necessity of strong wavelength absorption has been emphasised as a means of limiting formation of photooxidation products and thermal damage [13]. Other studies indicate problems in establishing appropriate laser parameters to prevent thermal and collateral damage to original painting materials, which may be attributable to thermal diffusion at nanosecond pulse widths [14,15].

Ultrashort pulses (<10 picoseconds) applied to the removal of biological tissue in surgical applications have been demonstrated to be capable of precise material removal while eliminating thermal and collateral damage, because picosecond pulse widths limit thermal diffusion [16]. Based on these results the potential of ultrashort laser pulses using a femtosecond laser for discrete glue-paste removal has been investigated in this study.

Ultrashort laser pulses, though extremely precise in ablating materials, do not discriminate between different materials. Ablation occurs via laser induced optical breakdown, initiated by multiphoton ionisation [17]. Removal of a coating from a substrate which is to be unaffected requires identifying the interface between the coating and substrate and guiding the laser precisely over the area targeted for ablation with a means of controlling the ablation depth. Glue-paste is applied in a semi-liquid state to the permeable canvas, and the interface between the two is indistinct. Compression between the original canvas and lining also gives the glue-paste an uneven topography, with large quantities of adhesive lying in the weave interstices. As with all ablation techniques, use of a sensitive monitoring technique to control the extent of glue-paste removal is therefore a fundamental requirement. Fluorescence Lifetime Imaging (FLIM) has been used for this application because it offers the possibility to integrate the optics of both the ablation and monitoring system into one sensor head and thus on-line monitoring during ablation is possible.

3. Ablation of Glue-paste from Canvas

3.1 Experimental Set-up

Experiments were carried out in the Physics Department at Imperial College of Science, Technology and Medicine (ICSTM). A commercial mode-locked Ti:Sapphire laser system and Cr:LiSAF amplifier built at ICSTM were used. This system delivered 830 nm, 10 ps pulses at 1 kHz. Pulses were focused onto the sample with a working distance of 3.0 mm. The experimental setup is shown in Fig. 1. The energy to the sample was calculated by measuring the energy of the subsidiary beam split by a beamsplitter of known ratio. A neutral density filter, placed in the beam path, allowed continuous variation of the energy to the sample from 0-10 μJ. A fluence of 0.032 $\mu J/\mu m^2$ (3200 mJ/cm^2) was maintained during the experiment.

3.2 Sample Preparation

The lining canvas was peeled away from the commercially primed canvas to expose the glue-paste film. In isolated areas glue-paste deposits came away with the lining, revealing areas of the reverse of the commercially primed canvas and small deposits of priming in the weave interstices. The lining canvas was discarded. Small rectangular samples (up to 0.5cm^2) were cut from the canvas. Each was selected to include an area of intact glue-paste film and an area of the exposed canvas in an approximately 50:50 ratio. In ablation tests the laser beam focus was guided over the sample surface to cross both such areas, in order to assess any laser-induced effects on the different sample materials. The canvas and glue-paste samples were mounted priming-side down on glass microscope slides.

3.3 Experimental Method

The microscope slide was fixed vertically to the XYZ translation stage and tilt-adjusted until perpendicular to the beam axis. The computer-controlled Z axis and the manual XY axis had resolutions of 0.1μm and 20 μm respectively. To achieve selective ablation of the glue-paste layer, measurements were made of the thickness of the glue-paste layer above the canvas tops, and of the total thickness of the canvas and priming using a stereo microscope micrometer. Based on theoretical calculation of the depth of focus and focal spot size, a series of thirteen increasingly deep lines were ablated across the sample. For each line the focal position was lowered incrementally in 10μm steps.

3.4 Results of Ablation Experiments

The sample was examined using a stereo microscope. Evenly spaced parallel lines were clearly visible along the length of the sample, corresponding to scans 2-13 of the test where material had been removed. All lines were intermittent and of varying length because of the extremely uneven topography of the sample surface. Where the focus passed through a void such as a weave interstice, there was no visible interaction with sample materials. Measurement of the depths of ablation lines was undertaken using the focusing micrometer on the stereo microscope. In order to accommodate the extreme variation in the depth and the discontinuity along each scan line, sample materials (glue-paste, canvas and priming) were examined individually. Measurements were made along each scan line of the depth at which the focus had entered a given material, and the deepest it had penetrated in that material (Figs. 2a, b & c). The data gave an indication of the confocal parameters of the beam and the depth within the sample at which ablation had occurred.

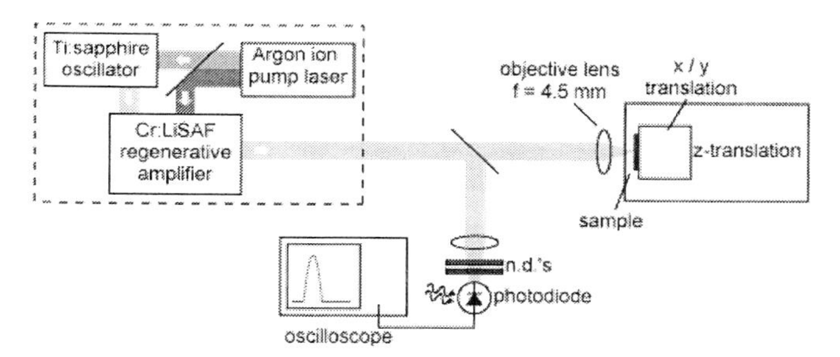

Figure 1. Experimental Configuration for Ablation

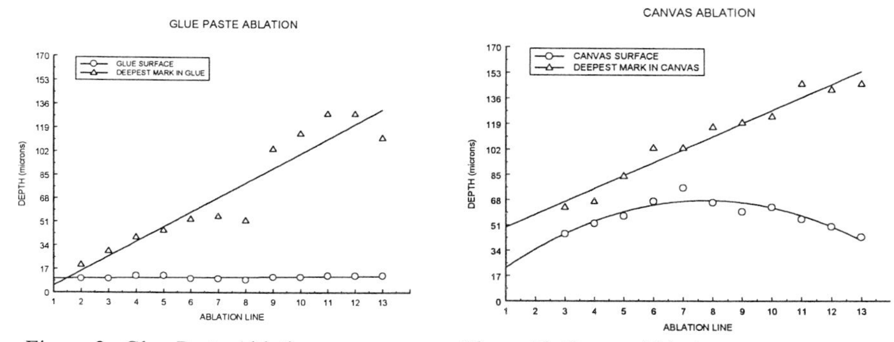

Figure 2a Glue Paste Ablation

Figure 2b Canvas Ablation

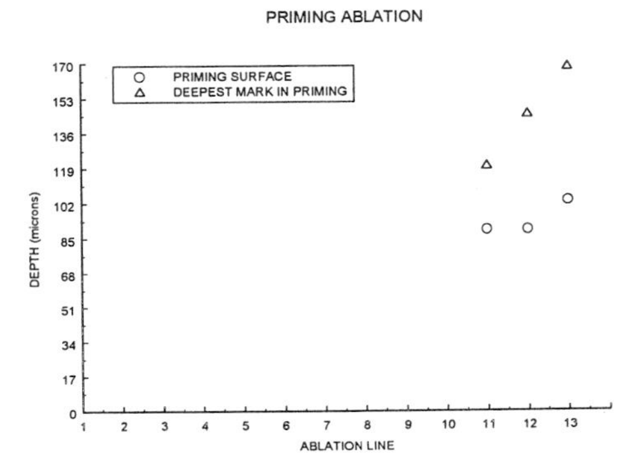

Figure 2c Priming Ablation

As expected marks in the glue-paste at lines 2, 3 and 4 were 10 μm, 20 μm and 30 μm deep, respectively, and are significant because they correspond to the repositioning of the focus. This shows the potential precision of the laser system, given correct guiding of the beam. This is essential given that canvas was also ablated in all lines after no. 2 and also some areas of priming where these areas were exposed to the beam focus. This confirms that this type of ablation is not dependent on wavelength absorption, emphasizing the need for a monitoring technique which could accommodate such an irregular interface. The irregularity of the increase in depth for each successive line is mainly due to low resolution and repeatability in the XYZ translation stages resulting in the delivery of uneven numbers of pulses per unit area as the beam was scanned across the sample. All lines were 5μm wide, because of the Gaussian beam profile only the central area is at or above the ablation threshold. Using a lens with appropriate confocal parameters the ablation depth could be tightly restricted, and the considerable precision already demonstrated in this study might be enhanced.

Microscopic examination of marks in glue-paste and canvas revealed no apparent changes to surrounding material which might have indicated thermal or mechanical damage. A black discoloration was visible in areas of priming, within the material along the beam path and in the area immediately surrounding it. This could not be accounted for, and appears to be a laser-induced alteration in the material suggesting that in this material, effects may not have been confined to material removal in the specific area of the beam focus. External constraints prevented repeat FTIR analysis of sample materials after irradiation, which might have enabled characterisation of any laser-induced changes.

4. Conclusions

Femtosecond lasers show potential as precision tools for the controlled removal of glue-paste in the delining treatment without causing thermal damage. The uneven topography of both the glue-paste and canvas and the likely presence of quantities of glue-paste within the sized canvas fibres present problems in the reversal of a glue-paste lining which emphasise the need for a sensitive, integrated, on-line monitoring technique. An experimental feasibility study has shown that FLIM would be the most appropriate technique for this application [18]. Recent reports indicate that highly reactive radicals may be produced during ablation, suggesting that lasers could contribute to the accelerated ageing of materials. It was not possible to undertake FTIR analysis of samples after exposure to the laser, and work in progress at the Tate Gallery is focusing on characterising the effects of irradiation on the sample materials. At present the laser system investigated is expensive and confined to the laboratory, requiring high level expertise to run and maintain. However, it is under development for use in medicine and costs are expected to fall, suggesting that it might become a viable option as a conservation tool.

Acknowledgements

This work was made possible by Dr Paul French of the Femtosecond Optics Group in the Physics Department at Imperial College who generously gave us access to his laser facilities.

References

[1] Hontzopoulos E., "Excimer Laser in Art Restoration", *Proc SPIE Vol 1810 (1992), pp748-751*

[2] Shekede L., "Lasers: A Preliminary Study of Their Potential For The Cleaning and Uncovering of Wall Paintings", *Diploma Project 1994, Courtauld Institute of Art, Wall Paintings Dept. (unpublished) pp69*

[3] Fotakis C., et al, "Lasers for Art's Sake!", *Optics and Photonics News, May 1995, pp30-35*

[4] Fotakis C., et al, "Lasers in Art Conservation", *1995 AIC Paintings Speciality Group Postprints, June 1995, pp36-42*

[5] Real W., et al, "Use of KrF Excimer Laser for Cleaning Fragile and Problematic Paint Surfaces", *ICOM Conference 1996, Preprints, pp303-309*

[6] Giorgiou S., et al, "Excimer Laser Restoration of Painted Artworks: Procedures, Mechanisms and Effects", *Accepted 1997 for publication in Applied Surface Science*

[7] Anglos D., et al, "Laser Diagnostics of Painted Artworks: Laser-Induced Breakdown Spectroscopy in Pigment Identification", *Applied Spect. Vol 51, 7 (1997) 1025-1030*

[8] Anglos D., et al, "Laser-Induced Fluorescence in Artwork Diagnostics: An Application in Pigment Analysis", *Applied Spectroscopy Vol 50, 10 (1996) pp1331-1334*

[9] Giorgiou S., et al, op. Cit.

[10] Hontzopoulos E., et al, op. Cit.

[11] Morgan N., "Excimer Lasers Restore Fourteenth Century Icons", *Opto and Laser Europe 7 (Sept 1993) pp36-37*

[12] Fotakis C., et al, AIC Postprints 1995

[13] Giorgiou S., et al, op. Cit.
[14] Shekede L., op. Cit
[15] Real W., et al, op. Cit.
[16] Neev J., et al, "Ultrashort Laser Pulses for Hard Tissue Ablation", *IEEEE Journal of Selected Topics in Quantum Electronics Vol 2, No 4, Dec 1996*
[17] Liu X. and Mourou G, "Ultrashort Laser Pulses Tackle Precision Machining", *Laser Focus World, Aug 1997, pp102-104*
[18] Young C., Shepard J., "The Application of FLIM in the Monitoring of Laser Cleaning of Paintings", submitted to *ICOM-CC 12th Triennial Meeting, Lyon, Sept., 1999*

UV-laser Ablation of Polymerized Resin Layers and Possible Oxidation Processes in Oil-Based Painting Media

V. Zafiropulos*, A. Galyfianaki, S. Boyatzis[1], A. Fostiridou[2], E. Ioakimoglou[2]
Foundation for Research and Technology-Hellas (FO.R.T.H.),
Institute of Electronic Structure & Laser, P.O. Box 1527, 71110 Heraklion, Greece
*Tel.: (+30) 81 391-485, -315, Fax: -305, -318, E-mail: zafir@iesl.forth.gr
[1]National Center for Scientific Research "Demokritos",
Institute of Physical Chemistry, Agia Paraskevi, 15310 Attiki, Greece
[2]Technical Educational Institute of Athens
Dept. of the Conservation of Antiquities and Works of Art, 12210 Egaleo, Greece

Abstract: The use of short-pulse lasers for ablating polymerized resin layers has been examined in detail. In particular examples of KrF excimer laser ablation of varnishes, artificially or physically aged, are presented. Ablation rate data, operational processes and critical experimental parameters are discussed. The aim is to develop a methodology for finding the optimal laser parameters, under which, no harm is done to the pigments below the usually unknown varnish layers. For this, Gas Chromatography coupled to mass analysis (GC-MS) has been used to assess the photochemical oxidation of the binding medium when different laser parameters are used. The combined results from the above studies lead to an understanding of UV-laser ablation in polymerized resins in relation to possible photo-oxidation processes.

1 Introduction

Modern technology has been playing a crucial role in the preservation of cultural heritage. In particular, laser based techniques have been used over the last 20 years in a variety of art conservation applications, ranging from diagnostics [1] to the removal of different types of encrustation and layers [1-4]. In contrast, laser applications for the conservation of painted artworks are more recent and rare [5-7]. This is a particularly demanding application due to the high sensitivity of paints and binding media upon exposure to light, requiring a good knowledge of the material and laser parameters involved. In addition to laser cleaning, intense interest also exists for the development and application of laser based analytical techniques as tools for artwork diagnostics. Such techniques can be performed in situ and are essential to artwork conservation and restoration since they provide important information regarding the physical and chemical structure of artworks. Laser based analytical techniques, such as Laser Induced Breakdown Spectroscopy (LIBS), can also be used for the on-line monitoring of the cleaning process to safeguard against potential damage [5-8].

Cleaning is important in maintaining the aesthetic aspects of the artwork and prolonging its lifetime by removing destructive environmental pollutants from the surface or the support material. There are many instances of demanding cleaning applications in which the use of lasers has been proved more effective than the traditional restoration techniques. Laser cleaning is based on the removal of well-defined layer of surface varnish, overpaint or encrustation under fully controlled conditions. The understanding of the laser ablation processes operative in each case studied is the key for a successful treatment. In the case of paintings, aged, contaminated or overpaint surface layers can be removed by UV-laser ablation with a sub-micron resolution. Characteristic ablation rate data as a function of laser fluence will be presented for various types of layers. From such data the ablation efficiency curves are derived and the optimum laser fluence can be determined for each material. The role of every major mechanism has been addressed in relation to possible photo-oxidation processes initiated by laser radiation. For this, laser ablation is combined with Gas Chromatography coupled to Mass Spectroscopy (GC-MS) analysis to study the oxidative cleavage and production of low molecular weight volatile products as the end-results of possible oxidative processes.

2 General Considerations

A deeper understanding of the physical processes and mechanisms involved into the interaction of laser light with matter is essential for the proper treatment of Artworks. The knowledge of important laser parameters such as the optimum laser beam focusing (laser energy fluence or just fluence), laser wavelength, pulse duration, pulse repetition rate etc. is absolutely necessary for the reliable use of lasers in the conservation of diverse materials. For the proper understanding of the importance of these parameters on the laser ablation mechanisms operating in different cases, one has to distinguish the materials in two discrete classes. The first class consists of materials whose main substance is polymerized organic matter, as for instance aged varnish, rigid overpaint layers or even organic dirt that has become hard and stiff with time. The second class includes mostly inorganic encrustation layers on marble, stone, stained glass, metals etc. originated generally from environmental attack. This separation is made based on the different ablation mechanisms operating on these two types of surfaces. In the present work we concentrate on the former class.

Material removal by laser ablation can be mainly explained by three major mechanisms acting simultaneously, but contributing in a varying extent to the amount of ejected material. These include thermal interactions, photomechanical and photochemical effects. The effects however, that these different processes can have on the future state of the restored artwork, is of major importance. Although in principle all these processes may operate simultaneously, photomechanical effects become important at high laser fluence or/and when dealing with artworks having considerable defects (detachments, cracks, voids, etc.). Both thermal and photomechanical effects are favored when short pulsed infrared lasers are used, e.g. the use of Nd:YAG laser for the removal of black encrustation from marble/stone sculpture. Photochemical effects, on the other hand, resulting in covalent bond cleavage mainly followed by small fragment formation [9] may be dominant with short pulsed UV-lasers (short wavelength excimer lasers and 4th-5th harmonic of a Q-switched Nd:YAG laser), when ablating polymers, resins, overpaintings or other polymerized materials.

Thermal effects are primarily determined by the thermal diffusivity parameters characteristic of the material processed and the laser pulse duration [10]. In particular, for nanosecond pulsed excimer lasers (at 193 or 248 nm) the heat affected zone for typical polymerized materials (such as aged varnishes used in paintings) is estimated to be comparable to the optical penetration depth. The result is that at low laser fluence and low pulse-repetition rates the temperature increase is confined within a very small proportion of the remaining resin. On the other hand, a laser operating at the visible or infrared part of the spectrum acts as a heat source and in most cases the evaporation of surface layers is the main cause of material loss [2, 3].

Photomechanical effects [11], for example thermoelastic stresses and shock wave formation, may also be important during laser ablation. Their direct contribution in laser ablation becomes significant when (i) the ablated material is mostly of inorganic nature such as marble, stone or thick overpaintings (ii) there are delaminations that can lead to detachment/ejection of flakes and (iii) high laser fluence is employed. In such cases a large percentage of the material comes out as small particles or flakes. When removing aged varnish on the other hand, the shock wave produced seems to have a negligible effect especially when working at low laser energy densities. Of course, this applies to the contribution of the photomechanical effects in the ablation process itself. The possible long-term consequence of the shock wave is a different issue that is currently under investigation.

Photochemical processes are mostly responsible for the laser ablation for organic materials when low energy densities of pulsed UV lasers are used [12]. The absorbed photons result in bond-breakage in which the products have a larger volume than the solid ablated volume. The subsequent ablation due to this sudden volume increase produces small fragments, basically two- or three-atom molecules, radicals, atoms or ions [9]. In parallel to the ablation process itself, the free radical formation as well as any excess of photons penetrating into the binding medium could result in a variety of unwanted effects, leading to possible long-term deterioration of the work. To address the importance of these processes during laser cleaning, laser ablation is combined with GC-MS analysis. This work is described in section 4.

3 Ablation of Polymerized Materials

Common problems encountered in painting conservation have to do with aging and/or photochemical degradation of the varnish layer, accumulation of various kinds of pollutants on the surface, or even overpaints on the original painting that must be removed. Cleaning of the artwork has to be done in a way that will guarantee the integrity of the original work.

Laser cleaning is based on the removal of well-defined layer of surface varnish, overpainting or stiff organic dirt under fully controlled conditions. In the case of paintings, aged/contaminated surface layers of 0.1 μm up to tens of μm in thickness can be removed by UV-laser ablation. It becomes evident that cleaning by UV-laser ablation can be a highly selective process, capable of leaving a thin protective layer of clean varnish over the painted area.

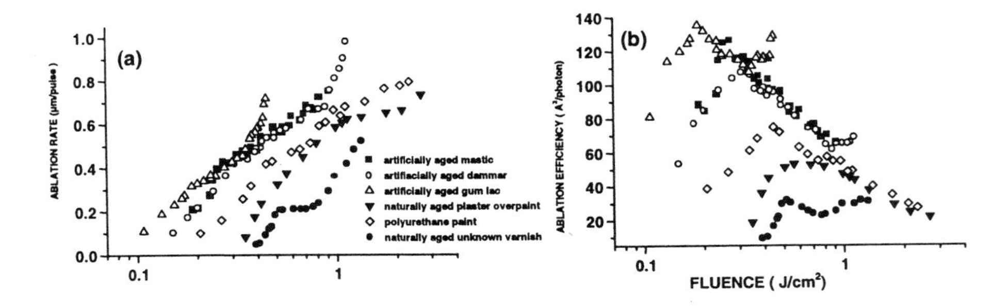

Fig. 1. Plots of the mean ablation rate (a) and mean ablation efficiency (b) versus the incident KrF excimer laser fluence. The uncertainty in measurements is estimated to be <10%. The laser fluence is measured by dividing the laser pulse energy by the affected area of a PVC sheet irradiated by one laser pulse.

In the case of overpaintings on the other hand, the material to be removed consists of (a) the medium used to prepare the overpaint. The unsaturated oil molecules (with double/triple covalent bonds) are usually polymerized with time, forming a hard three-dimensional lattice and (b) the pigment particles, which are trapped within this lattice. Owing to the bond cleavage during the UV-laser ablation, these particles are released while a fraction of them is also further ablated producing a characteristic plume emission. The existence of the pigment particles as well as their random size result in an ablation process less selective, from a topological point of view, than in the case of varnish removal.

Characteristic ablation rate curves as a function of laser fluence obtained from model and real samples using KrF excimer laser (emitting at 248 nm) are shown in Fig. 1. The six representative polymerized materials presented here are three types of aged varnish (mastic, dammar and gum lac), a multi-layer white plaster overpainting, a naturally aged unknown varnish and a black polyurethane film for comparison. The way such measurements are made, as well as the dependence of ablation rate on laser wavelength and pulse duration are described elsewhere [9].

The ablation rate curves of Fig. 1a reveal experimental data on the laser ablation of the six polymerized materials mentioned above. The data show the mean removed depth per laser pulse as a function of the laser energy fluence in units of J/cm^2. It can be seen that the resolution attained by the laser can be as low as 0.1 μm per pulse, which is simply impossible when using other techniques. Fig. 1a also demonstrates the attainable definite control of varnish removal, that is, it is possible to completely adjust the depth of each layer removal by appropriately either focusing the laser beam or varying the working energy of the laser. The axis of the laser fluence has intentionally a logarithmic scale to demonstrate the linear response of the ablation rate to the logarithm of the laser energy fluence, at low fluence values. This linearity can be explained by the Beer-Lambert law for light absorption from matter and the operation of the photochemical mechanism (photo-induced bond-cleavage) in the UV-laser ablation process [12]. For every polymerized material this linearity is always strictly localized just above the ablation threshold of this material and within a narrow range of laser fluence.

The ablation efficiency curves (Fig. 1b) provide the optimum laser fluence for treating each material. Here the word 'optimum' refers to the most efficient and at the same time least damaging laser fluence (see next section), although for many restorers 'optimum' may always be the laser fluence with the highest resolution. The efficiency is a quite significant parameter in cases where large surfaces or thick varnish layers are encountered, and therefore, the goal is to remove the maximum possible volume of material per available photon. The data presented in Fig. 1b indicate for instance, that the optimum laser fluence for the ablation of aged dammar and gum lac is about 0.3 and 0.2 J/cm^2 respectively, while for the white plaster it is about 0.65 J/cm^2. Another important observation in Fig 1b is the similar behavior between the six experimental trends. As the energy fluence increases, the efficiency starts rising until it reaches a maximum. This maximum corresponds to the point where there is saturation in the light absorption within the first layers of the surface. At this optimum laser fluence, the excited chromophores [13] (light absorbing sites within the 3-D macromolecular lattice of the polymerized substance) have reached a maximum density. For even higher fluence values, the excess light is mostly converted into heat and the efficiency drops. Especially in the case of the three varnishes, as we further increase the energy fluence and above a certain value (0.9, 0.35 and 0.8 J/cm^2 for dammar, gum lac and naturally aged varnish respectively), the ablation efficiency starts to increase for a second time. This phenomenon may be attributed to the fact that, for strong laser focusing the temperature of the surface increases rapidly above the sublimation-point and therefore a lot of material is lost through evaporation. At this high-energy fluence range the generated shock waves become also important, especially when the superficial layers present delaminations and entire flakes of varnish could be ejected. For varnish removal, it is not recommended to work at such high laser energy densities.

4 GC-MS Analysis

In parallel to the ablation process itself, the free radical formation as well as any excess of photons penetrating into the binding medium could result in a variety of unwanted effects, leading to possible long-term deterioration of the work. To address the importance of these possible processes during laser cleaning, laser ablation is combined with Gas Chromatography coupled with Mass Spectroscopy (GC-MS) analysis. The goal was to detect possible oxidation products initiated by laser radiation.

The presence of metal salts in many known pigments used in painting induces a potential catalytic factor in the oxidation of binding media, some of them accelerating the thermal and/or photochemical discoloration and degradation processes of oils. Such a pigment is mercury (II) sulfide (cinnabar) that is known to be photo-chemically active in catalyzing similar processes [14]. The process of degradation of linseed oil medium under oxidative conditions involves a simultaneous polymerization and de-polymerization reaction. It is generally accepted that unsaturated fatty carbon chains undergo rapid auto-oxidation. Especially for linoleate and linolenate, with two and three double bonds correspondingly, the rates for

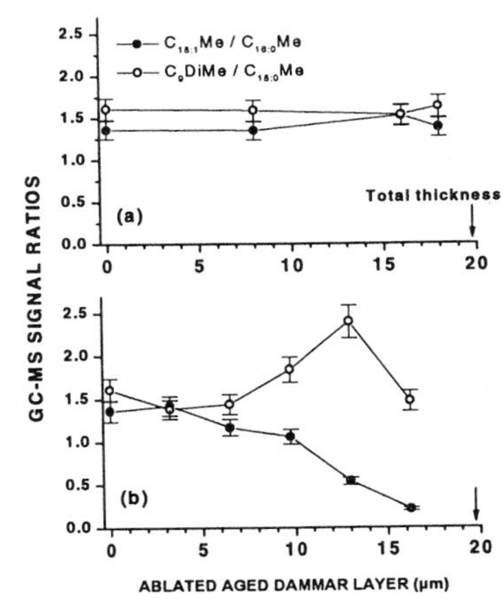

Fig. 2. Plots of GC-MS signal ratios for varying thickness of ablated material. (a) corresponds to a fluence of 0.3 and (b) to 0.8 J/cm^2.

oxidative degradation are even higher. This has been attributed to the pentadiene-containing chain of these esters which, in the presence of an initiator, form the stabilized pentadienyl radical which, after rearrangement of the C=C bonds, with the action of oxygen produces the thermally and photochemically unstable peroxides [15]. These finally result in yellow conjugated products, up to nine carbon atoms. The same time, cross-linking occurs as a basic consequence of the presence of radicals.

Oxidative cleavage and production of low molecular weight volatile products as the end-results of the possible oxidative processes, is routinely investigated by gas chromatography [16]. After transesterification of all possible glyceryl esters with hydrochloric methanol, a mixture of lower molecular weight methyl esters of all present fatty acids is produced and analyzed. Such a gas chromatogram from an aged linseed oil medium reveals that (a) long unsaturated fatty chains (e.g. oleate) are decreased due to oxidation, (b) saturated fatty chains are basically unchanged and (c) a number of oxidation products, such as dimethyl esters of azelaic (nonadioic) acid - not present in the starting material - is formed. The quantities of esterified fatty acid residues such as methyl palmitate (C16:0), stearate (C18:0), oleate (C18:1), as well as monomethyl and dimethyl azelate (C9) have been previously used as a measure of various oxidative routes in oils [16]. Oleate is slowly oxidized, while azelate is produced as an oxidation product, thus enabling an evaluation of the aging process. Therefore, low C18:1/C18:0 or C18:1/C16:0 and high C9/C16:0 or C9/C18:0 ratios suggest a strongly oxidized medium.

Based on the above, model samples of cinnabar pigment in linseed oil covered with dammar varnish (~ 20 μm thickness) were made and artificially aged using UV lamp and elevated temperature. Then for each sample a predetermined thickness of varnish was removed by using selected KrF excimer laser densities based on the ablation efficiency curve in Fig. 1b. Finally, the remaining pigment/medium of each sample was analyzed by GC-MS. Gas chromatography (column DB-5MS, 30m), at 9 psi, temperature range 90°-250° C (2° C/min) was performed with a mass detector (Micromass Platform II, calibrated with heptacosa reference, and tuned at 264 and 314 m/z of the same compound).

In general, the results have shown absence of oxidation products when at least a thin varnish layer is left intact and the optimum laser fluence is used. On the contrary, when all the varnish is totally removed and the laser directly irradiates the pigment the analysis of the remaining pigment reveals various oxidation products. In more detail, Fig. 2 presents the C18:1/C16:0 and C9/C18:0 ratios from the GC-MS signal for different values of laser ablation thickness (thickness of removed layer). Apart from the wavelength of the laser, the most important parameter in this type of experiments is the selection of the optimal laser fluence. Using the argumentation described in section 3 (also see Fig. 1b), the optimal fluence for the ablation of dammar is 0.3 J/cm^2, while a fluence of 0.8 J/cm^2 is considered rather high owing to saturation in the light absorption within the first layers of the surface. Figs. 2a and 2b correspond to the results using these two values of energy fluence. As it can be observed in Fig. 2a, there is no evidence of further oxidation (due to the laser radiation) compared to the reference sample, even when the remaining varnish layer is only 2 μm. This is owing to the use of all absorbed photons towards the ablation of material (absence of saturation); therefore the binding medium remains unaffected by the laser radiation. On the contrary, when using high laser fluence the concentration of the sensitive to oxidation oleic component is decreased as the thickness of the removed layer increases (see Fig. 2b). At the same time the concentration of the C9 oxidation product increases (the exception of the last point is owing to the further oxidation of C9 to lower molecular weight products). This behavior can be explained by the presence of saturation in the absorption, with a result, the remaining varnish layer acts like a filter to the penetrating towards the binding medium photons. As this filter gets thinner, the light that affects the binding medium is more.

5 Conclusions

The implementation of modern laser technology in the field of art conservation has been proven to have exciting possibilities and prospects both in cleaning as well as diagnostic applications. A successful laser-assisted removal of unwanted surface layers from various types of artworks is dependent on the appropriate choice of laser parameters, as well as on the application of on-line monitoring techniques as described elsewhere [5]. In any case, however, extreme care should be taken for the optimization of the laser parameters prior to the final laser application, so that to ensure the absence any unwanted effects on the artwork induced by laser irradiation.

Ablation rate studies have been proven to be essential for a deeper understanding of the various mechanisms involved in laser ablation of complex materials. The separation of the materials into two mayor categories (polymerized organic matter and inorganic encrustation layers) facilitates this understanding. For the former class of surface layers studied here, the ablation efficiency versus laser fluence diagrams provide essential information on the 'optimal' laser fluence for removing these layers. In the removal of aged varnish from sensitive pigments, this method of finding the optimal fluence was tested by analyzing the oxidation stage of the remaining binding medium using Gas Chromatography coupled to Mass Analysis (GC-MS). Further analytical work is in progress using sub-picosecond laser pulses in removing naturally aged varnish layers.

6 Acknowledgments

This work is partially supported by the "Laser Technology in the Conservation of Artworks" project N°640 of the EPET II Programme (Greece), from the EU Structural Funds and by the Ultraviolet Laser Facility operating at FO.R.T.H. under the TMR Program (DGXII, ERBFMGECT 950021) of the EU. We especially thank Michael Doulgeridis from the National Gallery of Athens for his helpful advice.

References

1. *Restauratorenblätter, Sonderband – Lacona I, Lasers in the Conservation of Artworks*, Eds. E. König, W. Kautek (Verlag Mayer & Comp., Vienna, 1997).
2. *Laser Cleaning in Conservation: an Introduction*, Ed. M. Cooper (Butterworth Heinemann, Oxford, 1998).
3. J.F. Asmus, IEEE Circuits and Devices Magazine, 6 (March 1986).
4. P.V. Maravelaki, V. Zafiropulos, V. Kylikoglou, M. Kalaitzaki and C. Fotakis, Spectrochimica Acta B 52, 41 (1997).
5. V. Zafiropulos and C. Fotakis, Chapter 6 in Ref. 2.
6. C. Fotakis, V. Zafiropulos, D. Anglos, S. Georgiou, P.V. Maravelaki, A. Fostiridou and M. Doulgeridis, in *The Interface between Science and Conservation,* Ed. S. Bradley (The Trustees of the British Museum, 1997) p. 83.
7. S. Georgiou, V. Zafiropulos, D. Anglos, C. Balas, V. Tornari and C. Fotakis, Applied Surface Science 127-129, 738 (1998).
8. I. Gobernado-Mitre, A. C. Prieto, V. Zafiropulos, Y. Spetsidou, and C. Fotakis, Applied Spectroscopy 51, 1125 (1997).
9. V. Zafiropulos, J. Petrakis and C. Fotakis, Optical and Quantum Electronics 27, 1359 (1995).
10. M. Himmelbauer, E. Arenholz and D. Bäuerle, Applied Physics A 63, 87 (1996).
11. B.P. Fairand and A.H. Clauer, Journal of Applied Physics 50, 1497 (1979).
12. R. Srinivasan and B. Braren, Chemical Reviews 89, 1303 (1989).
13. G.H. Pettit and R. Sauerbrey, Applied Physics A 56, 51 (1993).
14. F. Rasti, Studies in Conservation 25, 145 (1980).
15. N. Porter, Accounts of Chemical Research 19, 262 (1986).
16. J.S. Mills and R. White, *The Organic Chemistry of Museum Objects* (Butterworths, London, 1987) and references therein.

Vibration Monitoring by TV-Holography — a Diagnostic Tool in the Conservation of Historical Murals

Thomas Fricke-Begemann, Gerd Gülker, Klaus D. Hinsch, Holger Joost
Department of Physics, Carl von Ossietzky University Oldenburg
P.O.B 2503, D-26111 Oldenburg, Germany

Abstract. A TV-holography system is designed to monitor detached plaster areas in historical murals. Defects in the wall respond to sound irradiation by vibrations. These are detected optically. A feature of the system is the enhanced sensitivity obtained by reference wave modulation. Experimental results obtained at a historical site show very good agreement with data gained by the traditional percussion method.

1 Introduction

Deterioration of historical works of art, usually the result of a century-long impact by man and climate, has been accelerated by the pollution of recent decades. In historical murals, for example, dilapidation of plaster or paint layers from the supporting wall is driven by the crystallization of salts which may lead to partial detachments or even a complete loss of sections in the mural [1]. In the preservation of such delicate artwork it is important to assess the stability of the mechanical bond between plaster and wall and to identify any debonded regions. A commonly used technique is the so-called percussion method: a restorer knocks gently with his finger against the test section and estimates the quality of the bond from its acoustical response. This method shows some crucial disadvantages. When an exploration of large wall sections is required, as in the selection of regions for immediate remedy or in the control of conservatory steps, the procedure is very cumbersome and time consuming. In addition, expensive scaffolding is usually required. Finally, the method is intrusive because the finger contact may affect the delicate layer of paint.
TV-holography, also termed electronic speckle pattern interferometry (ESPI), has been introduced successfully for displacement and deformation investigations in historical buildings and monuments [2]. Portable equipment was developed that operates directly at the monument and allows nonintrusive and remote long-term measurements. So far, mostly basic studies have been made, contributing to the general understanding of deterioration processes; the introduction as a general tool in everyday practical work, however, is still overdue. Monitoring of mural detachment promises to be an essential step towards everyday routine application.

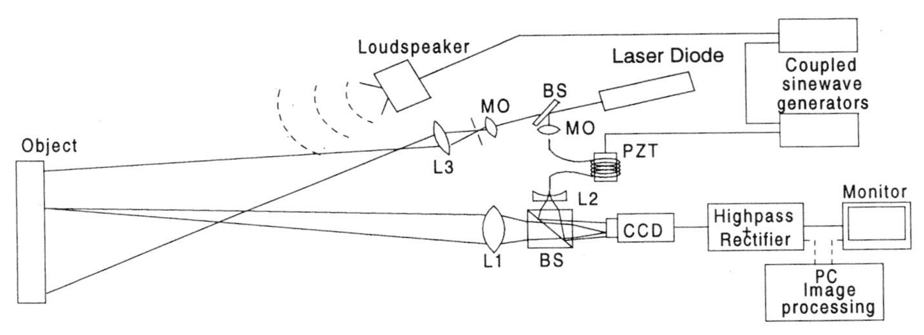

Fig. 1. Experimental setup for vibration monitoring (microscope objective MO, piezoelectric cylinder PZT, lens L, beam splitter BS

2 Measuring Method and Experimental Setup

In a first step, any debonded regions in the mural are excited to minute vibrations by exposing the wall to sound from a loudspeaker. To ensure considerate handling of the delicate artwork and avoid acoustical annoyance of the environment, minimum sound pressure levels are a prerequisite. Thus, in the second step, the resulting vibration amplitudes of some ten nanometers only are recorded with an analog TV-holography system of enhanced sensitivity (Fig.1). As known from literature such sensitivity can be achieved by modulating the path length of the reference beam of the interferometer at a frequency corresponding to the vibration frequency of the object [3,4]. In our system the reference beam is guided by a monomode optical fiber, a section of which is wrapped around a piezoelectric cylinder which is driven to periodically stretch the fiber. The video signal is processed by an analogue circuit, displayed on a monitor, recorded on videotape or stored in a computer memory.
As in conventional TV-holography, this new method offers full-field and video real-time capability. By using a laser diode light source and a CCD-camera pickup a very small and compact setup is obtained. For practical purposes, a frequency shift of a few Hz between reference signal and acoustical excitation provides a flickering intensity signal wherever there are vibrations. This frees from absolute intensity estimation, difficult in objects of varying background intensity, and attracts the attention even of an unscilled observer. Such considerations are essential for the successful introduction of a high-tech tool into a traditional field.

3 Testing the Method: Medieval Funeral Chapel

Performance and features of the new system were originally investigated in a test wall with artificially produced detachments. In this paper, we demonstrate results that were obtained on medieval murals in the small funeral chapel *St. Just* at

Fig. 2. Test area from mural in funeral chapel *St. Just* at *Kamenz*, Saxony, width about 3m (Photo curtesy R. Berg, HHFBK, Dresden)

Kamenz in Saxony. Here, the apse is covered with a fresco originating from the 14th century. Probably due to high amounts of salt and humidity inside the walls, parts of the mural show pronounced dilapidation. In 1996 the adherence of the plaster to the wall was examined in the whole apse by the percussion method. Thus, a complete map of percussion exploration of the fresco is available. On this basis, a thorough comparision could be carried out between the results obtained by the new acoustical-optical method and those from the traditional method. Several areas in the church were investigated including portions located about 8 meters high at a vaulted part of the ceiling of the chorus. These, too, were studied with the optical setup and the loudspeaker mounted on two stable tripods on the church floor, showing the great advantage of remote detection.
Here, we present results from a vertical portion in the apse wall that measures

Fig. 3. Exploration map of plaster detachments of the test area in Fig.2. Vertical hatching: acousto-optical method; 45° hatching: traditional percussion method

about 2x3 m^2 (Fig. 2). With a laser output power of about 150 mW an area of about one to two square meters, depending on the reflectivity, could be observed simultaneously. Thus, the test sections were studied by subareas and the measured results patched together for a final exploration map. The frequencies of the acoustical excitation and of the reference beam modulation were swept from 30 Hz to 500 Hz to locate debonded areas from resonances. At sound pressure levels of about 85 dB, the induced vibration amplitudes were about 20 nm. The resulting electronically filtered TV frames and one unprocessed original image of the test area were recorded on a video tape for later evaluation. For the final map regions were considered loose whenever they were found to vibrate anywhere during the frequency sweep. Fig. 3 shows both the results from the manual percussion method and the remotely operated optical technique, distinguished by the type of hatching. Obviously, in most places the results agree very well.

4 Conclusions

Video holography has proved a powerful restauratory tool for the mapping of loose plaster regions in murals. Only by continuous cooperation with expert restorers it was possible to guarantee consideration of all essential conservational aspects. Further development towards an automatic evaluation of data comparable to the well established detachment maps is necessary to gain general acceptance of the method in everyday practical monument diagnostics.

The investigations were done in close cooperation with R. Möller and R. Berg, 'Hochschule für bildende Künste' Dresden (HBFK). The present work is funded by the German Federal Foundation for the Environment (Deutsche Bundesstiftung Umwelt).

References

1. Mora, L. Mora, P. Phillipot, *Conservation of Wall Paintings*, Butterworths, London, 1984.
2. Gülker, G.; Helmers, H.; Hinsch, K.D.; Meinlschmidt, P.; Wolff, K., J. Opt. Las. Eng, Vol. 24, 183-213 (1996).
3. Løkberg, O.J., J. Acoust. Soc. Am., 75 (1984), 1783-1791.
4. Høgmoen, K., Pedersen, H.M., J. Opt. Soc. Am. 67(11), 1578-1583 (1977).

Algorithms for Pigment Identification

Mónica Breitman
Dep. de Matemática Aplicada y Telemática, E.T.S.E.T.B.
Módulo C3, Campus Nord
c/ Gran Capitán s/n, 08071 Barcelona Tel 401 60 02, Fax 401 59 81
E-mail: monica@mat.upc.es

Abstract. In the study of pigments in art works, we want to identify which pigments were mixed by the artists. The technique used is the Raman spectroscopy. We designed a specific database and we developed some algorithms for processing and understanding the original information

1 Introduction

Raman spectroscopy allows us to obtain characteristic spectra of each material [1][2]. In the artwork, it is possible to identify different pigments with theirs respective Raman spectra [3].

The position of the peaks in the spectra gives enough information for recognition of the different pigments. No two different pigments with all their peaks exist in the same frequency.

In this paper, we present a specific spectra database and a manager of spectra. The latter needed to be developed for working with the spectra taken in the laboratory and for storing only the basic information. This program has some algorithms that recognise the pigments in a mixture. We have different techniques for spectra treatments, particularly in those cases where the mixed pigments do not behave as in the expected way.

2 Data Base and Spectra Manager

Although the parameters used for pigment identification are amplitude and frequency, we also decided to recover bandwidth. In this way, we are able to define with accuracy where a Raman band exists and where does not. To take the three parameters we created an algorithm that converts the real spectra into ones where each Raman band is represented by a lorentzian. At this point, the data can be exported to the database before the signal pass for several filters modified by the users. There exists a peak detector. In each case it is possible to change the limits of amplitude and frequency to define where a peak is.

On some occasions we wished to find only those pigments that coincide with the biggest in the mixture or with the two biggest and so on. In this case we introduced a variable P, which changes the number peaks to be taken in account.

Furthermore, it was discovered that some pigments have a non-linear behaviour in mixtures. We then introduce a mechanism for making sums of different pigments imported from the database, in different conditions, which is called arithmetic sample. The idea is to compare the physical sample obtained in the laboratory with the arithmetic one. We are attempting to classify of all the pigments according to their behaviour in mixtures, taking into account how much the concentrations vary from the theoretic value.

3 Experiments

Some mixtures of the Lead White and the Litergirio in different concentrations were made. Here we show what happens when they are mixed together in equal (50%-50%) quantities. Figure 1 is the real spectra and figure 2 is the arithmetic sum.

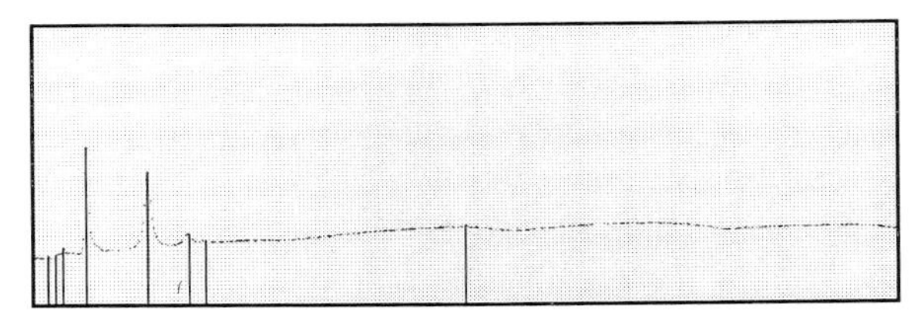

Fig. 1 . Real Spectra

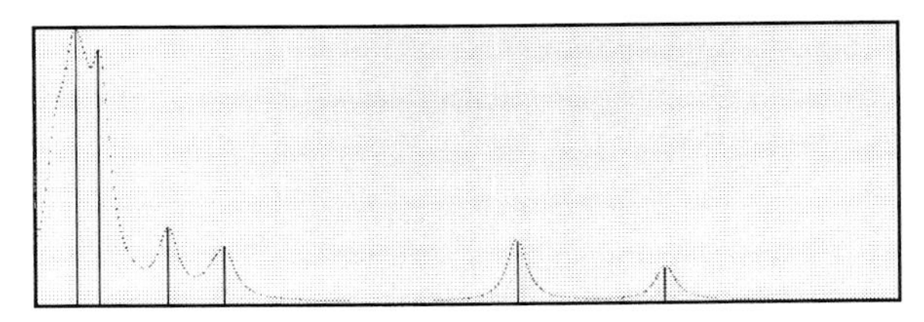

Fig. 2. Arithmetic Spectra

Our programs were used to study the differences. In the first screen (figure 3) we see the spectra of the physical sample; in the window we can see the identification. The last screen (figure 4) shows the same for arithmetic sample. It is obvious that most important peak of the Lead White has been lost. Observing the spectra, we may suppose that the proportion of the Litergirio is great than that of Lead White.

4 Conclusions

At the present time, there are an important number of pigments in our database. They are stored in an economic way and can easily be recovered.

Identifications that have been carried in pure samples are proof that algorithms are working well. As far as the mixtures are concerned, when the pigments are not behaving strangely, mixed pigments can be identified with their more important bands. Cases like those in the example can provide new information about the mixtures.

The causes of this are at the present being studied, and it is reasonable to suppose that relationship exists between the reflection and absorption coefficients [4][5].

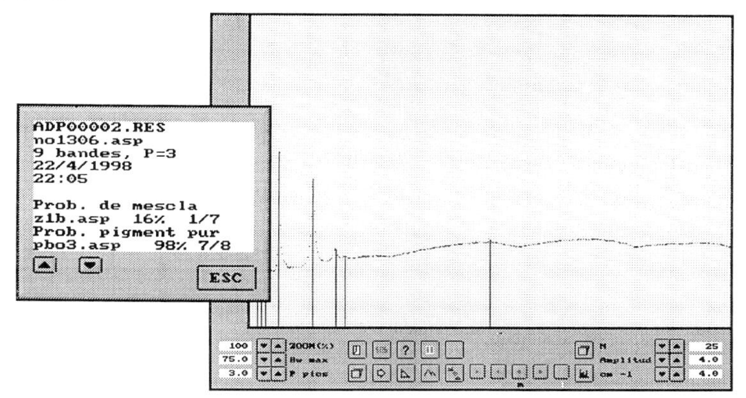

Fig. 3. Identification of the real sum

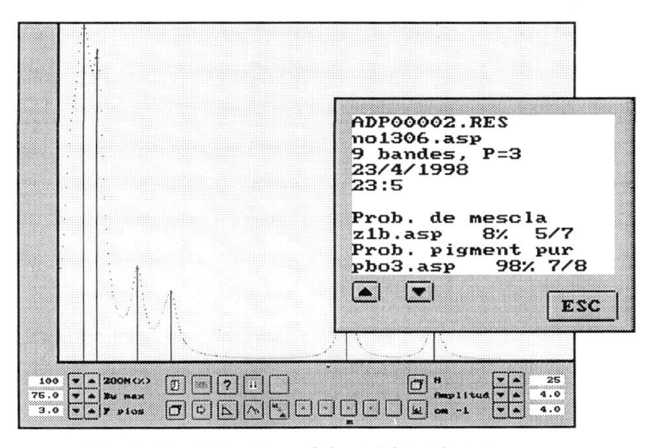

Fig. 4. Identification of the arithmetic sum

References

[1] S.P. Best, R.J.H. Clark, R. Withnall, Endeavour, vol 16, N°2 pp 66-73, 1992

[2] M. Ley, M.D. Brisant, M. Rudler. Analusis, vol 21 N°6 pp 47-49, 1993.

[3] M.J.Manzaneda, J.M. Yúfera, S. Ruiz Moreno, M.Soneira, P.Morillo, M.Breitman, Raman Spectroscopy for Pigments Analysis, Restauratornbläter, vol 16, 1996

[4] J.Reichman, Determination of Absorption and Scattering Coefficients for Nonhomogeneous Media.1:Theory, Applied Optics, vol.12, N°8, pp 1816-1823, 197

[5] W.G.Egan, T.Hilgeman, J.Reichman, Determination of Absorption and Scattering Coefficients for Nonhomogeneous Media.1:Experiment, Applied Optics, vol.12, N°8, pp 1811-1815, 1973

Investigations of Handwritten Manuscripts by Means of Optical Correlation Methods

Mathias Senoner[1], Sven Krueger[1], Guenther Wernicke[1], Nazif Demoli[2], and Hartmut Gruber[1]

[1] Humboldt-Universitaet zu Berlin, Institut fuer Physik, Labor fuer Kohaerenzoptik
[2] Institute of Physics at the University of Zagreb

Pattern recognition with coherent optical correlators was applied to different objects, among them handwritten documents [1,2]. A lot of investigations deal with character recognition, but only a few papers have analysed the relation between manuscript and writer. We used this method to analyse the origin of unknown manuscripts, which are of interest in music history. Collections of handwritten sources sometimes contain different versions of unknown origin forming compositions or arrangements for different instruments. Especially the distinction between the composer's autographs and handwritten copies of other writers is important in this regard. In this paper we present an analysis of manuscripts which are of interest in connection with the complete edition of works of Arnold Schoenberg (1874 -1951), known as a pioneer in music of the 20th century.

We try to determine the writer of one of these manuscripts by coherent optical filtering. The optical coherent 4f-correlator [3] is able to realize convolution operations on two-dimensional patterns $(f(\ ))$, $(h(\ ))$ via the fourier transform: $\mathcal{F}\{f(\ ) * h(\ )\} = \mathcal{F}\{f(\ )\}\mathcal{F}\{(h(\ )\}$. In the frequency plane complex-valued filter functions will be applied to the image spectrum to receive the correlation output.

We used this method on isolated signs from different manuscripts. We decided to compare treble clefs because their shape is sufficiently complex and therefore characteristic of the writer. Furthermore, they appear frequently on every page of the analysed manuscripts.

After scanning the signs from copies of the handwritten documents they were scaled, positioned and then the five staves were removed (fig.1). These isolated signs were colour inverted (white on black background) and normalized with respect to their energy (defined as the sum of the squared amplitudes of all pixels in the image).

For the identification of unknown objects it is usual to work with averaged filters (synthetic discriminant functions - SDF) which characterize a class of objects. Our first approach to the problem was to correlate single objects of different classes, because this method gives more detailed information than using the SDF.

We determined the similarity between two signs by the digital simulation of a coherent 4f-correlator with the classical matched filter [4] , which modulates amplitude and phase. We used this filter because only in this case the correlation is symmetrical with respect to an exchange of input scene and filter source. This symmetry corresponds to the usual understanding of similarity between two objects.

fig.1

The first sign (input scene) was placed in the input plane whereas the second sign (filter source) was fourier transformed, phase inverted and then placed in the frequency plane. The maximum of intensity (squared amplitude) in the correlation plane was taken as a measure of similarity between the two signs (fig.2). These correlation values were normalized with respect to the autocorrelation values (taken as 100%), which are the same for all signs due to the energetic normalization of input scenes and filter sources.

In a first step we analysed the similarity of the prepared treble clefs within a set of 14 signs (reference set) from a manuscript ("15 poems from The book of the hanging gardens by Stefan George", Opus 15 by Arnold Schoenberg) which was written by Arnold Schoenberg without any doubt. The results are given in a correlation table for the 14 treble clefs (tab.1). The similarity varies between 13.2% and 46.6% of the autocorrelation and the mean values for the columns range from 19.2% to 25.9%. The total mean value of 22.8% characterizes the similarity within the whole set.

In a second step we analysed a set of treble clefs written by the Schoenberg pupils Alban Berg and Felix Greissle (false class), because their shapes are more similar to Arnold Schoenbergs treble clefs than those of other pupils and co-workers. For ten signs of this class we calculated the correlation with the 14 signs of the reference set. The correlation values varies between 8.7% and 26.6%, whereas the mean values for the columns (characterizing one element of the false class) vary between 12.3% and 19.9%. These values overlap with the correlation values within the reference set.

Based on the comparison of reference set and false class we created a criterion to attribute unknown treble clefs to Arnold Schoenberg, which excludes all signs from the false class. Because of the strong variation of the treble clefs within the reference set we based the criterion on the maximum

matched filter correlation of treble clefs with a simulated 4f-correlator

the treble clefs are normalised to the same energy
the correlation values are the maximum intensities in the correlation plane,
they are normalized with respect to the autocorrelation (100%)
the treble clefs are taken from an autograph of "15 poems from The book of the hanging gardens by Stefan George", Opus 15 by Arnold Schönberg

46,6%
13,2%

filter source \ input scene	26_1	26_2	26_4	26_5	26_7	26_8	26_10	26_11	27_1	27_2	27_6	27_7	27_9	27_10	
26_1	100.0	19.8	22.5	24.7	31.5	14.0	19.3	20.6	24.8	19.0	26.5	29.0	18.5	25.6	
26_2	19.8	100.0	**13.2**	**46.6**	26.3	27.2	17.0	19.7	18.7	14.8	23.2	27.7	35.5	25.8	
26_4	22.5	**13.2**	100.0	16.8	22.9	14.7	28.0	17.1	16.8	23.9	17.6	16.6	17.2	22.2	
26_5	24.7	**46.6**	16.8	100.0	25.6	29.2	21.4	17.0	17.4	21.9	23.9	19.9	31.1	30.9	
26_7	31.5	26.3	22.9	25.6	100.0	26.6	17.7	21.0	22.1	19.6	29.4	18.2	29.5	32.0	
26_8	14.0	27.2	14.7	29.2	26.6	100.0	17.8	14.2	13.8	18.0	20.2	19.5	23.7	32.6	
26_10	19.3	17.0	28.0	21.4	17.7	17.8	100.0	17.0	25.2	21.4	17.2	27.8	22.8	15.9	
26_11	20.6	19.7	17.1	17.0	21.0	14.2	17.0	100.0	22.2	19.1	26.2	28.4	16.1	18.7	
27_1	24.8	18.7	16.8	17.4	22.1	13.8	25.2	22.2	100.0	20.0	29.1	25.6	21.0	16.3	
27_2	19.0	14.8	23.9	21.9	19.6	18.0	21.4	19.1	20.0	100.0	20.4	23.0	25.0	17.0	
27_6	26.5	23.2	17.6	23.9	29.4	20.2	17.2	26.2	29.1	20.4	100.0	36.6	31.4	24.8	
27_7	29.0	27.7	16.6	19.9	18.2	19.5	27.8	28.4	25.6	23.0	36.6	100.0	36.0	19.5	
27_9	18.5	35.5	17.2	31.1	29.5	23.7	22.8	16.1	21.0	25.0	31.4	36.0	100.0	29.5	
27_10	25.6	25.8	22.2	30.9	32.0	32.6	15.9	18.7	16.3	17.0	24.8	19.5	29.5	100.0	
mean value (without autocorrelation)	22.8	24.3	19.2	25.1	24.8	20.9	20.7	19.8	21.0	20.2	25.1	25.2	25.9	23.9	**22.8**

tab.1

correlation of a column K_{max} and not its mean value. This corresponds to the demand that at least one sign of the reference set is sufficiently similar to the sign of unknown origin.

In the presented case the maximum correlation value of the false class is 26.6 % and therefore we attribute all treble clefs with $K_{max} \geq 28\%$ to the true class. With this threshold 13 of the 14 signs of the reference set and no sign from Greissle and Berg are elements of the true class.

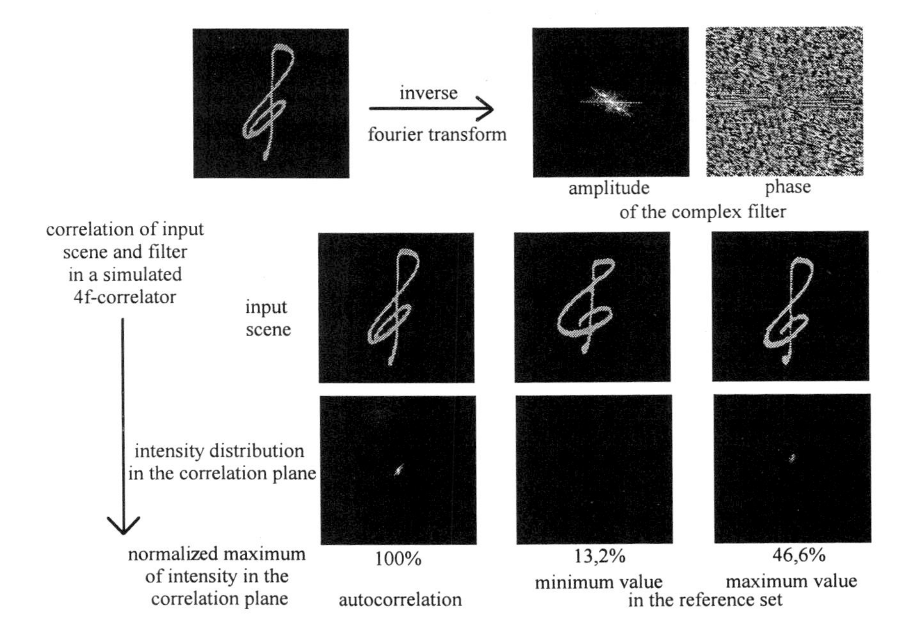

fig.2

In the third step we analysed a manuscript of unknown origin (part of the first violin in the monodram "expectation", opus 17 by Arnold Schoenberg). We calculated the correlation values of 13 treble clefs from two pages of this manuscript with the 14 treble clefs of the reference set. The cross correlation table yields values between 10.2% and 37.7%, and its maximum value is exceeded by only one correlation value within the reference set. The mean values (characterizing one sign of the unknown class) vary between 15.0% and 24.2%. By application of the threshold $K_{max} \geq 28\%$ we attribute 10 of the 13 signs to the true class.

From these facts we conclude that the manuscript of unknown origin is an autograph of Arnold Schoenberg.The presented results show that coherent optical filtering leads to a quantitative and objective measure describing the similarity of handwritten signs.

References

1. T.S.Wilkinson, D.A.Pender, J.W.Goodman (1991) , Use of synthetic discriminant functions for handwritten verification, Applied Optics 30, 3345
2. N.Demoli (1996) in, Fourth International Conference on OWLS IV , D.Dirksen and G.von Bally (Eds.), Muenster, Germany, 161
3. A.Vander Lught (1964) IEEE Trans. Inform.Theory, IT-10, 139
4. V.Kumar, L.Hassebrook (1990), Performance measures for correlation filters, Applied Optics 29, 2997

Optical Fibers for the Cultural Heritage II: The Monitoring of Lighting in Museum Environments

R. Falciai *, A.G. Mignani *, C. Trono *, B. Tiribilli §
* IROE-CNR, Dept. Optoelectronics and Photonics, Via Panciatichi 64, I-50127 Firenze, Italy, ph. +39-055-42351, fax +39-055-4379569,
e-mail: falciai@iroe.fi.cnr.it
§ Istituto Nazionale di Ottica, Largo Enrico Fermi 6, I-50122 Firenze
ph. +39-055-23081, fax +39-055-2337755

Abstract. An optical fiber sensor for monitoring lighting in museum environments is presented. It is an extrinsic, intensity-modulated type sensor that uses a photochromic material as transducer. The prototype is operative at the Uffizi Gallery of Florence.

1 Motivation

Light is one of the most important factors enabling visitors to fully enjoy the visual aspect of art. However, all organic materials such as paper, textiles, resins and leathers are damaged by excessive or unsuitable lighting, because of the light-induced chemical reaction. A typical reaction is photo-oxidation, which is the formation of free radicals caused by photon absorption and subsequent combination of the radicals with oxygen molecules. Photo-oxidation is a cumulative and irreversible effect; consequently, artistic material remains permanently damaged by the wrong lighting.

The guidelines for correct illumination suggest a maximum illuminance of 50 lux for highly delicate materials such as watercolors, paper and textiles, while many robust materials, such as tempera paintings and wood, can support an illuminance of 150 lux. As far as the fraction of UV light is concerned, all materials should be illuminated by means of 75 μW/lm at maximum[1].

In this paper, an optical fiber sensor is presented which is capable of measuring lighting in museum environments. Optical fibers are used in combination with a photochromic transducer which operates as sensitive element. Calibration of the sensor is provided together with the on-line monitoring of daily lighting.

The low environmental impact offered by optical fibers, and their optimized integration with fiber optic communications networks, which are being used always more often in modern museums and galleries, make optical fiber sensors attractive for application in safeguarding the cultural heritage.

2 Working Principle

The optical fiber sensor for monitoring lighting is an extrinsic, intensity-modulated type sensor. The sensitive element is a photochromic material made of fulgides [2,3] which are immobilized on a polymeric matrix. The photochromic transducer is coloured by UV-lighting, and is bleached by VIS-lighting. The absorption spectra of the photochromic material when exposed to fractions of UV in the 60-100 μW/lm range are shown in Figure 1. The temporal behaviour of fulgide colouring when exposed to both UV and VIS lighting can be expressed as

$$y(t) = -\text{asympt} + A \cdot e^{-m \cdot t} \qquad (1)$$

where A is a constant depending on the starting state of the fulgide, and asympt and m are proportional to

$$\text{asympt} \propto \frac{UV(\mu W/m^2)}{UV(\mu W/m^2) + VIS\,(lux)} \quad ; \quad m \propto UV(\mu W/m^2) + VIS(lux) \qquad (2)$$

In practice, the asymptotic value, asympt, represents the attainment of the photostationary state, while the time decay constant, m, represents the combined action of UV and VIS lighting. Consequently, measurement of both asympt and m makes it possible to obtain the fraction of UV, together with the individual UV and VIS light levels.

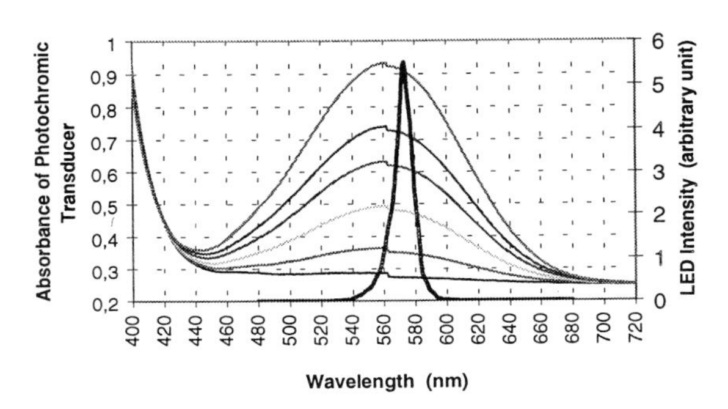

Figure 1 Absorption spectra of the photochromic transducer when exposed to fractions of UV in the 60-1000 μW/lm range, and emission spectrum of LED for sensor interrogation

3 Sensor Operation

The probe consists of a 0.17-mm thick layer of fulgides aligned between of an optical fiber link coupled to the interrogation electro-optic unit. Multimode optical fibers 200-μm core diameter were used. The electro-optic unit makes use of commercially-available LED and PIN as source and detector. The LED spectrum is shown in Figure 1, superimposed on the fulgide absorption spectra. The electronics is interfaced to a portable PC equipped with a specifically-developed software for both calibration and measurement functions.

Sensor calibration was performed by simultaneously exposing the transducer to intensity-tuned tungsten and mercury lamps, while UV and VIS light levels were measured by means of a reference photometer. Figure 2, 3 and 4 show the calibration curves of the optical fiber sensor. The response time of the sensor, which depends on the UV and VIS light intensity, is in the 20-30 minute range for the first attainment of the photostationary state, while it is of the order of few minutes when the sensor continuously follows the variations of the daily lighting.

In order to test the optical fiber sensor during normal operating conditions, both the reference photometer and the fiber sensor were exposed to natural lighting for 24 hours while their outputs were recorded. The agreement between their responses was encouraging, suggesting the testing of the optical fiber sensor in a real museum environment.

At present, the sensor prototype is operative at the Uffizi Gallery of Florence, in the room housing the paintings of Antonio del Pollaiolo. That room has a south-west orientation, with two big windows, the curtains of which are manually operated by the museum custodians. Figure 5 shows a typical response curve of both fiber optic and reference sensor recorded in the Gallery. Since both sensors are placed at the height of the paintings being monitored, in front of which people pass, the reference photometer output, which is rather fast, is highly fluctuating.

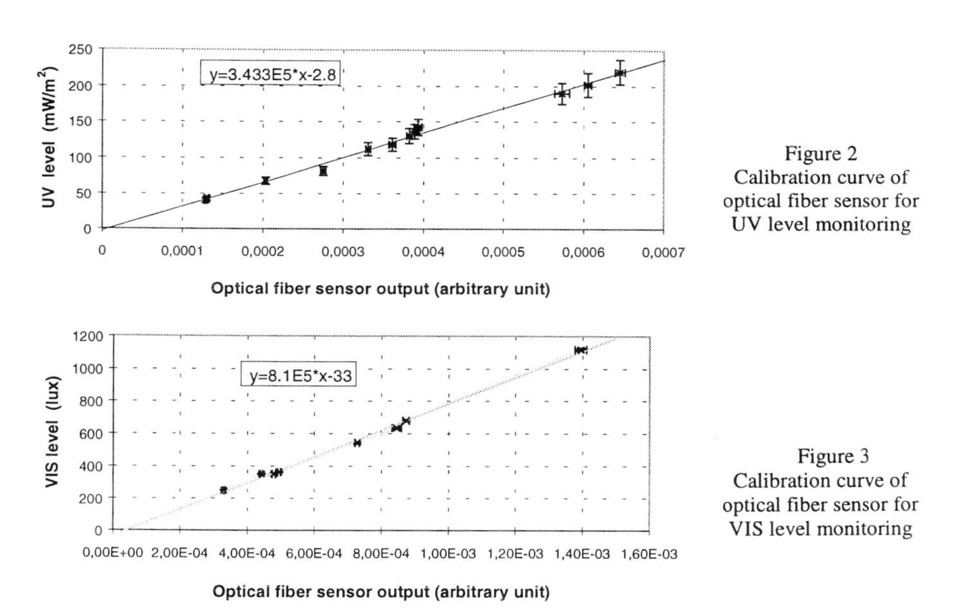

Figure 2
Calibration curve of optical fiber sensor for UV level monitoring

Figure 3
Calibration curve of optical fiber sensor for VIS level monitoring

On the contrary, the long response time of the fiber optic sensor, which could be a drawback for other photometric applications, in this case gives more stable and significant results. It is to be noted that, when the Gallery is closed to the public, emergency lighting is operative. The level of emergency lamps is too low to be detected by the photometer, while fiber optic sensor records the small UV fraction.

4 Remarks

The fiber optic sensor for monitoring of lighting in museum environments has demonstrated its functionality in real operational conditions. Another application of the sensor for safeguarding the cultural heritage, is for lighting monitoring during artwork restoration. In fact, restoring procedures are performed under high illumination, although for a short time compared to the life of the artwork. Since there are regulations also on excessive lighting for short times, continuous monitoring can indicate when the exposure is becoming too excessive. The low cost of the technology involved, indicated as a further development of the sensor an implementation as a battery-powered unit to be permanently installed where critical conditions are feared.

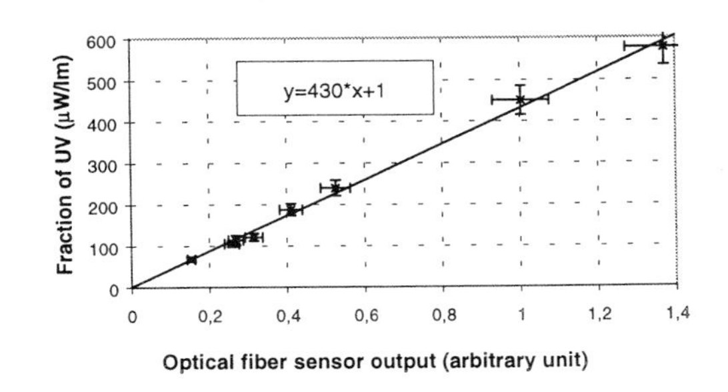

Figure 4 Calibration curve of optical fiber sensor for monitoring the fraction of UV

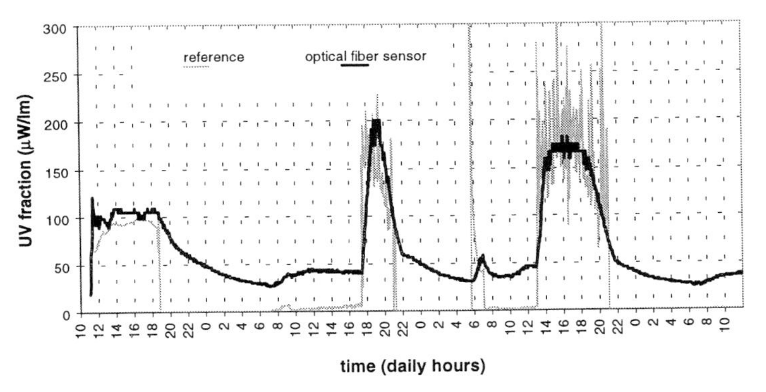

Figure 5 Continuous monitoring of UV fraction during 3 days at the Uffizi Gallery of Florence

Acknowledgements

This work has been partially funded by the National Research Council of Italy, under the *"Cultural Heritage"* Special Project.

References

1. G. Thomson, *The Museum Environment*, Butterworth-Heinemann Ltd., Oxford UK, 2nd Edition, 1986.

2. H.G. Heller, "Photochromic heterocyclic fulgides: part 7", *J. Chem. Soc. Perkin II*, 1992, pp. 591-608.

3. H.G. Heller, "Fulgides and related systems" in *CRC Handbook of Organic Photochemistry and Photobiology*, Boca Raton FL, 1995, pp. 174-191.

Fast and Precise Determination of Painted Artwork Composition by Laser Induced Plasma Spectroscopy

Alessandro Ciucci[1], Vincenzo Palleschi[1], Simone Rastelli[1], Azenio Salvetti[1], Elisabetta Tognoni[1], Roberta Fantoni[2], and Ilaria Borgia[3]

[1] Istituto di Fisica Atomica e Molecolare del CNR
Via del Giardino, 7 - 56124 Pisa (ITALY)
WWW: http://www.ifam.pi.cnr.it/libs.htm
[2] ENEA, Dip. Innovazione, Frascati (ITALY)
[3] Dipartimento di Chimica dell'Università di Perugia (ITALY)

Abstract. In this paper we present recent results of application of a new technique of Laser Induced Plasma Spectral analysis, developed and patented by IFAM-CNR, to the study of painted artworks composition. Thanks to this new approach, a fast and precise determination of paintings composition can be obtained without need for reference samples or calibration curves.

1 Introduction

In recent years, a great deal of interest has been devoted to laser applications to Cultural Heritage study, both in the fields of artwork conservation [1],[2] and analysis [3],[4],[5]. Among them, Laser Induced Plasma Spectroscopy seems to be one of the most promising approach; a number of applications have been suggested, ranging from automatic control of the laser cleaning process [2] to paintings analysis for study and authentication [4]. However, LIPS applications to artwork conservation and analysis were limited, up to now, to qualitative or semi-quantitative information on material composition. The main limitation of LIPS technique was related to the so-called 'matrix effect', i.e. to the strong dependence of LIPS spectra on relatively small variation of the material composition. This effect was particularly difficult to deal with in analytical application to paintings composition, because of the great variability of pigments and binders used even within the same piece of artwork. For that reason, the general belief was that LIPS, despite of its intrinsic speed and ease, could not be used for obtaining precise quantitative analysis of paintings composition. As a matter of fact, no attempt was made to give even semi-quantitative estimations of paintings composition with LIPS. This scenario dramatically changed after the introduction of a new technique of Calibration Free LIPS spectra analysis, recently developed and patented by the IFAM group. Thanks to this new approach, it's now possible to obtain precise quantitative information on composition of materials of interest in

Cultural Heritage conservation, without need for reference samples or calibration curves.

2 Experimental Set-Up

At IFAM, since several years is operating a LIPS prototype, widely described elsewhere [6].

The IFAM-LIPS prototype has been successfully used for pollutants detection in atmosphere, in water and soils, and for metallic alloys analysis (gold caratage). The prototype was also operated 'in field' for power plants pollution detection. Because of the high power density of the focused laser beam, a small portion of the material under study (around 1 μg per shot) is instantaneously ionised. The optical radiation emitted by the resulting microplasma is then spectrally analysed in order to get information on material composition.

2.1 CF-LIPS Analysis of Spectra

The method used for the analysis of LIPS spectra has been described in detail in ref. [7], therefore we will just present here the basic characteristics of our approach. Assuming the plasma in Local Thermodynamic Equilibrium (LTE), the LIPS line integral intensity corresponding to the transition between two levels E_k and E_i can be expressed as:

$$I_\lambda^{ki} = FCA_{ki}\frac{g_k e^{-\frac{}{}}}{U(T)} \tag{1}$$

Where λ is the wavelength of the transition, C is the concentration of the emitting atomic specie, A_{ki} is the transition probability for the given line, g_k is the k level degeneracy and F is a constant value to be determined after normalisation of the specie concentrations. U(T) is the partition function for the emitting specie. Taking the logarithm of eq. 1 and substituting, we obtain a linear relationship between the parameters which can be graphically represented as a 'Boltzmann plot'. According to eq. 1, the slope of the plot is related to the specie temperature, while the intercept is proportional, via the F constant factor, to the logarithm of the specie concentration. The F factor is determined by normalising at one the species concentrations.

3 Experimental Results

In order to demonstrate the application of CF-LIPS method to paintings analysis, we determined the elemental composition of five artificial pigments (titanium white, zinc white, cadmium yellow, cadmium red, and ultramarine

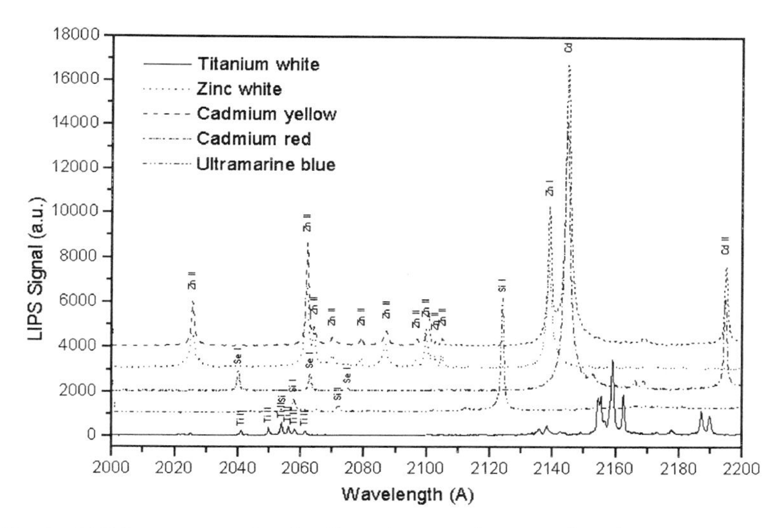

Fig. 1. LIPS spectrum for the five pigments here analysed

blue). A typical section of spectrum is shown in fig. 1 for the five samples; the emission lines of the characteristic components are well recognisable, making simple the qualitative identification of the different pigments. However, the peculiar feature of the CF-LIPS method goes beyond the simple qualitative identification. Using the calibration-free method above described, is possible to obtain a precise quantitative determination of painting composition without need for any reference sample or calibration curve. The relevant results are shown in fig. 2 It is clear from the figure that several elements are present in the pigments, besides those expected according to their chemical base composition. The proportions of additional elements in the pigments are determined almost uniquely by the specific manufacture procedures. The quantitative information obtained by CF-LIPS method might thus be usefully exploited for painted artworks authentication and datation.

4 Conclusions

The results shown in this paper are the first quantitative determination of paintings elemental composition by Laser Induced Plasma Spectroscopy. The intrinsic speed of the calibration free technique (the whole process of data collection and analysis takes a few minutes) and the precision of the calibration free technique (of the order of a few parts percent) propose the CF-LIPS as a viable method for paintings analysis. At present time, the group at IFAM is

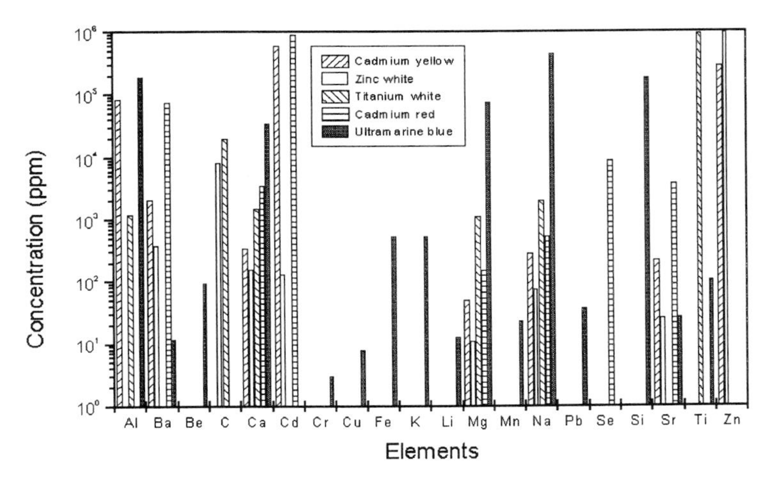

Fig. 2. Pigments elemental composition, as determined by CF-LIPS method

collaborating to the realisation of a commercial CF-LIPS system for Cultural Heritage conservation and analysis applications.

References

1. Maravelaki, P. V., Zafiropulos, V,, Kilikoglou, V., Kalaitzaki, V., Fotakis, C. (1997) Spectrochimica Acta Part B 52 41–53.
2. a) Ciucci, A., Palleschi, V., Salvetti, A., Rastelli, S., Tognoni, E., Towards the 'smart' laser. Spectroscopic control of the laser cleaning process (1995) IFAM Internal Report B01LS/96
   b) Gobernado-Mitre, I., Prieto, A. C., Zafiropulos, V., Spetsidou, Y., Fotakis, C. (1997) Appl. Spectroscopy 51 1125–1129
3. Weibring, P., Andersson, M., Cecchi, G., Edner, H., Johansson, J., Pantani, L., Raimondi, V., Sundner, B., Svanberg, S. (1997) SPIE 3222 372–383
4. Anglos, D., Couris, S., Fotakis, C., Appl. Spectroscopy (1997) 51 1025–1030
5. Anglos, D., Solomidou, M., Zergioti, I., Zafiropulos, V., Papazoglou, T., Fotakis, C., Appl. Spectroscopy (1996) 50 1331–1337
6. Arca, G., Ciucci, A., Palleschi, V., Rastelli, S., Tognoni, E., Appl.Spectroscopy (1997) 51 1102–1108
7. Ciucci, A., Palleschi, V., Rastelli, S., Salvetti, A., Tognoni, E., Proceedings of XXV European Conference on Laser Interaction with Matter (ECLIM), Formia (ITALY) (1998)

Application of Factor Analysis and Multivariate Curve Resolution Techniques for the Separation and Identification of Raman Spectra from Pigments on Artworks

Llorenç Coma, Manuel J. Manzaneda and Sergio Ruiz-Moreno
Universitat Politècnica de Catalunya (UPC)
Sor Eulalia de Anzizu, s/n, Mod. D5, Campus Nord UPC, 08034 Barcelona, Spain.
E-mail: llorenc@arrakis.es

Abstract. The objective of this communication is to show the application of Multivariate Curve Resolution techniques (MVCR) for the separation, identification and quantification of Raman spectra of mixtures of pictorial materials. Its application to real samples, stratigraphies, and artworks analyzed will be shown.

1 Introduction to Raman Spectroscopy and Pigment Identification

Raman Spectroscopy has shown to have excellent features for identifying pictorial materials (pigments, varnishes, binders, etc.) [1] due to its capability of obtaining information from a microsample in a non-destructive way and to the possibility of application *in situ* with the aid of optical fibre probes and portable Raman spectrometers.
The analysis and interpretation of this information encompasses many disciplines of Art and Sciences. The analysis of the items under study is of valuable significance only if it answers questions about them, and only can be done with interdisciplinary partners. Identification of pigments [2] can help historians and curators to take care of the conservation with the maximum accuracy and respect to the original artwork. Besides, authenticating and dating is useful to discard fakes or to assess the authoring of artworks.

2 Data Treatment and Processing using MVCR and FA

Processing and mixture resolving techniques has been exhaustively reported on the literature in various fields of application over the past years [3]. Factor Analysis (FA) [3] or more in particular, Multivariate Curve Resolution (MVCR) [4,5] are chemometric disciplines that comprises several techniques for

- Establishing theoretical models on the data.
- Extracting the maximum amount of information from the data.
- Designing the experiments in order to facilitate the extraction of information.

The factor analytical model for Raman data is already established [3]. Therefore, the main purposes of FA and MCVR techniques are to extract information from spectroscopic data and to plan experimental procedures to allow this extraction.
The analysis of pigments in artworks is not straightforward. Different aspects of the procedure must be treated with care to get satisfactory results. For the analyst, the choice of the best procedure to resolve the information is not usually easy.
When a pigment data base is available, specially good results in determining and quantifying basic pigments are obtained using Target Factor Analysis (TFA) [6].
Methods of Evolving Factor Analysis (EFA) and Window Factor Analysis (WFA) [7] deserve special consideration. They use not only the information contained in the spectra of the samples but also the information from the order in which the spectra have been taken. This fact makes them especially useful when dealing with stratigraphies and with spectra recorded along spatial colour gradation. Rank Annihilation Evolving Factor Analysis (RAEFA) [8], like TFA, take as a data the spectra from a database, and is specially useful if our objective is quantifying the components.
Besides, three-dimensional methods [9] provide the so-called *second order advantage* for separating spectra without ambiguity. New experimental procedures will be used for appraising these methods applied to Raman spectroscopy.

3 Analytical Protocol

The research objectives and current results in the development and implementation of such procedure can be summarized in:

- Prior to the analysis of the artwork, information about the artist, like reported palette of colours, pigments available in the period, dating, etc. must be on hand. The more a priori information, the better the final results.
- Sampling of the work of art is important. Selective areas with similar colour tones must be chosen to identify individual pigments. Moreover, several samples must be recorded in each area in a spatial direction and samples must be classified by their position on the artwork. Some MVCR and FA techniques are sensible to that spatial information and the correct use of them improve their results
- Microphotography is also useful when available. Visual inspection of the area under study can reveal the presence of varnishes or artifacts that can degrade the performance of the methods
- Digital processing filters are useful to eliminate interferences and noise if correctly used. The popular Savitzky-Golay filter is discarded in front of maximum entropy methods, statistical inference, optimum low-pass filtering, etc., which give better results without introducing artifacts in the spectra recorded

In such conditions, coherent results are obtained and information cleared. The application of optimum pre-processing techniques, FA and MVCR methods are displayed in figures 1. In figure 1a raw Raman spectra are displayed. Pigments are sought by their typical Raman lines positions. Figure 1b shows the results after MVCR processing, and it is clear the underlying mixture of 3 components and a straightforward identification. By means of MVCR is possible (under study) to know the composition of the mixtures.

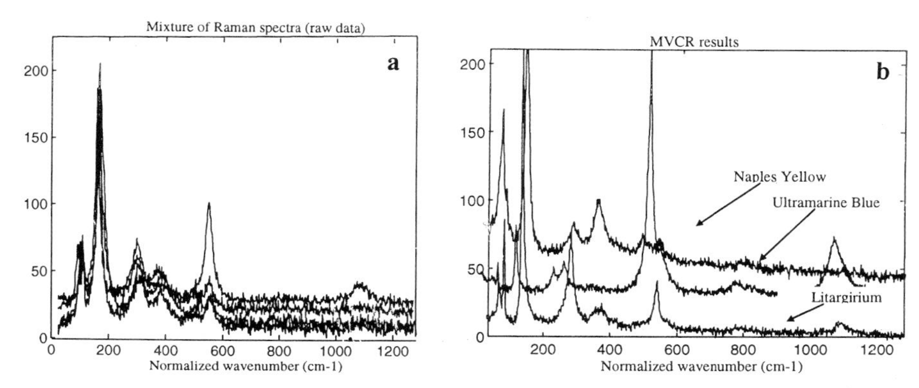

Fig. 1. The experimental raw data (in a) makes difficult to identify the individual Raman Spectra. MVCR techniques help the analyst, separating the original compounds (in b).

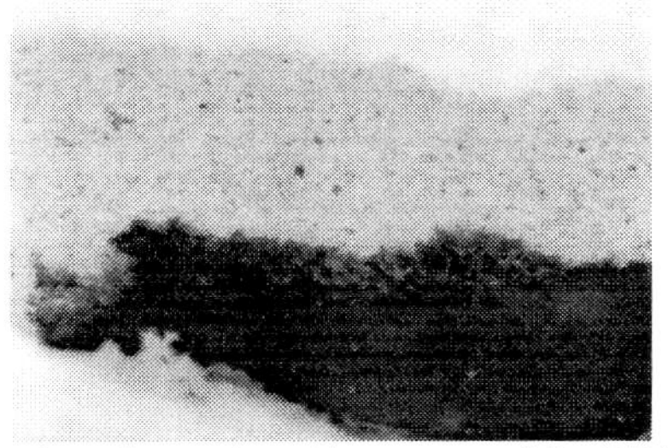

Fig. 2. Stratigraphy analyzed by means of the EFA method.

4 Application Example: Study of a Real Stratigraphy

To assess the techniques descried above, it was prepared on purpose a stratigraphy with four compounds to be analyzed by the EFA method. In figure 2 can be seen a zoomed photography of the item under study. Twenty Raman spectra where taken along an imaginary line that crossed all the layers that conforms the stratigraphy. The results of the application of the EFA method show clearly the information about the compounds resolved (see figure 3): not only the individual Raman spectra for identification (qualitative analysis) but also the concentration profiles (quantitative analysis).

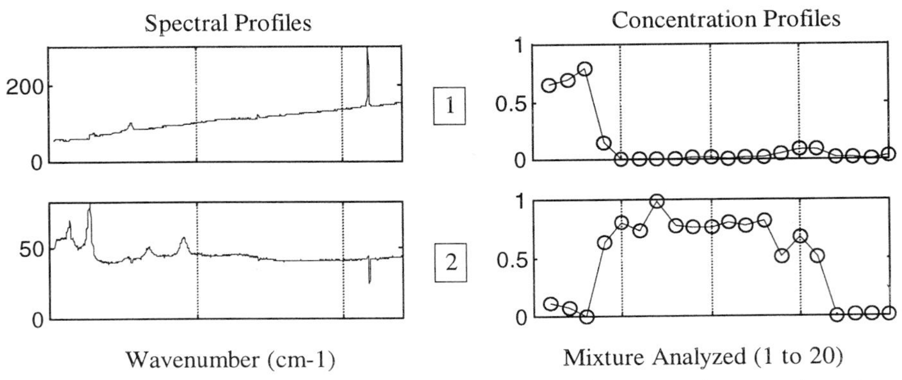

Fig. 3. EFA results showing the spectral (qualitative) and concentration (quantitative) profiles. Only shown Creta White or chalk (1) and Naples Yellow (2) for brevity.

5 Conclusions

Care must be taken in the processing available techniques to enhance information in Raman spectroscopy if correct results are sought. The steps and the techniques to such end are summarized. Future research will improve the results and the applicability to artworks in measures *in situ*.

References

1 S. P. Best et al., *Identification by Raman microscopy and visible reflectance spectroscopy of pigments on an icelandic manuscript*, Studies in Conservation, 40, 31-40 (1995).

2 M. J. Manzaneda, S. Ruiz-Moreno, M. Breitman and A. Tuldrá, *Separation and Identification of Raman Spectra for the Recognition of Pictorial Pigments.* International Series on Optics Within Life Sciences, vol IV (Springer Verlag, ISBN 3-540-63280-8, 1996)

3 E. R.Malinowski, *Factor Analysis in Chemistry*, (John Wiley and Sons, New York, 1992).

4 P. J. Gemperline, *Mixture analysis using factor analysis II: self modeling curve resolution* J. Chemom., 1-13 (1990).

5 R. Tauler, B. R. Kowalski and S. Fleming , *Multivariate Curve Resolution Applied to Spectral Data from Multiple Runs of an Industrial Process*, Anal. Chem., 65, 2040-2047 (1993).

6 A. Lorber, *Validation of Hypothesis on a Data Matrix by Target Factor Analysis*, Anal. Chem., 56, 1004-1010 (1984).

7 E. R. Malinowski, *Window Factor Analysis: Theoretical Derivation and Application to Flow Injection Analysis Data*, J. Chemom., 6, 29-40 (1992).

8 H. Gampp, M. Maeder, C. J. Meyer, A. D. Zuberbühler, *Quantification of a Known Component in a Unknown Mixture*, Anal. Chim. Acta, 193, 287-293 (1987).

9 E. Sánchez and B. R. Kowalski, *Tensorial Resolution: A Direct Trilinear Decomposition*, J. Chemom., 4, 29-45 (1990).

Automatic Acquisition and Evaluation of Optically Achieved Range Data of Medical and Archaeological Samples

Dieter Dirksen[1], Gert von Bally[1] and Friedhelm Bollmann[2]

[1] Laboratory of Biophysics, Institute of Experimental Audiology, University of Muenster, E-mail: dirksen@gabor.uni-muenster.de
[2] Dept. of Prosthetic Dentistry, University of Muenster, Germany

Abstract. A computer controlled scanning device for 360 degree acquisition of 3D coordinates using the technique of phase measuring profilometry is presented. Automatization during acquisition and evaluation of the data is a major requirement. Solutions for a robust digital surface reconstruction as well as filtering techniques for noise reduction adopted to the case of depth images are discussed. Applications in the fields of medical diagnosis and archaeological documentation are presented.

1 Introduction

Photogrammetry is a well established technique for the optical acquisition of 3D coordinates. In conjunction with digital image processing and active projecting techniques like phase measuring profilometry [1], its possible field of application greatly increases, since topometric sensors are now able to digitize far more coordinate points (10^4 to 10^6) automatically and thus allow a digital reconstruction of the object surface. With a computer controlled setup data from different viewpoints can be acquired, integrated, and processed automatically. Various types of diagnostic tools may then be employed for numerical evaluation and comparison of samples.

In the current case, this technique is applied to the problems of digitizing and comparing several features of series of dental plaster casts, as well as for the three-dimensional investigation of archaeological inscriptions with the goal of enhancing readability by digital filtering. An important side effect is that the digitized samples represent a new form of documentation with the additional advantage of transferability by Internet.

2 Experimental Methods

The basis of the system for 3D coordinate measurements is a topometric sensor head consisting of two CCD cameras, and a fringe projector fixed on an adjustable-height rail via a tilting device. An electro-mechanic precision positioning system, consisting of a translation table, a turntable, and a tilting table, allows automatized recording from different viewpoints, which together cover a 360 degree angle area.

During the measuring process a sequence of four phase shifted quasi-sinusoidal fringe patterns is projected onto the object and registered as stereo images by the CCD cameras. After calculating the phase distribution [2] the stereoscopic images are evaluated by photogrammetric techniques and a 3D coordinate is calculated for each valid pixel. The achievable measurement

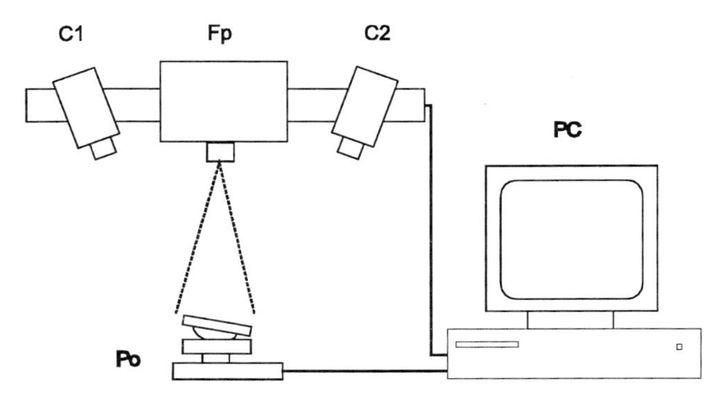

Fig. 1. Experimental setup for 360 degree 3D data aquisition
Fp: fringe projector, C_1, C_2: CCD-cameras, Po: positioning unit, PC: personal computer

accuracy depends on the triangulation angle (angle between the cameras), the image field size, as well as on the number of camera pixels. In the present case the height resolution - with a triangulation angle of 40 degrees and an image diagonal of approximately 20 cm - is situated at < 50 µm, the lateral at approx. 200 µm. If necessary, a higher resolution can be achieved by reducing the image field.

Using the obtained coordinate points (*point cloud*) the object surface is reconstructed by triangulation, i.e. by covering it with a grid of triangles. As the investigated objects are quite complex, an approach is used which does not make special assumptions about the surface structure and iteratively connects neighbouring points, only controlled by preset angular and distance constraints. It allows automatic processing — including the computation of range images – of large numbers of samples.

3 Medical applications

Dental plaster casts of the human jaw were digitized to quantify changes during orthodontical treatment and to aquire anatomical 3D reference data. The setup was equipped with a mounting plate that allows a fast and precise transfer of their anatomical orientation, adjusted in a dental articulator, to the topometric setup. Several hundred samples were measured automatically, integrating 7 different views for each. After calculating range images, cusp positions were interactively marked (Fig. 2(a)) — whereby the exact postions were determined numerically — and stored in a data base [3]. From this data arbitrary samples may then be selected and compared (Fig. 2(b)) or investigated by statistical routines.

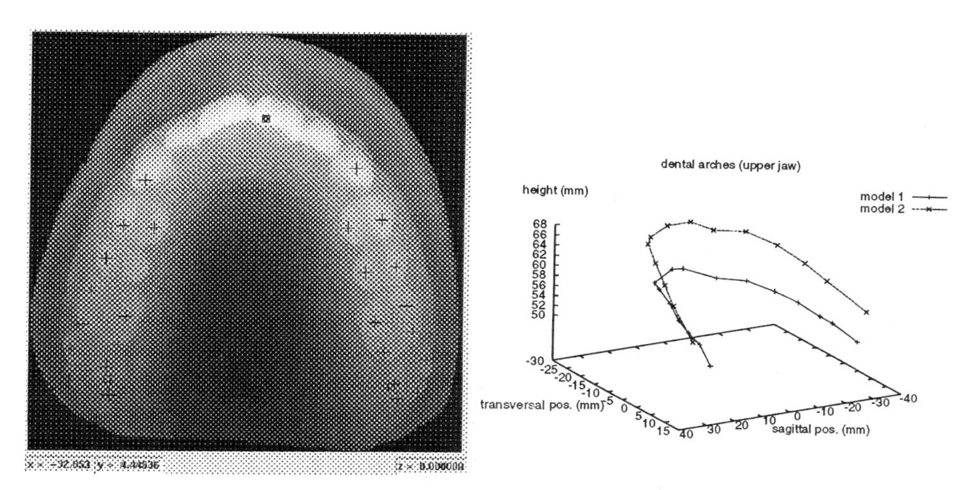

Fig. 2. False color range image (reproduced in gray-level form) of dental plaster cast with marked cusps (a) and comparison of 3D cusp positions of two dental arches (b)

4 Archaeological applications

3D measuring techniques have the potential to form an important supplement to the classical tools for investigation and documentation of inscriptions on archaeological or historical samples such as cuneiform tablets [4]. Handling of the original 3D data sets (e.g. point clouds or triangle sets) is a problem as they require vast amounts of memory (typically 10 to 100 MB for sufficient resolution) and computational capacity. Especially in the case of inscriptions this may be simplified by computing 2D range images, as the relevant information usually is located on nearly planar surfaces. Unfortunately, the objects are not precisely flat and thus, the information of small scale inscriptions is almost invisible in the false color representation of the range image (Fig. 3(a)). One solution is first to subtract a bias height obtained by smoothing, but with height distributions this approach often fails as height steps and edges lead to strong artefacts that may spoil the result. Thus, a range limited 3D smoothing filter was developed that does not include all pixels within a given two-dimensional range (with respect to the image plane) but only those pixel for which $(x_i - x_0)^2 + (y_i - y_0)^2 + (z_i - z_0)^2 < d^2$, where the $z_{i,0}$ represent (physical) height coordinates, $x_{i,0}, y_{i,0}$ are pixel indices transformed to physical coordinates and d is the choosen range of the smoothing operator. With this form of processing, subtraction leads to a significantly improved reproduction of small details as is shown in Fig. 3(b). Applications that have been investigated are the identification of characters that were unreadable before and, as a step towards an internet based documentation, the virtual reproduction of tablets as high resolution 2D range images, mapped onto low resolution 3D reconstructions of the tablets body.

Fig. 3. False color range images (reproduced in gray-level form) of a cuneiform tablet: (a) original range image, (b) range image with subtracted bias

This work was suported by the Deutsche Forschungsgemeinschaft and by the Ministry of Education, Research, Science and Technology of the Federal Republic of Germany.

References

1. Kraus, K. (1997) Photogrammetry Vol. 2. Duemmler, Bonn
2. Creath K. (1988) Phase measurement interferometry techniques. Prog. Opt. 26:349-393
3. Dirksen D., Diederichs S., Runte C., von Bally G., Bollmannn F. Three-dimensional Acquisition and Visualization of Dental Arch Features from Optically Digitized Models. Journal of Orofacial Orthopedics (in print)
4. Dirksen D., Kozlov, Y., von Bally G. (1997) Cuneiform surface reconstruction by optical profilometry. in: D. Dirksen, G. von Bally (Eds) Optics Within Life Sciences (OWLS IV): Optical Technologies in the Humanities. Springer, Heidelberg, 257-259

Surface Modification of Wood by Laser Irradiation

Hendrik Wust*, Michael Panzner*, Günter Wiedemann*, Kai Henneberg**, Lars Pöppel**, Thomas Wittke*
*Fraunhofer Institut für Werkstoff- und Strahltechnik, Winterbergstraße 28, D-01277 Dresden, Germany
**Fakultät Maschinenwesen, Institut für Holz- und Faserwerkstofftechnik, Technische Universität Dresden, Mommsenstraße 13, D-01069 Dresden, Germany
E-mail: wust@iws.fhg.de

Abstract. This contribution presents first results of thorough systematic investigations concerning the effect of laser irradiation on wood. It has been found that the structure of a surface layer can be modified in a definite way by suitably choosing the irradiation parameters.
For technical applications, thorough knowledge about the laser ablation process is necessary. Results of ablation experiments by excimer lasers, Nd:YAG-lasers, and TEA-CO_2 lasers on surfaces of different wood types and cut orientations are shown. The process of ablation was observed by a high speed camera system and optical spectroscopy. The influence of the experimental parameters are demonstrated by SEM images and measurement of the ablation rate depending on energy density.

1 Introduction

Wood is the most important domestic raw material. Because of its renewing resources, balanced material properties and ecological advantages, its importance is even increasing.

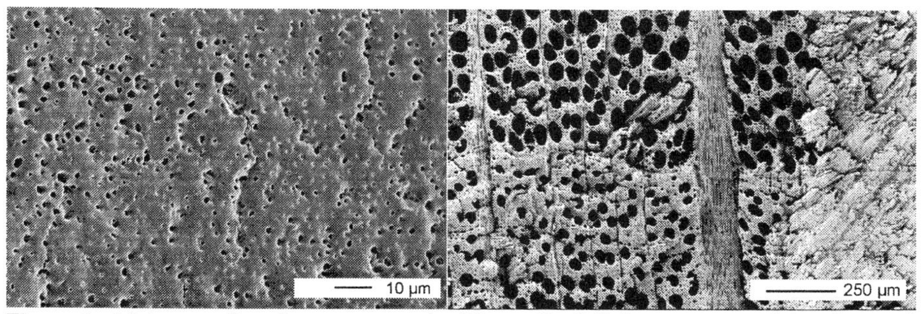

Figure 1: Molten surface layer generated with CO_2-laser.

Figure 2: Cellular structure of beech wood laid bare with XeCl-excimer-laser

In the past the use of this natural material was largely made obsolete by industrially produced materials suitable for novel products and manufacturing techniques. Along with the decreasing use, wood research slowed down. Until recently, laser application in wood technology has been restricted to a few techniques of cutting, perforating, and engraving of wood and wood products, also enhancing the visibility of the grain of veneered wood. In view of this state of art it

is highly justified to apply contemporary methods of material science to wood for better understanding of its structure and diverse properties in order to modify them with the aim of extended use.
This contribution presents first results of thorough systematic investigations concerning the effect of laser irradiation on wood. It has been found that the structure of a surface layer can be modified in a definite way by suitably choosing the irradiation parameters. The possible modifications cover the range from laying bare the cellular structure to sealing the surface by melting and solidification of a surface layer.

2 Experimental

Orech [1] measured the absorption spectrum of wood. Because of the absorption minimum at 1000 nm the lowest ablation per pulse is expected for Nd:YAG-lasers. Contrary to this, an effective ablation can be predicted for the UV- and IR - spectral range considering an absorption of > 80 % in this spectral ranges. So the investigations were concentrated on experiments in UV and IR ranges.

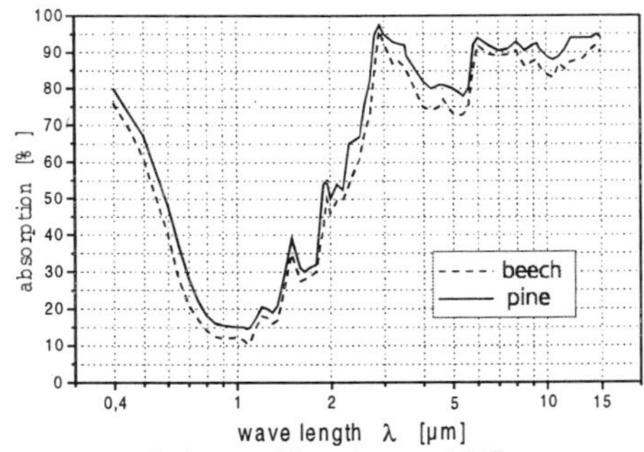

Figure 3: Absorption spectrum of pine and beech wood [1]

The small effect of 1064 nm laser light on wood was shown by simple ablation experiments with a Nd:YAG-laser (NY82S-10 Continuum,1064 nm, 8 ns, 1.8 J) which was focused in a way that several mean energy values were obtained. No significant ablation was found for energy densities below 100 Jcm^{-2}.
Regarding future technical applications, a XeCl-excimer laser (XP2020, Siemens , 308 nm, 40 ns) and a CO2-TEA-Laser (Uranit, Urenco, 10.6 μm, 1.2 μs) were used for ablation experiments on a coniferous and a deciduous wood (pine and beech).
Surfaces which were oriented according to the principal cut orientations (cross - across the trunk, quarter - radial, slash - tangential to the growth rings) have been investigated.
The excimer laser ablation process was in-situ observed by a high speed camera system (IRO-Image Intensifier Camera, PCO-Computer Optics).
The reached ablation was measured by a perthometer (S3P/PRK, Feinprüf Perthen GmbH).

3 Results

As found for artificial polymers like polyimide, an energy density of 1 Jcm^{-2} at 308 nm is already sufficient for ablation of wood of any species, orientation and moisture. The inner structure of the wood is exposed to the view after removal of the squeezed surface layer by laser ablation. Similar results were found for pine, although the microstructure is different.
The ablation depth strongly depends on the local structure of the wood. The ablation rate is higher for the thin-walled, spring wood cells and lower for the thick-walled, late wood cells (see Fig. 4).

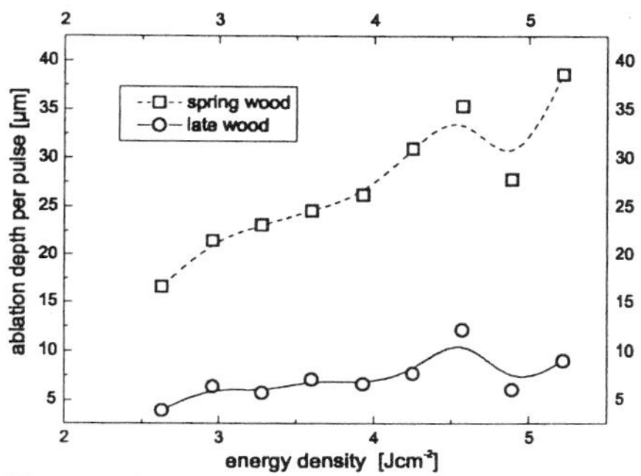

Figure 4: Ablation rate of wood versus energy density, XeCl-excimer laser, beech, averaging of spring and late wood, water content 10 %

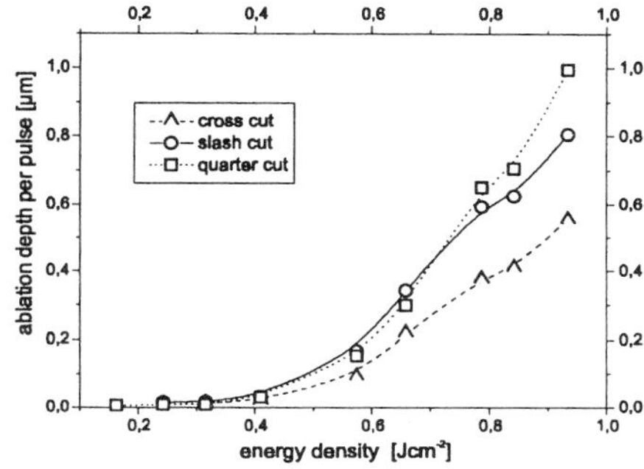

Figure 5: Ablation rate of wood versus energy density, CO_2-TEA-laser, pine, water content 10%

Lowest ablation depth per pulse was observed for cross cuts (Fig. 5), as expected from the anisotropy of the cellular structure: On the cross cut a large part of the incident power gets lost in the deep interior of the tube-like cells without effect. This anisotropy is more pronounced for coniferous wood.
Changing from 308 nm to 10.6 µm with 5-fold increase of energy density produces a 20-fold increase of ablation rate for pine (slash cut). For suitable parameter sets, damage-free ablation of the wood structure without carbonization is possible.
For example Fig. 5 shows ablation rates of pine for 10.6 µm laser wavelength. At low energy densities, carbonization and ablation have been observed. At higher energy density, carbonization gets suppressed while the ablation rate is high.
In situ high speed photographs of the excimer laser ablation process of wood show that three main stages of the process can be distinguished from the visible part of the emission spectrum of the plume: 1) absorption of the laser pulse and development of a thin light emitting plasma sheet on the surface (0...1 µs after laser pulse), 2) plasma expansion (1...200 µs after laser pulse), 3) burning of gases coming out of the wood surface because of chemical processes caused by thermal effects (200 µs...1 ms after laser pulse).
The existence and intensity of the third stage strongly depends on repetition rate and energy density. Distinct differences in the brightness were found for light emission of spring wood and late wood indicating differential intensity of the ablation process.

At 0,308 μm and 10,6 μm, melting was observed for all components of the wood structure (for example Fig. 6, 7). The thickness of the molten zone is in the range of 1 μm. It means that also for laser wave length in the UV range, ablation seems to be mainly caused by thermal processes, e.g. the excitation of phonons in the macromolecules of the wood components. The breaking of molecular bonds by direct interaction with a photon does not seem to be the dominating process. Ablation experiments on pure cellulose by parameters allowing an efficient ablation of wood (308 nm, 0,8 Jcm^{-2}) only causes a change of the cellulose color accompanied by melting of a thin surface layer. Probably other wood components, such as lignin which glues the cell walls made of cellulose, effectively absorb the laser light. This seems to be supported by observation: even in the presence of molten and foam-like expended cellulose from the cell walls, an open slit remains between the cells where the lignin had been (Fig. 7).

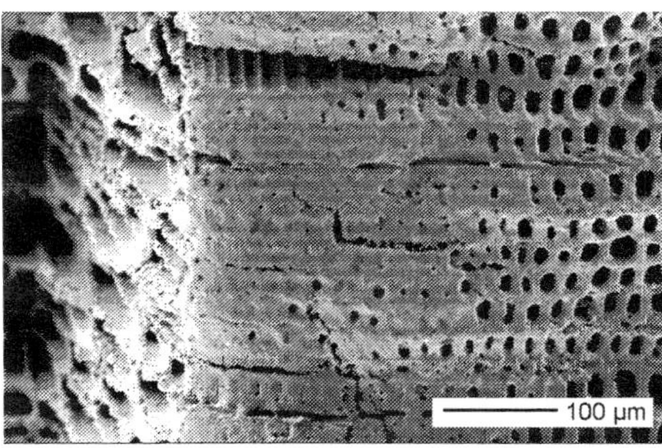

Figure 6: Melting of the ablated surface, pine, cross cut, CO_2-TEA-laser, 5.2 Jcm^{-2}, 1 Hz, 30 pulses

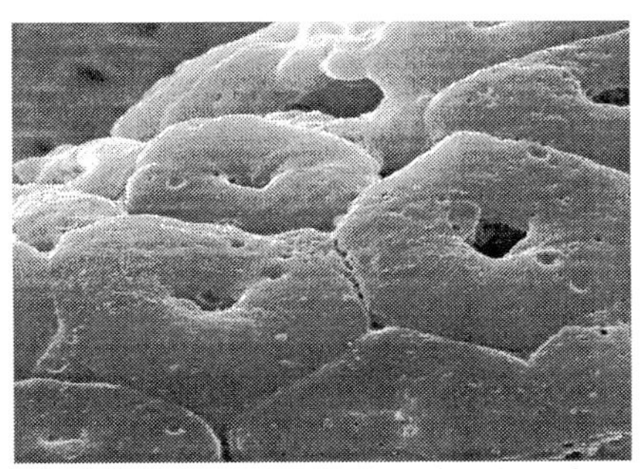

Figure 7: Melting of the ablated surface, beech, cross cut, XeCl-excimer laser, 0.8 Jcm^{-2}, 20 Hz, 100 Pulse

4 Summary

Laser ablation of wood without carbonization of the surface is possible for certain parameter sets. UV and IR-pulse lasers are suitable for this purpose because of the absorption properties of wood. For these laser types, melting of the cellulose in the μm range was observed, which supports the assumption of mainly thermal ablation at laser wavelengths above 300 nm.

References

1 Orech, T., Kleskenova, M.: Drevo, 30 (1975) 324

Decoration of Glass by Surface and Sub-surface Laser Engraving

Andreas Lenk, Thomas Witke
FhG IWS Dresden, Winterbergstraße 28, 01277 Dresden, Germany
E-mail: lenk@iws.fhg.de

Abstract. Unfortunately glass is highly transparent for the wavelength of 1.06 μm commonly used by marking systems. Using absorption via non-linear effects can be a solution of this problem. There are three conditions for an effective use of non-linear effects: a high repetition rate (500 Hz), TEM_{00} beam quality ($M^2 < 1.5$) and a high pulse power (> 0.5 MW).
The formation of laser induced plasma in the case of surface and sub-surface focusing regime was investigated by a special high speed framing microscope.

1 Introduction

Marking of glass is done with either TEA-CO_2-laser or excimer lasers. The pulse power of these lasers is high enough to induce an ablation process on relatively large areas. For this the mask ablation is the preferred marking technology.
The main disadvantage of mask ablation technology is its low flexibility. A change of pattern requires a change of mask. This precludes a programmable and individual marking. On the other hand, for non-transparent media the marking of pixel based or vector based images by means of a galvoscanner is state of the art. For this technology lasers with high beam quality and high pulse repetition rate like cw-pumped solid state lasers are required. Unfortunately glass is highly transparent for the commonly used wavelength of 1.06 μm. Using absorption via non-linear effects can be a solution of this problem.

2 Laser Technology

Power densities of 10^{10} W/cm^2 and more are necessary to use non-linear effects (self-focusing, multi-photon absorption, inverse Bremsstrahlung)[1,2].
In the case of focused Nd-YAG-laser ($\lambda = 1.06$ μm, $M^2 = 1.5$, d = 10 mm) with a focal length of 100 mm the focus diameter is nearly 20 μm. Thus a typical focus area of $5 \cdot 10^{-6}$ cm^2 requires a pulse power of more than 0.5 MW to achieve power densities above 10^{11} cm^2.
Two types of single stage TEM_{00}-mode solid state laser fulfil this condition: the mode-locked, q-switched, cw-lamp pumped laser and the q-switched, quasi-cw diode pumped laser. The results presented below are based on experiments carried out with these two laser types.

3 Sub-surface Engraving

In the focus of the laser beam within a transparent target, phenomena as plasma formation and subsequent microcracking occur. The formation of the laser induced plasma was investigated by a special high speed framing microscope.[3,4].

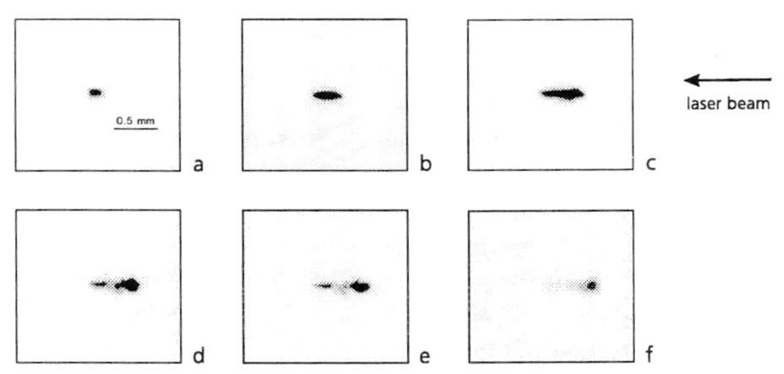

Fig 1 Sequence of 6 (a-f) high speed photos delay time 0, 100, 200, 300, 400, 500 ns, exposure time 100 ns

Fig. 1 a-d shows a sequence of 6 frames. Because of delay time = exposure time = 100 ns the complete development of laser induced plasma for 600 ns was recorded. It is visible that the plasma seems to travel in direction of incident laser beam. Shortly after plasma ignition the irradiated front of the plasma absorbs the laser energy. This fact results in an increased plasma formation at the front of the plasma and in cooling at the rear of the plasma. Therefore a moving light spot was

Fig. 2 3D sub-surface engraving using a CAD-file of the „Dresdner Frauenkirche" (cube size 60 mm)

observed. Moreover, this effect is an explanation for the anisotropic, cigar like shape of the micro-crack structure.

Obviously, it is possible to compose 2D- and 3D-objects by programmed relative motion between laser beam and target. This idea was presented by Russian inventors[5] as early as 1970. Admittedly, applications like subsurface marking of lead crystal for decoration or anti-counterfeiting marking of sterile glasscontainers are only economical with high repetition rates and with correspondingly high speed of relative motion between laser beam and target. For example the processing time of the 3D-image of the „Dresdener Frauenkirche“ is about 2 min (Fig. 2).

4 Micro Laser Plasma Sputtering

Another field of application could be a laser induced sputtering process. This sputtering regime is carried out by focusing the laser beam at the surface with power densities in the range of 10^{10}-10^{11} W/cm^2. In this range the laser induced plasma has such a high electron density soon after the onset of the pulse that the target is nearly completely shielded from the laser radiation. Therefore the further ablation is done by hot plasma ions. The ions acquire their high kinetic energy from the electron gas by collision processes. The electron gas itself is heated by the laser radiation via inverse Bremsstrahlung[6].

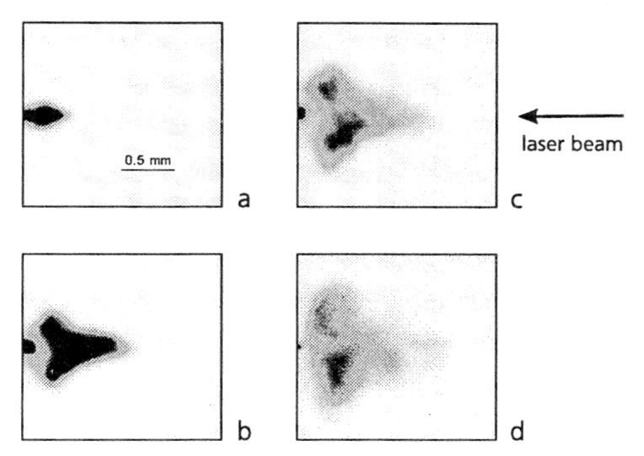

Fig. 3 Sequence of 4 high speed photos (a-d) delay time 0, 100, 200,300 ns, exposure time 100 ns

The formation of the laser induced plasma emanating from the beam focus on the surface was recorded by the high speed framing microscope (Fig. 3). The relatively long plasma plume indicates that shortly after the dielectric breakdown at the target surface an additional air breakdown is induced by the emitted plasma particle. This air breakdown results in the perfect shielding of the target from the incident laser beam.

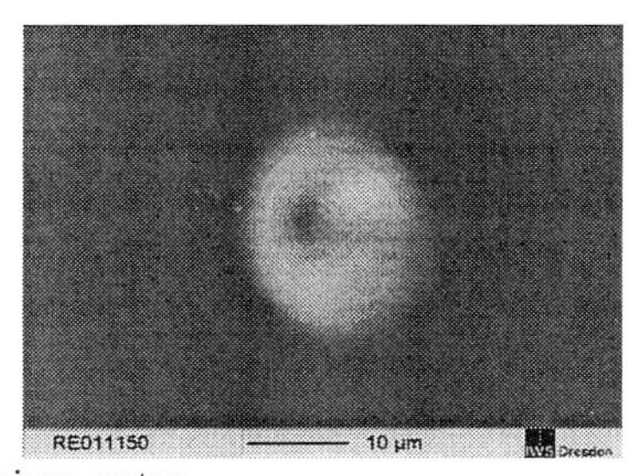

Fig. 4 SEM image of micro crater

The plasma sputtering regime has two important advantages. First, the ablation area is limited to a diameter of about 20 µm. The depth of the ablation crater was determined by interference microscopy. It amounts to some micrometers. Second, the crater and its vicinity are absolutely free of cracks (Fig. 4). What seems to be a blurred image at first sight is actually a very smooth crater surface, properly focused. Moreover, by means by light microscopy between crossed linear polarisers, we did not find any indication of stress birefringence in the vicinity of the ablation crater.

Micro laser plasma sputtering could be applied as pixel based micro marking of glass surfaces, structuring of special scattering surfaces and non-slip finishing of glass flooring. Just like in the case of subsurface engraving repetition rates up to 500 Hz and more guarantee efficient production with short processing time.

5 Summary

The use of non-linear effects makes the precision machining of transparent materials like glass with solid state lasers feasible. There are a variety of potential applications for the surface and the sub-surface focusing regime. Micro laser plasma sputtering in particular allows a marking of glass surfaces without any damage.

References

1 W. Koechner, „Solid-State Laser Engineering“, chapter 11, pp. 540-558,Springer-Verlag New York Berlin Heidelberg, 1988
2 In, Lawrence Livermore National Laboratory Laser Program, Annual Report UCRL-509021-82 1982 p.15
3 P. Siemroth, T. Witke, Th. Schülke, A. Lenk, *„Short-Time Investigation of Laser and Arc Assisted Deposition Processes“*, Surface and Coating technology 68/69 (1994) pp. 713-718
4 A. Lenk, B. Schultrich, T. Witke, *Diagnostics of laser ablation and laser induced plasmas*, Applied Surface Science 106 (1996) pp. 473-477
5 W.W. Agadshanow et. al. *„Production technology for decorative articles“* , Patent SU 321422
6 H. Hora, *„Nonlineare Plasma Dynamics at Laser Irradiation“*, Lecture Notes in Physics 102 (1979) Springer Verlag, Berlin Heidelberg New York

Experimental Studies on Black Encrusted Sandstone Cleaning by Various UV Wavelengths

S. Klein[2], V. Zafiropulos[1], T. Stratoudaki[1], J. Hildenhagen[2], K. Dickmann[2], and Th. Lehmkuhl[3]

[1] Foundation for Research and Technology Hellas (FORTH), Institute of Electronic Structure and Laser (IESL), P.O.Box 1527, Heraklion 71110, Crete, Greece
[2] Laserzentrum Fachhochschule Münster (LFM), 48565 Steinfurt, Germany
[3] Dipl. Restaurator, 48565 Steinfurt, Germany

Abstract. Laser cleaning studies were performed on black crusted sandstone samples from the Dresden *Zwinger* by employing the KrF wavelength (248nm), the 3rd harmonic of the Nd:YAG (355nm) and its fundamental wavelength (1064nm). The elemental composition and the depth of the crust and the underlying stone were determined by a combination of ablation rate studies with laser induced breakdown spectroscopy (LIBS). Using two UV wavelengths and the traditional IR-cleaning wavelength on the same sample it was possible to achieve an adequate comparison for cleaning sandstone. The different LIBS spectra of the black crust and the stone as a reference were used for on-line monitoring the cleaning process.

1 Introduction

Laser radiation is established as a tool to remove surface contaminants from a wide variety of substrates [1,2]. Lasers were introduced to artwork conservation by J. Asmus et al. [3,4]. Since then the technique of laser cleaning has been established and is mainly applied on marble [5], limestone [6], paintings [7] and other materials. Within the frame of this experimental study a type of sandstone called *Elbsandstein* from the Dresden *Zwinger* was used to perform a comparative study of the interaction of different laser wavelengths. The stone has a superficial black crust of 200 to 300 μm thickness. In order to remove this crust most efficiently and without harming the original structure, three wavelenghts were used. The Excimer wavelength of 248 nm, the third harmonic wavelength of the Nd:YAG at 355 nm and its fundamental wavelength at 1064 nm. Additionally laser-induced breakdown spectroscopy (LIBS) was used to determine the trend of the elemental composition when removing the crust and reaching the stone surface. The LIBS technique [8] is based on the spectroscopic analysis from the plasma when a pulsed laser is focused onto a sample. LIBS is an established method and was applied under different topics [9–12]. LIBS was also previously applied for analysis for laser cleaning [5]. The information about the different compositions of superficial crust and the underlying sandstone could be used for on-line cleaning [6].

2 Experimental

A diagram of the experimental arrangement consisting of the laser cleaning system and the LIBS technique is shown in Fig. 1. First LIBS was used to

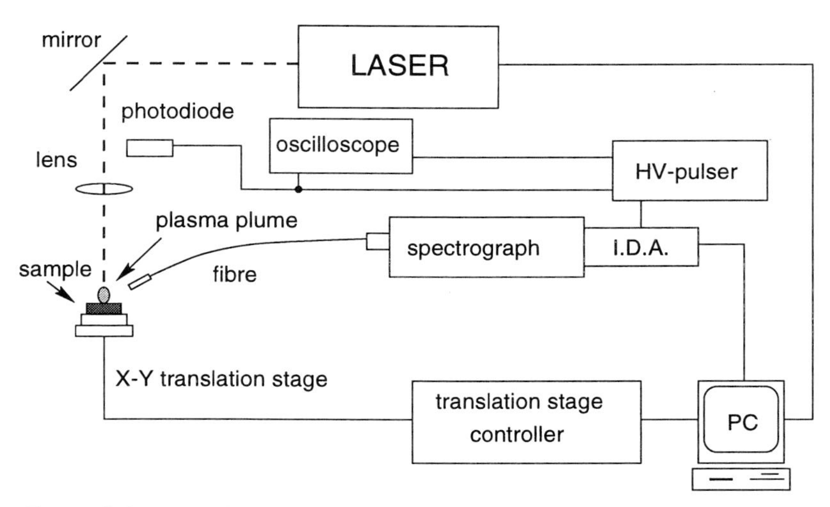

Fig. 1. Schematic diagram of the experimental setup

determine the elemental composition of the black crust and the underlying stone. Pulses of 30 ns duration and 240 mJ energy of a KrF Excimer (Lambda Physik) emitting ultraviolet (λ = 248 nm) were focused onto the target with a cylindrical lens with a focal length of f = 300 mm. The plasma emission was monitored 0.5 μs after the laser pulse, which was detected by a fast photodiode. A quartz fibre was used to guide the emitted light to the detection system, consisting of a spectrograph (PTI Model 01-001 AD), which is equipped with two interchangeable gratings of 300 and 1200 grooves/mm (wavelength coverage of 270 and 70 nm) and an optical multichannel analyser (EG&G PAR OMA III 1420 UV) with an intensified photodiode array detector. The typical laser fluence was about 6 J/cm^2.

3 Results and Discussion

The intense plasma that is generated when the pulsed laser beam is focused onto the sandstone surface and the subsequent emission spectra of the excited atoms/ions were used for elemental analysis. Emission spectra of the black crust and the sandstone were recorded in a spectral region from 245 to 650 nm. It was observed that the major differences were recorded in the spectral region from 245 to 320 nm. Therefore only this region is shown by Fig. 2. The

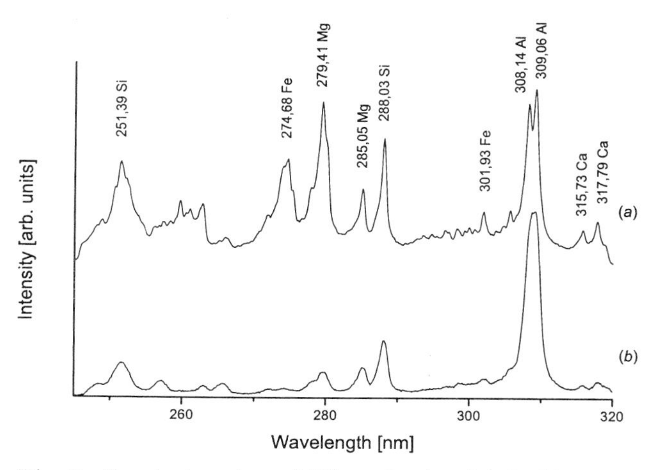

Fig. 2. Spectral region of Elbsandstein. (**a**) Inside encrustation layer after three pulses (**b**) Bulk sandstone

assignment of the atomic lines is based on [13]. The main difference of the black crust and this type of sandstone is the major presence of Fe within the crust and the change to a minor Fe content within the sandstone. Al, Si and Mg could be found in both layers. It should also be mentioned that Ti and Na where also observed in the crust and in the stone by analysing the spectral region of 320 to 650nm. With these results we designed a closed loop control based on LIBS and demonstrated *automatic* cleaning of sandstone with strong encrustation. In a second objective we applied different wavelengths to clean sandstone with corroded surface. Fig. 3 shows a laser cleaned area. The third harmonic wavelength of the Nd:YAG with λ = 355 nm was applied for this

Fig. 3. Laser cleaned area of Elbsandstein with 355 nm

experiment. The black crust was irradiated with a fluence of 1,8 J/cm^2 and 50 pulses. In contrast, for the fundamental wavelength of the Nd:YAG (1064 nm) the effect of *yellowing* was observed as shown by Fig. 4. As for the 355 nm the sandstone cleaned by 248 nm showed no discoloration.

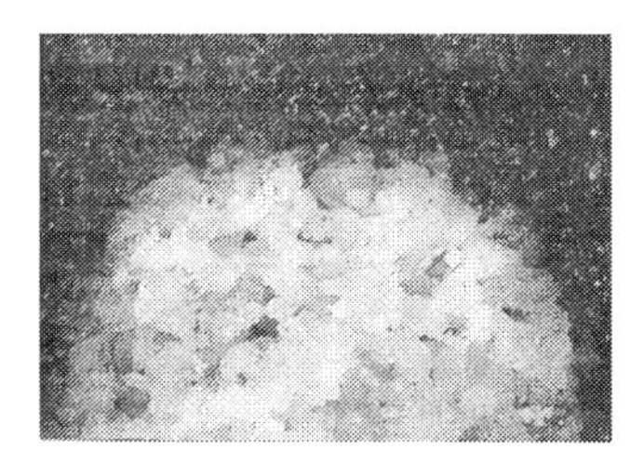

Fig. 4. Laser cleaned area of Elbsandstein with 1064 nm (due to the black and white picture the mentioned effect is not obvious)

References

1. Zapka W., Ziemlich W. et al., Appl. Phys. Lett. **58**, 2217 (1991)
2. Lee S. J., Imen K. et al., J. Appl. Phys. **74**, 7044 (1993)
3. Asmus J. F., Studies in Conservation **18**, 49 (1973)
4. Asmus J. F., Newton H. T. et al., J. Vac. Sci. Tech. 12, 1352 (1975)
5. Maravelaki P. V., Zafiropulos V. et al., Spectrochimic Acta Part B 52, 41 (1997)
6. Gobernado-Mitre I., Prieto A. C. et al Appl. Spec. **51**, 8 (1997)
7. Georgiou S., Zafiropulos V et al. Appl. Surf. Scie. 5048 (1998)
8. Radziemski L. J., Cremers D. A. (1989) Laser Induced Plasmas and Applications, Marcel Dekker, New York
9. Anderson D. R., McLeod C. W. et al. (1995) Depth Profile Studies Using Laser-Induced Plasma Emission Spectroscopy. Appl. Spec. **49**, 6: 691-701
10. Barbini R., Colao F. et al. (1997) Semi-Quntitative Time Resolved LIBS Measurements. Appl. Phys. B, 101-107
11. Yamamoto K. Y., Cremers D. A. et al. (1996) Detection of Metals in thr Environment Using a Portable Laser-Induced Breakdown Spectroscopy Instrument. Appl. Spec. **50**, 2: 222-232
12. Arca G., Ciucci A. et al. (1997) Trace Element Analysis in Water by the Laser-Induced Breakdown Spectroscopic Technique. Appl. Spec. **51**, 8: 1102-1105
13. Tables of spectral lines. IFI/Plenum Data Corporation, New York/London (1970)

Laser Induced Breakdown Spectroscopy in the Analysis of Pigments in Painted Artworks. A Database of Pigments and Spectra

T. Stratoudaki, D. Xenakis, V. Zafiropulos and D. Anglos
Foundation for Research and Technology-Hellas (FO.R.T.H), Institute of Electronic Structure & Laser, Laser and Applications Division, P.O. Box 1527, GR 71110, Heraklion, Greece (e-mail : anglos@iesl.forth.gr)

Abstract. The application of Laser Induced Breakdown Spectroscopy (LIBS) in the identification of pigments, used in paintings, is outlined. Selected examples from the analysis of pigments in model samples and in painted artworks are given. A spectroscopic database including the analytical spectroscopic information has been developed and used to assist in pigment analysis.

1 Introduction

Analysis of pigments in paintings, is of major significance in art conservation, as it can lead to charecterisation of materials which is important for the dating and authentication as well as for possible conservation or restoration of the artwork. Several techniques such as Optical and Scanning Electron Microscopy (SEM), X-ray fluorescence (XRF), Fourier-Transform Infrared (FT-IR), Laser Induced Fluorescence (LIF) or Raman spectroscopy [1-5] have been employed in the identification of pigments. In this work, we present examples from the application of Laser Induced Breakdown Spectroscopy (LIBS) in the analysis of pigments in paintings [6]. In addition, the structure and function of a database designed to assist in the rapid interpretation and analysis of the LIBS spectral data is described.

LIBS is an atomic emission spectroscopic technique for elemental analysis of materials. Focusing of an intense nanosecond laser pulse on the surface of the sample, results in plasma formation, which upon cooling emits radiation (fluorescence from excited atoms and/or ions) characteristic of the elements contained in the sample. LIBS features high sensitivity and selectivity and can be performed in situ -on the artwork itself- thereby eliminating the need for sample removal, which can be damaging to the artwork. In a LIBS analysis, only a minute amount of sample is consumed in the process of atomisation (achieved by tightly focusing a pulsed laser beam on the surface of the sample), hence the technique is characterised as microdestructive.

2 Experimental

The instrumentation required for performing LIBS experiments has been described elsewhere [6]. In brief, the main components of the experimental setup include a nanosecond pulsed laser, appropriate beam delivery optics and the spectrum acquisition system. In a typical LIBS experiment, a nanosecond Q-Switched Nd:YAG laser operating at its fundamental (1064 nm) or harmonic frequencies (532nm, 355nm) is employed. The laser beam is focused on the sample surface. A single laser pulse of energy ranging from 2 to 20 mJ/pulse is used for each measurement, producing strong emission signals. Because of the value and sensitivity of the objects studied, work is deliberately carried out with lower than usual power density values (0.2-2 GW/cm^2) without of course sacrificing spectral quality. The light emitted from the plasma plume is collected with an optical fiber into a 20 cm focal length spectrograph (PTI model 01-002AD) equipped with two holographic gratings of 1200 and 300 lines/mm, for high and medium spectral resolution measurements respectively. The spectrum is recorded with an Optical Multichannel Analyzer (OMA III system, EG&G PARC model 1406 with an intensified photodiode array detector, EG&G PARC model 1420UV) permitting fast gating with adjustable delay and gate width.

Pigment Name	Chemical Composition / Formula	Identified Elements
Lead White	$Pb(OH)_2 \cdot 2PbCO_3$	Pb
Titanium White	TiO_2	Ti
Zinc White	ZnO	Zn
Lithopone	$ZnS \cdot BaSO_4$	Ba, Zn, (Ca)
Cadmium Yellow Lemon	$Cd(Zn)S \cdot BaSO_4$	Cd, Zn, Ba
Chrome Yellow	$PbCrO_4$ (PbO)	Cr, Pb (Ca)
Cobalt Yellow	$2K_3(Co(NO_2)_6) \cdot 3H_2O$	Co
French Ochre	$Fe_2O_3 \cdot nH_2O, SiO_2, Al_2O_3$	Fe, Si, Al
Cadmium Red	$CdS_xSe_{(1-x)}$	Cd
Cinnabar	HgS	Hg
Burnt Sienna (Pompeii Red)	Fe_2O_3, Al_2O_3	Fe, Al
Minium	Pb_3O_4 $(PbO \cdot Pb_2O_3)$	Pb
Ultramarine	$Na_7Al_6Si_6O_{24}S_3$	Al, Si, Na
Cobalt Blue	$CoO \cdot Al_2O_3$	Co. Al, Na
Prussian Blue	$Fe_4[Fe(CN)_6]_3 \cdot nH_2O$	Fe, (Ca)
Azurite (Azure Blue)	$2CuCO_3 \cdot Cu(OH)_2$	Cu, (Si)
Cobalt Green deep	$CoO \cdot Cr_2O_3$	Co, Cr
Malachite	$CuCO_3 \cdot Cu(OH)_2$	Cu, (Si)
Chrome Green dark	Cr_2O_3 $(PbCrO_4)$	Cr, Pb, Ba
Viridian Green	$Cr_2O_3 \cdot 2H_2O$	Cr

Table 1. List of selected pure pigments analysed by LIBS and elements identified

3 Results

3.1 Pure Pigments

A systematic investigation of an extended variety of pigments has been undertaken, corresponding LIBS spectra have been obtained, and elements have been identified based on their characteristic emission lines. Table I shows several pigments analysed and the elements identified. Analytical spectral lines have been selected which are appropriate for the determination of the elements contained in the pigments. This information has been organised and compiled in a database which can be searched according to criteria set by the user, giving the possibility to identify pigments or obtain information on specific elements and their spectral lines. The database fields include pigment name, colour and chemical structure, spectral analytical lines, LIBS spectra in a picture format, and experimental parameters used to obtain the corresponding spectra. For example the user may provide with one or more peak wavelengths from a LIBS spectrum and ask the database to suggest element and/or possible pigment. Alternatively the database can be searched for spectra of individual pigments and help the user to select the appropriate spectral range and resolution and experimental conditions for performing measurements.

3.2 Case Studies

Several byzantine icons were examined during this study in order to identify pigments used and selected results are presented.

On a late 16th century (egg tempera on wood) icon, portraying Virgin Mary holding Jesus, we examined white, green and red paint areas. As expected, based on the approximate date the icon was made, lead was identified in the white pigment suggesting that lead white was the pigment of choice. In the green paint, emission lines from copper were identified suggesting that a copper based pigment most likely malachite had been used (Fig. 1). The LIBS spectrum from the red paint shows emission lines from iron, aluminium and calcium indicating the use of an iron based red pigment (sienna, hematite).

A mid 19th century icon (egg tempera on silver foil) of Saint Nicholas was also examined. On this icon, it was found that the originally used white pigment was lead white. However, when a different area was examined, characteristic emission from zinc was obvious in the first LIBS spectrum on that point, indicating the use of zinc white. Furthermore, upon delivering three more pulses to the same point and recording the corresponding LIBS spectra, lead was detected, indicating that the zinc white layer was some kind of overpaint on the original lead white paint (Fig. 2). Finally, and after forty pulses on the same point only calcium was present indicating that the preparation layer (most likely calcite or gypsum) had been reached. This case, demonstrates the capability of LIBS to effectively carry out a depth profile analysis of the successive paint layers. The obvious principle behind this feature is that during the analysis each laser pulse removes from the surface a small amount of material and therefore the following pulse probes always new fraction of the paint layer slightly deeper than the previous one. As a result, successive LIBS spectra on the same point reveal the stratigraphy of the paint layers performing essentially an in situ cross section analysis.

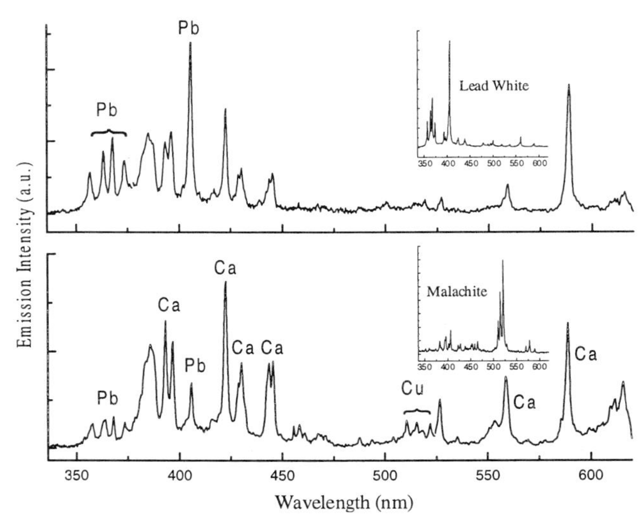

Fig. 1. LIBS spectra from white (top) and green (bottom) paint of byzantine icon. The insets show the corresponding spectra of the pure pigments lead white and malachite. Analytical line wavelengths (in nanometers) for lead: 357.27, 363.96, 368.35, 373.99, 405.78; and for copper: 510.55, 515.32, 521.82.

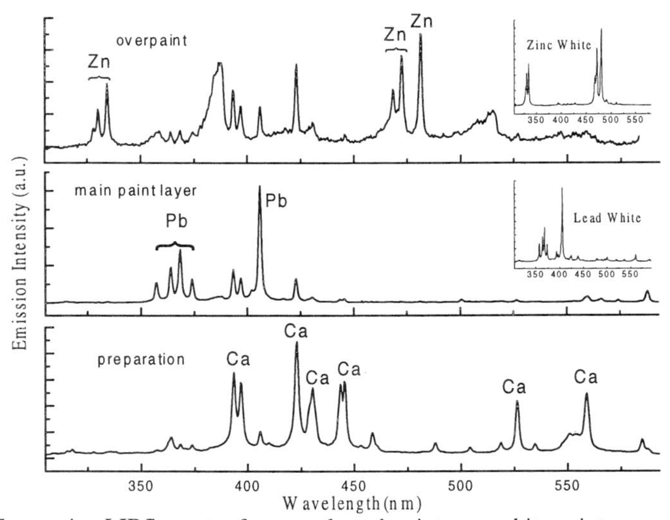

Fig. 2. Successive LIBS spectra from a selected point on a white paint area of byzantine icon revealing the thin overpaint (2nd pulse), the original paint (5th pulse) and the preparation (40th pulse). Each spectrum has resulted from a single laser pulse. Emission lines for zinc: 328.23, 330.26, 334.50, 468.01, 472.21, 481.05; lead: (see Fig. 1); and calcium: 393.37, 396.85, 422.67, 443.50, 445.48, 527.03, 558.20.

A 16th century icon, The Annunciation, was also examined and the results are shown in Figure 3. The red paint analysed, showed characteristic peaks of mercury proving that the pigment used was cinnabar (HgS). The LIBS spectrum was slightly more complicated when a yellow pigment was examined. Atomic emission lines due to lead, chromium and calcium were observed. These findings suggest that lead chromate (chrome yellow), has been used on this icon. A question is raised about the originality of this yellow paint, given that chrome yellow was introduced in the mid 18th century [7] while the icon is dated back to the 16th century.

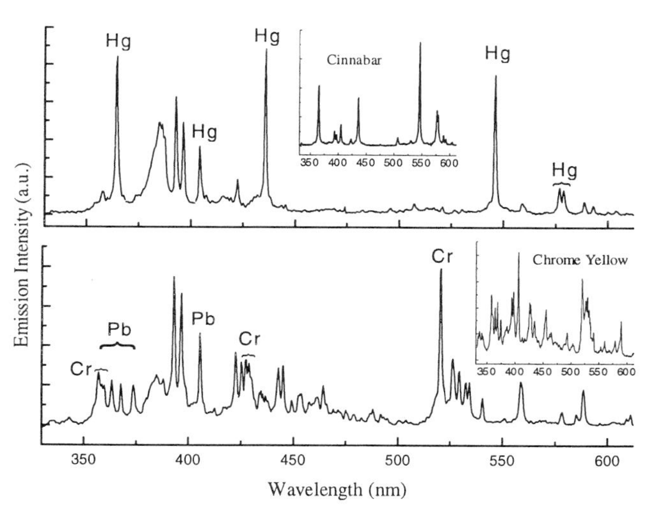

Fig. 3. LIBS spectra of red (top) and yellow (bottom) paint layers of byzantine icon. The insets show the spectra of the corresponding pigments, cinnabar and chrome yellow. Analytical line wavelengths (in nanometers) for mercury: 365.02, 404.66, 435.83, 546.07, 576.96, 579.07; for lead (see Fig.1) and for chromium: 357.87, 359.35, 360.53, 425.43, 427.48, 428.97, 520.45.

4 Conclusions

LIBS is a rapid elemental analysis technique having key analytical advantages with respect to its potential in the area of artwork diagnostics. It is a sensitive and selective technique, does not require any sample removal and thus can be performed in situ - on the artwork itself. It is practically non-destructive and provides nearly microscopic spatial resolution as well as depth profiling information revealing the stratigraphy of painted surfaces as shown in several examples.

Acknowledgements

The authors would like to thank Mr. M. Doulgeridis from the National Gallery of Athens and Mr. S. Stasinopoulos and Ms. M. Benaki from the Benaki Museum for offering the icons for analysis and for providing with valuable information on the materials used.

References

1. P. Mirti, Ann. Chim. **79**, 455, (1989).
2. S. E. Filippakis, B. Perdikatsis and K. Assimenos, Studies in Conservation **24**, 54, (1975).
3. T. Learner, Spectroscopy Europe, **8**, 14, (1996)
4. D. Anglos, M. Solomidou, I. Zergioti, V. Zafiropulos, T.G. Papazoglou, C. Fotakis, Appl. Spectrosc., **50**, 1331, (1996).
5. R.J.H. Clark, Chem. Soc. Rev. **24**, 187, (1995).
6. D. Anglos, S. Couris and C. Fotakis, Appl. Spectrosc. **51**, 1025-1030, (1997).
7. H. G. Friedstein, J. Chem. Educ. 58, 291, (1981).

Optical Fibers for the Cultural Heritage I: Picture Varnishes as Thermosensitive Fiber Cladding

A.G. Mignani, M. Bacci, C. Trono
IROE-CNR, Dept. Optoelectronics and Photonics
Via Panciatichi 64, I-50127 Firenze, Italy
ph. +39-055-42351, fax +39-055-410893, e-mail: mignani@iroe.fi.cnr.it

Abstract. Varnishes used for the protection of paintings are also key factors influencing the appearance of paintings, since they are able to provide more gloss and saturated colors. In order to give a stable appearance to a painting, varnishes must have stable or reversible optical characteristics (color, refractive index) even in the presence of different environmental conditions. This work describes how optical fibers can be used to monitor temperature effects on the varnish refractive index.

1 Introduction

Transparent varnishes have been always applied to oil and tempera paintings, and have been always regarded as protective layers. Actually, the effect of the varnish is twofold, not only on the safety of the painting, but also on the perception of the colors. In fact, the varnish substantially alters the appearance of paintings by creating a microscopically-smoother surface and by providing a refractive-index matching-medium between the air and the pigments [1]. The incident light better penetrates the paint layer that appears glossier, and the scattered white light is reduced thus giving more saturated colors [2,3]. Specialists say that varnishing paintings "bring out the colors" and "make them shiny"[1]. Because of the influence they have on the appearance of paintings, varnishes must exhibit stable or reversible optical characteristics, especially color and refractive index, even on ageing and in the presence of different environmental conditions.

This paper shows how optical fiber technology can be used for the real-time and continuous monitoring of the temperature effects on the varnish refractive index.

2 Working Principle

A plastic-cladding silica-core (PCS) optical fiber was used. A short length of the cladding was stripped off and replaced by a layer of varnish, as shown in Figure 1. Consequently, the light intensity guided by the optical fiber was dependent on the refractive index of the varnished fiber-section, and any temperature variation affecting the varnish refractive index resulted in light-intensity modulation.

Three varnishes made of the most common natural resins used in picture restoration were tested: dammar resin, amber in linseed-oil and gum mastic, having refractive indices of 1.539, 1.546, and 1.536, respectively[4]. Because of the high values of these refractive indeces, the light from the quartz core of the varnished section was radiated into the varnish layer, and was totally reflected at the varnish-air interface. Polystyrene optical fibers, that have a core refractive index of 1.59, should be much more appropriate for obtaining a standing wave also in the varnished fiber section.

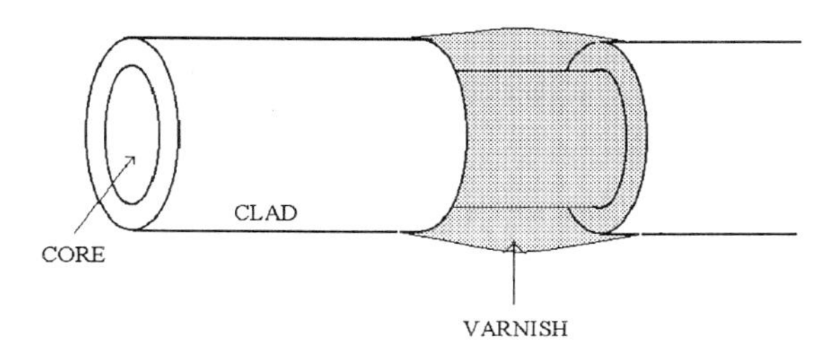

3 Experimental Results

The PCS optical fiber was a 3M-EOTec™-FP200LMT fiber with a core diameter of 200 μm. Lengths of fibers were stripped of the cladding at the center for a length of ≈15 mm using a razor blade; traces of plastic and grease were further removed using solvents (trichloroethylene and acetone). Each fiber was then fixed to a plastic support to prevent breakage and to be varnished and tested during thermal cycles.

The fiber optic test-unit is shown in Figure 2. The source was a LED emitting at 660 nm and the detectors were a couple of PINs. The LED was connected to the common arm of a 1x2 fiber optic coupler. One arm of the coupler, a_1, was directly connected to the PIN in order to check source stability (reference arm), while the other arm, a_2, was connected to the varnished fiber and then to the other PIN. A mode scrambler was used in arm a_2 to feed the varnished fiber by means of a stationary modal distribution. LED and PINs were interfaced to a PC by means of a Data Acquisition Processing (DAP) board providing LED driving current and modulation, together with signal detection and processing. The ratio between the varnished-fiber output and the reference-arm output was considered to be the sensor output.

The bare fiber-core was treated by spraying it with varnish, and the sensor output was continuously monitored for two days so as to check the varnish hardening-phase. Once hardened, the varnished fiber-section was heated by means of a hot plate positioned under the fiber while sensor output was being measured. The reference temperature was measured by means of a thermocouple the output of which was processed by the DAP. Several temperature cycles in the 20-65°C range were performed in order to check both thermal sensitivity and reversibility.

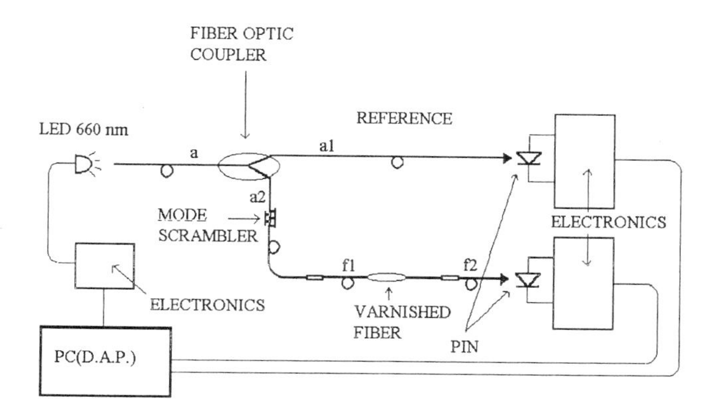

The temperature sensitivity of dammar, amber in linseed-oil and gum mastic is shown in Figures 3, 4, and 5, respectively. All three of these varnishes exhibited temperature sensitivity to some extent: thermal behaviour of the refractive index was fully reversible for gum mastic, irreversible for dammar, slightly sensitive and poorly reversible for amber.

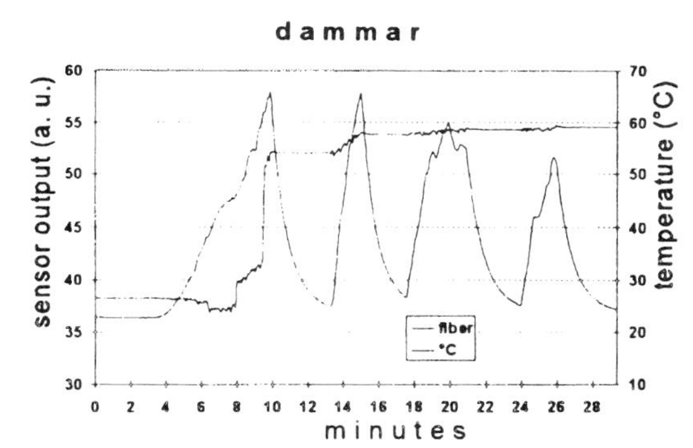

Figure 3. Optical fiber varnished using dammar: response to thermal cycles

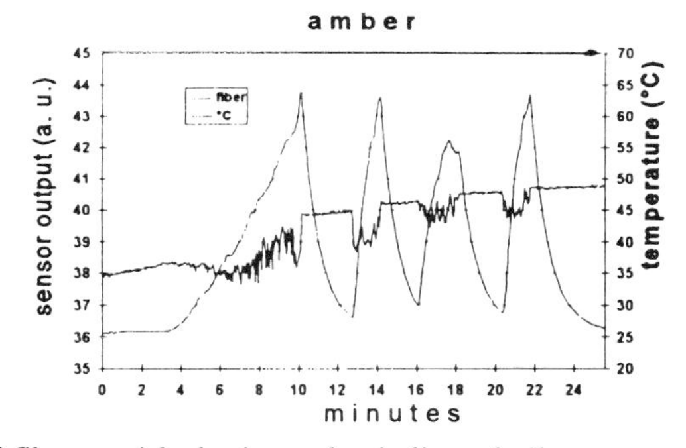

Figure 4. Optical fiber varnished using amber in linseed-oil: response to thermal cycles

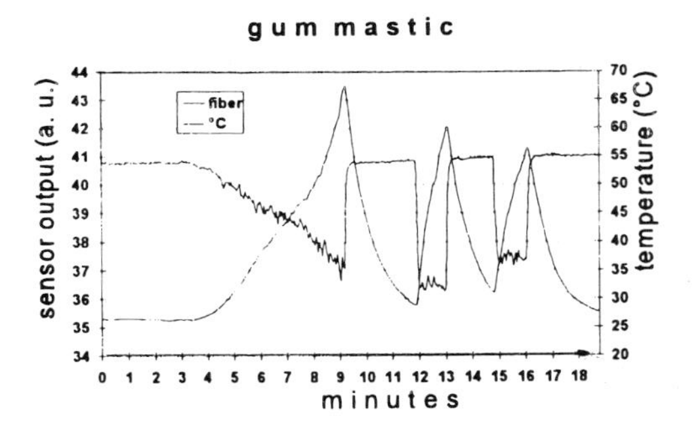

Figure 5 Optical Fiber varnished using gum mastic: response to thermal cycles

4 Remarks

The experimental results obtained suggest two different uses for the varnishes. When a thermally-stable refractive index is required for a stable appearance of a painting, the most suitable varnish is dammar, provided that a preliminary thermal cycle is carried out. On the contrary, if the monitoring of painting thermal excursions should be performed, the gum mastic varnish is the most suitable. In fact, thanks to its good temperature sensitivity and reversibility, gum mastic could also be considered as a transducer for the implementation of a temperature sensor to be permanently inlayed in the painting. By embedding the optical fiber in the painting together with the picture varnish, for example on a corner, continuous temperature monitoring could be possible, in order to prevent risk conditions that can arise when illuminating the painting with the use of lamps, as happens during television shots.

5 Acknowledgements

This work has been partially funded by the National Research Council of Italy, under the *"Cultural Heritage"* Special Project.

References

1. E. René de la Rie, "The influence of varnishes on the appearance of paintings", *Studies in Conservation* 32, pp. 1-13, 1987.
2. T.B. Brill, *Light, its Interaction with Art and Antiquities*, Plenum Press, New York, 1980.
3. R.S. Hunter, *The Measurement of Appearance*, John Wiley & Sons, New York, 1975.
4. E. René de la Rie, "Old Master Paintings", *Anaytical Chemistry* 61, pp. 1228A-1240A, 1989.

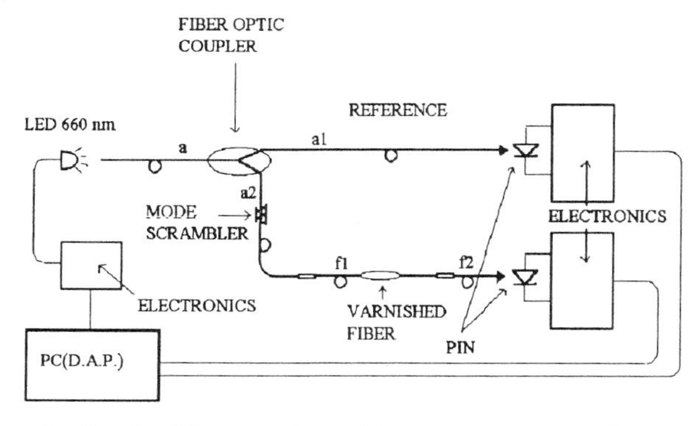

Figure 2 Sketch of the test-unit used for characterizing the varnished fiber

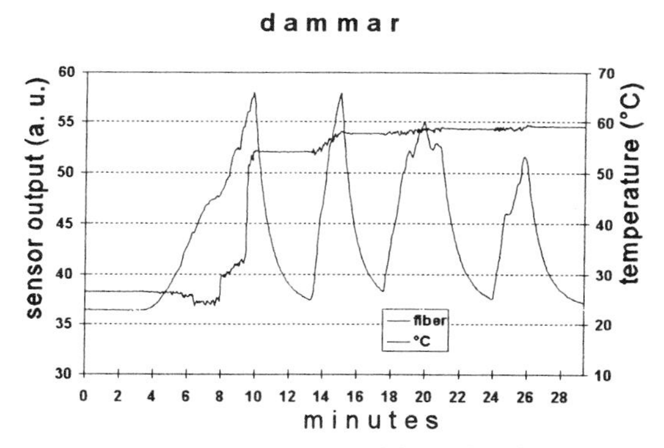

Figure 3 Optical fiber varnished using dammar: response to thermal cycles

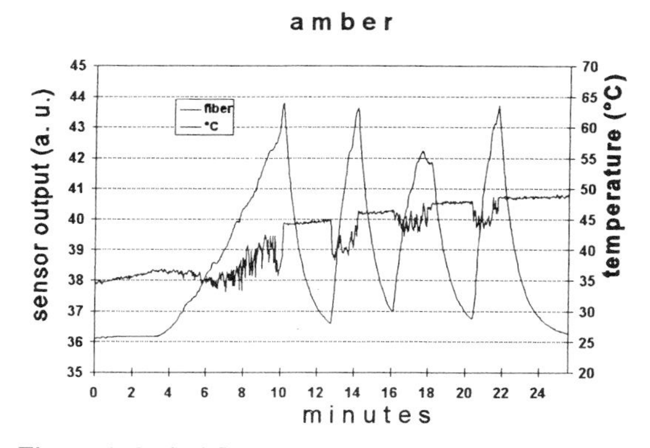

Figure 4 Optical fiber varnished using amber in linseed-oil: response to thermal cycles

Non-invasive Measurements of Damage of Frescoes Paintings and Icons by Laser Scanning Vibrometer: A Comparison of Different Exciters Used with Artificial Samples

Paolo Castellini, Enrico Esposito, Nicola Paone, Enrico P. Tomasini.
Dipartimento di Meccanica, Università di Ancona,
Via Brecce Bianche, 60131 Ancona - ITALY
Phone: +39-71-2204487, Fax: +39-71-2204813
e-mail:tomasini@mehp1.cineca.it

1 Introduction

Frescoes, icons and composite materials sho òw analogies in terms of defects, both present layer-to-layer detachments and delaminations and surface cracks; past experiences demonstrated that the study of surface vibrations could be used to locate defects position and size. At present a non-invasive diagnostic system is under development and the aim of this work is to propose and compare different kinds of structural exciter.

After initial measurement set-ups based on accelerometers and impact hammers a novel system based on laser vibrometers and acoustic stimulation of structures to allow full remote and contactless investigation of detachments and delaminations has been develop.

Three kinds of exciters are currently employed: horn loudspeakers, elliptic mirror acoustic focer and piezo actuators.

The first source is easy to implement and commercially available, while the second one has been developed in our laboratories to concentrate sound power in a very small area so to excite only localised defects of the structures and avoid annoying noises to be propagated around them. Piezo discs are cheap and efficient sources of mechanical energy and a very simple actuator has been constructed around one of them.

Measurements results on icons and frescoes samples obtained with these three devices will be compared and relative pros and cons examined.

2 Measurement Procedure

We have defined a general measurement procedure consisting of two different stages, leading to defects identification and characterisation. The first scan on the work of art is done by white noise excitation of the work of art and measuring the RMS value of surface vibration by the Scanning Laser Doppler Vibrometer

(SLDV): the result is a point by point map of the surface RMS velocity. This puts in evidence the detached areas very quickly because they show as higher velocity ones.

After defects localisation it is possible to investigate the associated spectrum by pointing the laser, for example, at the centre of the detached regions. Employing an FFT analyser, resonance frequencies are identified and subsequent scans are executed looking surface vibrations at these same single frequencies. Signal to noise ratio is greatly improved if compared with the one of a RMS scan although measurement time grows in a similar way and the defect shape is defined less clearly. Resonance frequencies are also needed if one wants to study a model of the work of art by FE analysis.

In Fig. 1 a comparison of the same sample examined by RMS and FFT analysis is reported (plate dimensions are 240x160x25 mm).

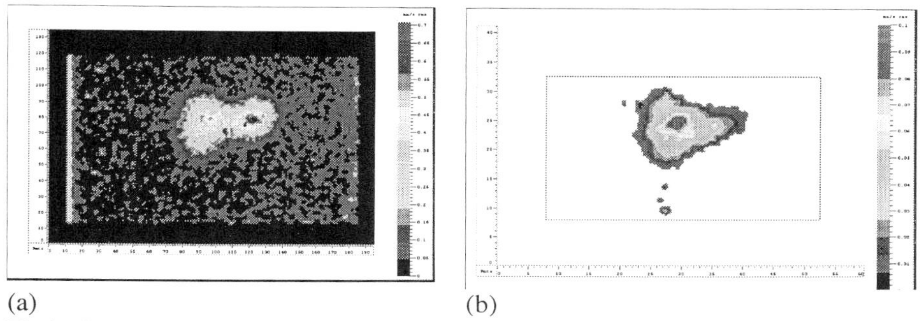

(a) (b)

Fig.1 - Plate III analysis (a) RMS analysis of plate vibration (b) same sample vibration map at 3338 Hz.

3 Artificial Excitation Devices

As previously mentioned we use three different excitation devices:

a) constant directivity horn loudspeakers;
b) elliptic acoustic mirror;
c) piezo actuators.

The basic idea behind type b) is to exploit the focusing property of elliptic mirrors to concentrate acoustic energy in a very small region of the object to be studied instead of dispersing it everywhere. The speaker is positioned in front of the mirror exactly in its first focus (f1) while the specimen is on the second focus (f2) of the ellipsoid: in this way sound waves emitted in f1 reach f2 exciting vibrations. The mirror was realised in our laboratory and is made of a sandwich of seven fibreglass layers. We fixed a laser pointer to the mirror to indicate the position of f2.

Type c) exciter was developed because there is a physical limit inherent in acoustical excitation, i.e. the great mismatch between sound waves and specimens (usually hard solids) impedances. Energy transmission is very low and only the very high sensibility of SLDV allows good measurements in this situation. Mechanical impedance of ceramic piezo composites (PZT) is much higher than that of a sound wave and gluing small disks to specimens greatly reinforces mechanical coupling. At present we use 10x0.5 mm PI GmbH PZT 151 disks, capable of exciting structures up to 2 MHz. Maximum voltage that can be applied is 1 kV and a 100 W custom designed amplifier to drive disks adequately has been acquired.

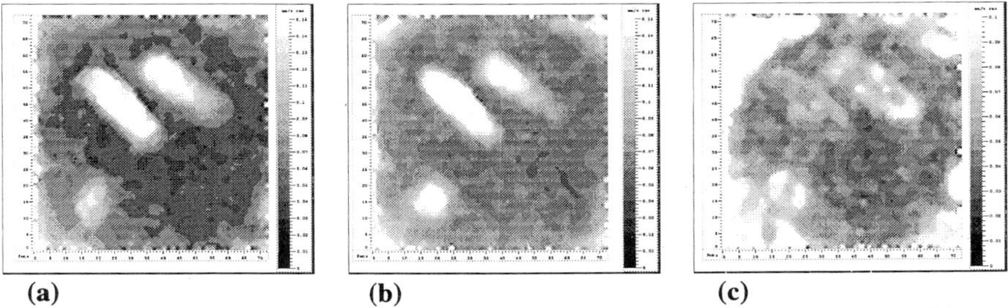

(a) (b) (c)

Fig. 2 – RMS analysis of fresco sample (40x40x5 cm) - (a) mirror source (b) horn speaker (c) piezo actuator.

4 Natural vs. Artificial Excitation Sources

We consider two natural excitation sources in our study

d) knocks on surfaces;

e) crowds and live music.

In the following table we compare amplitude and spectra of artificial and natural sources. Natural sources generally exhibit greater or comparable amplitudes and narrower frequency bands.

	Peak Press. (N/cm^2)	Max Freq. (kHz)
Mirror	**6.5E-04**	**10**
Horn	**6.5E-04**	**15**
Piezo	**0.5**	**2000**
Knocks	**30**	**0.5**
Live Music	**1.3E-04**	**3.5**

5 Measurements on Artificial Samples

We present the results of RMS measurements on a fresco, an icon and a composite sample (Fig. 2,3,4). It is evident that the piezo performs at its best on very small defects, while acoustic sources on medium to large ones. The latter allow a fully contactless diagnostic to be performed, too.

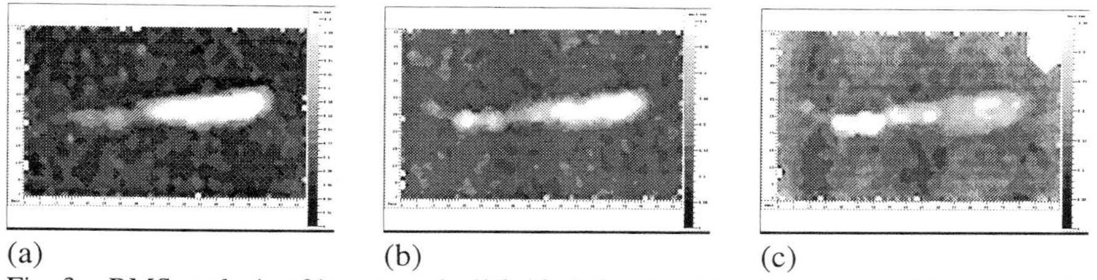

(a) (b) (c)

Fig. 3 – RMS analysis of icon sample (15x10x2.5 cm) - (a) mirror source (b) horn speaker (c) piezo actuator.

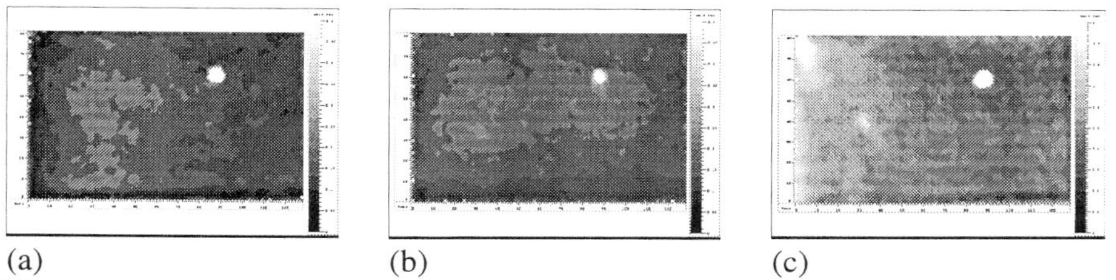

(a) (b) (c)

Fig. 4 – RMS vibration analysis of compo sample (30x20x0.2 cm) - (a) acoustic mirror (b) horn speaker (c) piezo actuator.

6 Concluding Remarks

We executed a number of tests on different samples using three excitation sources. The horn is capable of detecting from large to very small defects, irradiates sound over a wide angle to excite large surfaces at great distances but can be very noisy and annoying for people around.

The mirror overcomes this problem and can be used even in crowded rooms. We saw that it was not efficient only with very small defects but its main drawback could be the fixed focus distance.

The small piezo actuator we realised demonstrated a useful frequency range above some kHz and so it performed optimally for very small defects. The piezo must be applied to the object and this makes them not suitable, for example, for distant frescoes. Our future work is to improve the actuator performance by testing new materials to build it so to improve energy transfer towards the samples. In the following table we summarise the performance of the three exciters referred to defects dimensions and sample material.

7 Acknowledgements

Most of the activity has been financially supported by the EEC-Standard Measurement & Testing programme under contract Laserart-SMT4-CT96-2062.

Defect size ➔ Source ➘	Large (> 10 cm)	Medium (2< <10 cm)	Small (< 2cm)	Dep.ce on material (*)
Mirror	***Good***	**Good**	**Fair**	**Low**
Horn	**Good**	**Good**	**Fair**	**Low**
Piezo	**Poor**	**Poor**	**Good**	**Low**

The contribution of the following bodies in the field of art conservation is acknowledged:

Laboratoire de Recherche des Monuments Historiques (LRMH), Champs-sur-Marne, France, for the fresco samples (Dr. M. Stefanaggi).

Benaki Museum, Athens, Greece, for the icon samples (Dr. S. Stassinopoulos).

We would like also to thank Audiomatica Srl (Florence, Italy) for supplying the CLIO audio measurement system.

References

1. E.Esposito "*Ecospettrografia. Una tecnica per lo studio dello stato di conservazione delle opere d'arte*", Tesi di laurea, Università di Ancona, A.A. 1989-90.
2. W. D'Amrogio, A. Mannaioli, D. Del Vescovo "*Use of FRF measurements as a non-destructive tool to detect detachments of frescoes*", Proc. of the 12th International Modal Analysis Conference, pp. 1083-1088, Honolulu, 1994.
3. P.Castellini, N. Paone, E. P. Tomasini "*The Laser Doppler Vibrometer as an Instrument for Non-Intrusive Diagnostic of Works of Art: Application to Fresco Painting* ", Optics & Lasers in Engineering, Vol. 25, pp. 227-246, May 1996.
4. P.Castellini, N. Paone, E. P. Tomasini "*A Laser Based Measurement Technique for the Diagnostic of Detachments in Frescoes and Wooden Works of Art*", oral presentation, LACONA II, 2nd International Conference on Lasers in the Conservation of Artworks, Liverpool, 1997.
5. P.Castellini, E. Esposito, N. Paone, E. P. Tomasini "*Conservation of frescoes paintings and icons: noninvasive measurement of damage by a laser scanning vibrometer*", Proc. of the SPIE International Symposium on Nondestructive Evaluation Techniques for Aging Infrastructure &Manufacturing, SPIE Vol. 3396, San Antonio (Texas), 1998.

FTIR Imaging Spectroscopy for Organic Surface Analysis of Embedded Paint Cross-Sections

Ron M.A. Heeren, Jaap van der Weerd and Jaap J. Boon
FOM Institute for Atomic and molecular Physics, Kruislaan 407, 1098 SJ Amsterdam, The Netherlands.

Abstract: A description of the microscopic analysis of an embedded paint cross section with a novel FTIR imaging technique will be provided. Sample preparation turns out to be crucial for the success of the method used and some of its practical aspects for organic surface analytical techniques will be discussed. FTIR imaging analysis carried out with this novel technique is successfully employed to determine the binding medium type in individual paint layers.

1 Analysis of Embedded Paint Cross-Sections

The start of the priority program MOLART of the Dutch science foundation NWO, marked the beginning of a new consolidated effort to develop new methodologies and strategies in paintings research. The MOLART project focuses on the molecular aspects of ageing in painted art. These molecular processes occurring in the course of time underlie the chemical and physical changes in a painting. Especially the ageing mechanisms in the organic constituents of an aged paint formulation are poorly understood. We develop and employ new spectroscopic methodologies to study the organic surface chemistry of embedded paint cross-sections. In this paper, we will focus on reflection type Fourier Transform InfraRed imaging of these surfaces to study the functional group distribution.

Characterization of materials using infrared functional group mapping microscopy has been developed approximately one decade ago [1]. The infrared spectrum yields a set of absorption and/or reflection bands related to the chemical functional groups. Non spatially resolved FTIR microscopy is used as an analytical technique in conservation science since the early 70's. It has demonstrated its use in material identification of minute paint samples and reference materials [2,3,4]. Functional group mapping combines spatial and spectral information, identifying and localizing various components of the sample surface. Mapping larger surface areas with sufficient spatial resolution is extremely slow, especially if a reasonable spectral resolution and signal-to-noise ratio are required. A slit or diaphragm in one of the intermediate focal planes of the FTIR microscope defines the spatial resolution of a single measurement. The introduction of a small (in the order of the wavelength) aperture in the optical path potentially creates various instrumental artifacts such as interference and scattering. These effects tend to blur the functional group maps created in this manner. These considerations hamper the routine use of FTIR microspectroscopy in conservation science.

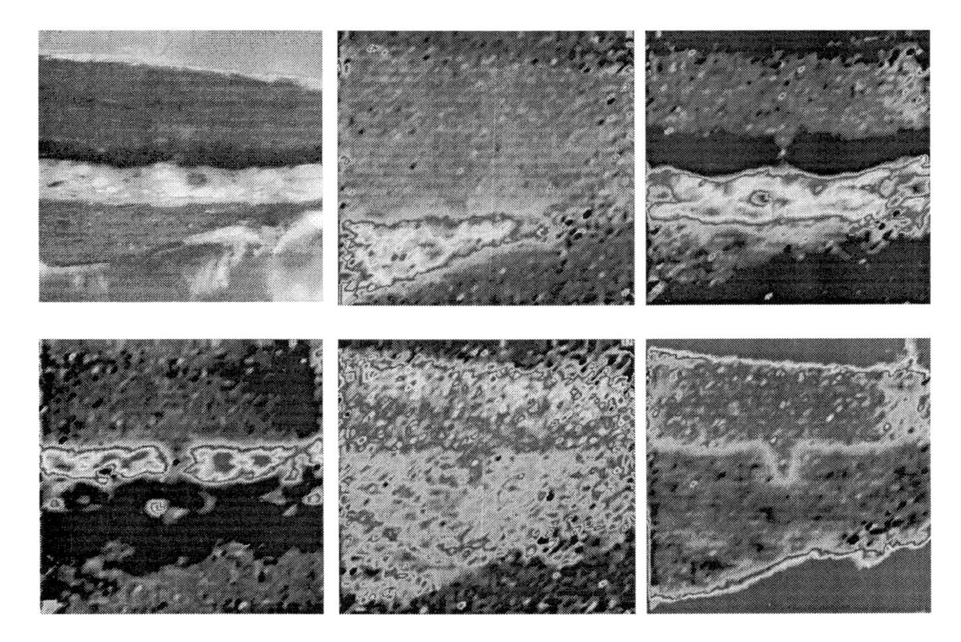

Figure 1. Images of an embedded cross-section. A) visual image B) FTIR reflection image at 1138 cm^{-1} C) FTIR reflection image at 1475 cm^{-1} D) FTIR reflection image at 1510 cm^{-1} E) FTIR reflection image at 1775 cm^{-1} and F) FTIR reflection image at 1750 cm^{-1}

Here, we will use a non-dispersive infrared imaging microspectrometer to examine the infrared radiation absorbed by an embedded paint cross-section without the need for elaborate mapping experiments. The experimental set-up consists of a Michelson-Morley type interferometer equipped with a step-scan mirror, a FTIR microscope and a 64x64 pixel Mercury-Cadmium-Telluride IR array camera. In step-scan operation, the moving mirror moves in discrete steps, and the mirror surface stays a fixed position for a variable period of time. An interferogram is build up from detector signals at incremental optical path differences (or mirror displacement). After acquisition of the interferogram, the signal is Fourier Transformed to yield the infrared absorption spectrum. This mode of operation opens up new possibilities for FTIR imaging [5,6,7] which can be employed in the examination of paint cross-sections. Briefly, at each individual "step" of the moving mirror an IR camera acquires and stores an image. After completion of the full scan a dataset is created which consists of a stack of images which is sometimes referred to as a hyperspectral data-cube. Each pixel stack in this data-cube represents an interferogram, which is subsequently Fourier Transformed to yield the corresponding IR spectrum. Each of the pixels of the IR camera is an individual detector. Hence, it is possible to perform a full multi-channel FTIR analysis of the image created by the microscope on the camera without the need for sample scanning. Note that the IR light coming through the

interferometer illuminates the sample, and absorption characteristics can be determined in transmission as well as reflection.
All reported results in the literature up to date have been achieved in transmission mode, in which a soap/water/lipid film or biological tissue sections have been examined. No such measurements seem to have been performed in microscopical reflection. The analysis of paint cross-sections in conservation science is one of the potential applications of reflective microscopical imaging. Here, it provides an instantaneous correlation of high-resolution spatial and spectral information of the embedded paint samples. The resulting functional maps can be readily correlated with the regular microscopical image. When the fingerprinting region of the infrared spectrum is imaged, varnish layers, binding media and certain pigments as well as the interaction between the various compounds can be visualized. This will lead to a better understanding of the materials used by the artist, the condition of the painting and insight into the aging process. In this paper we will present the first reflection FTIR imaging experiment on a paint sample carried out with this technique.

2 Results and Discussion

An example of a visual image taken with the FTIR microscope is presented in figure 1A. This sample has been embedded in Scandiplast ™ and hand ground on mesh 4000 polishing paper to reveal the layered structure in figure 1A. Indicated are the red grounding layer (1), a white paint layer (2), a dark black layer (3), a laminated varnish layer (4) and the embedding material (E). One of the particular questions regarding this cross-section was the nature of the binding medium in the dark layer 3. After reflectance FTIR imaging with a 8 cm-1 spectral resolution with symmetric interferogram acquisition, a selected set of results is displayed in figure 1B-F. Each of the panels B-F reveals a different functional group distribution, related to the main organic components of each paint layer. To obtain these images, the raw data was Fourier Transformed and subsequently ratio'ed against a Fourier transformed ZnSe background image set which has approximately the same reflectance (a few percent) as the paint cross-sections. The resulting spectral images were used as is without further work-up of the data. As no Kramers-Kronig transformation has been applied some specular reflection peaks appear in the spectrum. This also introduces a slight offset in the peak center wavelengths. Only the positive parts of these peaks were integrated to create the spectral image data presented in figure 1B-F. Figure 1 B is an image at 1138 cm-1 characteristic for a carbonate (C-O) absorption attributed to a strong lead or calcium carbonate containing grounding material. Figure 1C is the 1475 cm-1 image originating from the C-H absorption's in an oil containing binding medium. The most remarkable finding was the image displayed in figure 1D, which was the amide II band (1510 cm-1) indicating a protein-containing layer. This finding was confirmed by the observation of the amide I band in the same layer (image not shown). The laminted varnish layer showed up at the carbonyl wavelength of 1776 cm-1 (figure 1E) a small shoulder of the carbonyl absorption of the embedding material at 1750 cm-1 (figure 1F).Combined together the images 1B-F elucidate the binding medium of each different layer found in this cross-section .

3 Conclusions

We have demonstrated the use of reflectance FTIR imaging spectroscopy on embedded paint cross-sections. This method is successfully used to identify the nature of the binding medium in individual layers of a paint cross-section. This allows the rapid examination of the distribution of organic components on a complex surface, even with a low reflection coefficient. Next to binding medium analysis it potentially can be used to identify various organic and inorganic pigments as well as varnish types and gum containing glazes. FTIR imaging in reflection mode provides a unique combination of high resolution chemical contrast and rapid analysis.

Acknowledgements

The authors are indebted to all MOLART team members, past and present, for their continuous support and invaluable input. We would also like to acknoweledge H. Brammer of the Gemaelde Gallerie Kassel for his kind supply of the cross-section discussed in this paper. This research is performed with financial support of the priority program MOLART of the Nederlandse organisatie voor wetenschappelijk onderzoek (NWO) and the foundation for Fundamenteel Onderzoek der Materie (FOM), a subsidiary of NWO.

Reference

1. M.A. Harthcock and S.C. Atkin, Appl.Spect. 42 (1988) 449-455
2. M.R. Derrick, E.F. Doehne, A.E. Parker, D.C. Stulik, J. Am. Inst. Cons. 33 (1994) 171-184.
3. R.L.Feller, Science, 120 (1954) 1069-1070.
4. J. Pilc and R. White, Nat. Gal. Tech. Bull. 16 (1995) 73-84
5. P.J. Treado, I.W. Levin and E.N. Lewis, Appl. Spectrosc. 46 (1992) 1211
6. P.J. Treado, I.W. Levin and E.N. Lewis, Appl. Spectrosc. 48 (1994) 607.
7. E.N. Lewis, A.M. Gorbach, C. Marcott and I.W. Levin, Appl. Spectrosc. 50 (1996) 263-269.

Advanced Workstation for Controlled Laser Cleaning of Paintings

J.H. Scholten, J.M. Teule, V. Zafiropulos*, R.M.A. Heeren§
Art Innovation b.v., Westermaatsweg 11, 7556 BW Hengelo, Netherlands
email: Info@Art-Innovation.nl
*Foundation for Research and Technology-Hellas (Fo.R.T.H.), Laser and Application Division, Heraklion, Crete (Greece)
§FOM-institute for Atomic and Molecular Physics (AMOLF), Amsterdam (Netherlands)

1 Introduction

Any artwork conservation procedure relies on effective diagnostic techniques and controlled cleaning methods. Traditional methods for the removal of unwanted layers from an artwork comprise mechanical or chemical techniques, which often cause (unknown) damage to the object. New technologies can offer valuable tools to support conservators and restorers.

The Dutch company Art Innovation develops an innovative laser restoration tool for non-contact cleaning of painted artworks. Accurate beam manipulation techniques in combination with an on-line detection system make the system suitable for selective cleaning of delicate surfaces.

Pioneering work by researchers at the Fo.R.T.H.-institute (Crete) demonstrated that the laser cleaning technique can be applied on artworks in a safe and efficient way using Laser-Induced Breakdown Spectroscopy (LIBS) as a diagnostic tool to control the process. In contrast to conventional techniques, when using an excimer laser with a proper selection of laser parameters, material can be accurately removed with minimal damage to the underlying material. The utilisation of lasers obviates the use of various chemicals, and provides a method to remove layers untreatable using conventional methods. A laser beam can easily and accurately be manipulated, the cleaning process is controlled on-line and there is no mechanical contact with the artwork.

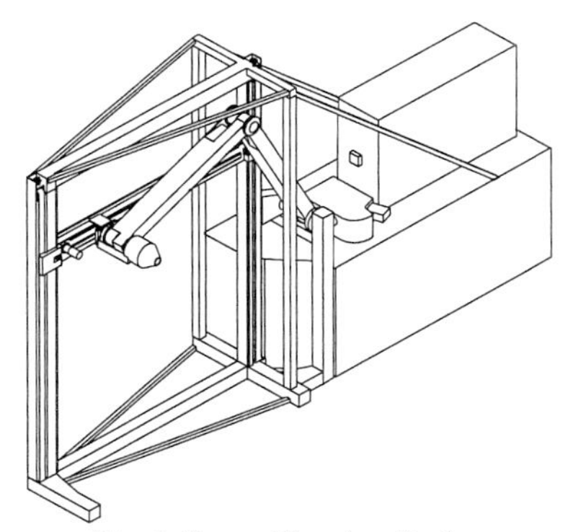

Fig. 1. Laser Cleaning Station

2 The System Concept

The Laser Cleaning Station basically consists of a "passive arm" which follows the movement prescribed by a separate XY-table. Advantages of this design:

- The rectangular laser spot and scanning protocol requires the laser spot to move in long straight lines. An XY-manipulator can accomplish this task easier than a motorised anthropomorphic arm.
- The separate driving systems of a XY-manipulator can easily be optimised according to the specified scanning motion (a large number of rapid, short steps in one direction followed by one side step at the end of a scan line).
- The XY-table supports the optical arm at the tip (wrist), where the actual scanning takes place. This enhances the stiffness and position accuracy of the system.

The optical arm employs mirrors in a configuration similar to that of a periscope. Since the mirrors are mounted as complementary pairs, the laser fluence profile and orientation of the spot are guaranteed during a rotation of the arm. The optical path is of constant length which simplifies the task of maintaining the necessary laser spot properties.

2.1 Accuracy and Cleaning Resolution

A combination of a multi-finger shutter and a cylindrical lens provides a resolution which is suitable for regular cleaning activities (varnish removal). When a higher resolution is preferred, a mask with a thin slit is inserted into the beam. The dimensions

of the single resulting spot are determined by the slit width and the focusing lens and can be adjusted to the wishes of the restorer (0.3-3 mm).
To ensure that the laser spot and scanning motion are within the specified accuracy of 0,04 mm, a mechatronic design approach is practised. The software controls and corrects the position of the head to compensate for the drift of the laser spot during movement of the optical arm.

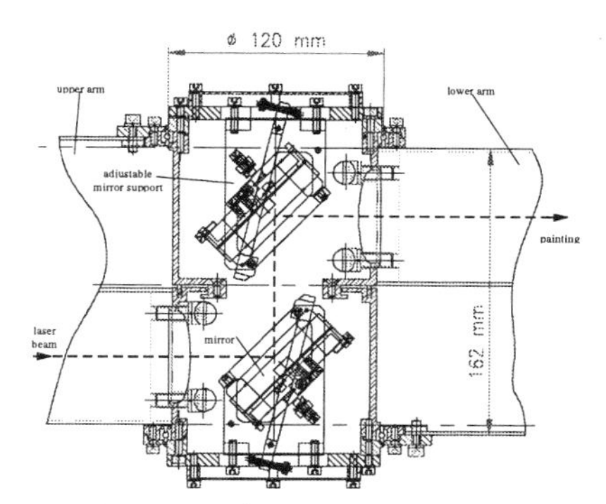

Fig. 2. Detail of elbow joint optical arm

2.2 System Control

Excimer laser cleaning of artworks is strictly a 'hands-off' operation using the LIBS control technique and a sophisticated scanning protocol, in order to properly apply the radiation to the painting surface.
For automatic control during the laser cleaning process, the LIBS spectra are continuously recorded in a pre-selected spectral region. The pre-selection is based on results of a preliminary study, conducted at the beginning of each project. A project specific data algorithm for process control is obtained, based on the peak intensity ratios characteristic for the treated material. Certain boundaries are defined. When these conditions are exceeded, the laser stops firing and is pointed at a new position, where the cleaning process starts again.

2.3 User Interface

The user interface plays an important role in simulating the "hands-on" treatment preferred by the restorer. An important aid in providing a feeling of control for the restorer is extensive visualisation of the cleaning progress. With a computer program similar to the Windows environment the user will be able to select different spot sizes on a computer display for treatment of a selected area of the painting.

3 Laser-Induced Breakdown Spectroscopy

The successful operation of the laser cleaning station relies heavily on the quality of the LIBS diagnostic control technique. Laser-Induced Breakdown Spectroscopy invloves the measurement of emission after illumination of the painting surface. The emitted light from the surface and plasma plume is projected onto an array of fibers connected to a spectrograph and ICCD-detector.

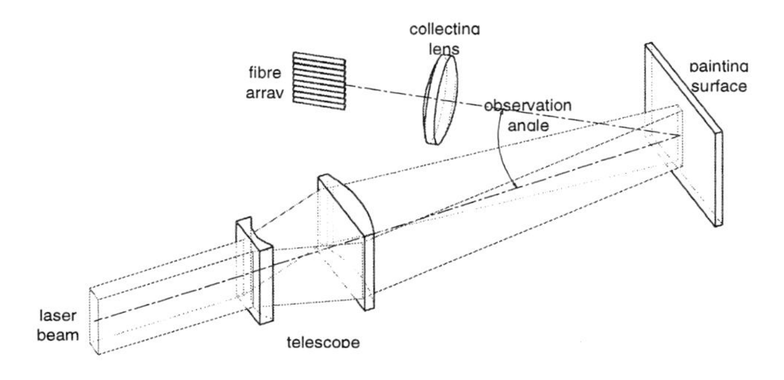

Fig. 4. LIBS configuration

3.1 Goal

- Verify the possibility of imaging the laser spot on a fiber array at a certain distance from the surface
- Determine the spatial resolution, observation angle and observation spot in case of a fiber array
- Determine an effective data algorithm to enhance the resolution

3.2 Experimental Procedure

LIBS spectra are collected around an interface between two distinct pigments. A single fiber is moved across the projec-ted image of the emitted light from different parts of the laser spot. Peaks of atomic emission, which have minimal mutual interference are

selected, and relative changes are monitored. The spatial distinguishing ability of the LIBS equipment is calculated from the ratio of the selected peaks for different positions of the fiber near the pigment transition.
A change in ratio of one order of magnitude is supposed to be a sufficient indication of the presence of another pigment. The spatial resolution is defined as the change in fiber position required to span one order of magnitude change in signal ratio.
The experiments are performed on a sample of lead within linseed oil next to titanium white in walnut oil. The latter shows a large background signal ("organic belt") caused by the broad emission from organic material.

Experimental parameters	
Spectrographic grating	300 lp/mm
Pulse delay	800 ns
Pulse width	460 ns
Pulse energy on surface	50mJ
Observation angle	40 °
Spot height	12 mm
Spot image	37,1 mm
Distance from laser exit	3,5 m
f (lens at 212 mm)	160 mm

Experimental setup	Res.	Impr.
crude setup	1.9mm	
substraction backgr.	1.6mm	16 %
addition of iris	1.1mm	40 %
iris&substr. Backgr.	0.6mm	67 %
iris,substr. bkgr., focussing & algorithm	0.4mm	80 %

3.3 Varnish Removal

Spectrum A and B show LIBS spectra obtained after irradiation of a naturally aged dammar sample, after 10, 14, 20 and 20, 25, 27 pulses, respectively.
In spectrum A a gradual decrease is observed in the CO and CN peaks, showing a change in CO/CN ratio going deeper into the material. After about 20 pulses the pigment layer is reached, and an increase in Pb and Ca peaks is observed (spectrum B).

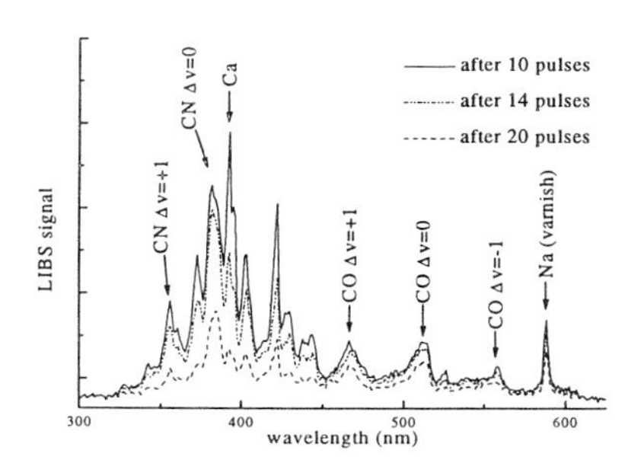

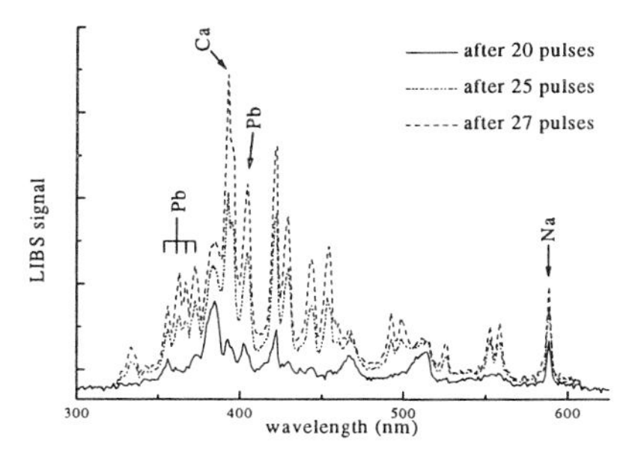

Illumination System for the Theodosius' Plate

Eusebio Bernabeu, Javier Alda, Angel García-Botella,
José A. Gómez-Pedrero, Enrique Olivera.
Departamento de Óptica. Universidad Complutense de Madrid.
Facultad de CC. Físicas. Ciudad Universitaria s/n 28040 Madrid Spain
Tel: (34) 913 94 45 55 Fax: (34) 913 94 46 74 e-mail: j.alda@fis.ucm.es

Abstract: The Theodosius' Plate constitutes an unique piece of the final period of the Roman Empire (IV century) in the Iberic peninsula. We have developed an illumination system that preserves the object without thermal effects and allows an obtimum viewing of the Plate. The system is based on a customized fibre optic bundle whose output ends are located along a line that is place laterally to the Theodosius's Plate. The illumination is almost at grazing incidence. This feature enhances the relief of the plate and improves its observation. Besides, the reflected light is not directed to the observer. The halogens lamps that act as the light source for the plate are located in a hidden space below the disk. This allows an easy maintenance without interaction with the space where the disk is located. At the same time the thermal emission of the lamps is avoided by using anti-caloric filters at the input window of the bundle of fibres.

1 Introduction and Historical Background

Theodosius "The Great" (379-395) was the last of the great Roman Emperors. He was the last governor of an unified Roman Empire, because when he died (in 395) he divided the Empire between his sons, Honorius and Arcadius being this division definitive. Theodosius' Plate is a silver disk with 73.5 cm diameter and 15.344 Kg. The Plate represents allegorically the Theodosius' government. Theodosius' Plate is one of the greater European ancient jewels. The Plate was made in 380 A.D. and it was lost for 14 centuries until its casual discovery in 1847 in Almendralejo (Spain). After its discovery, the Royal History Academy of Spain acquired the Plate. In these last years, the Plate has been carefully restored in order to expose it at the Royal History Academy of Spain.
For the public exposition of such archaeological treasury it has been necessary to develop a special illumination system. This system must meet certain requirements. First of all, it is necessary to avoid the incidence of infrared radiation over the plate surface, in this way non-desired thermal effects are eliminated. Second, the illuminance over the plate surface must be uniform in the vertical direction although it could be not uniform in the horizontal one. This illuminance distribution produces a pleasant sensation to the observer. Third, it is necessary to isolate the Plate in order to ensure the best conservation of it. This implies that all the maintenance operations must be done out of the showcase where the Plate is located. The last requirement is to illuminate the plate with almost grazing illumination, in order to show the beautiful relieves of the plate, avoiding at the same time undesired glares caused by specular reflections.

Figure 1: Image of the Theodosius' Plate (380 AD)

The Optics Department, UCM, has designed an illumination system for the Theodosius's Plate which matches all the requirements above mentioned. This system is based in four optical fibre arrays equipped with absorbing filters for the infrared radiation. The fibre arrays are illuminated by means of four halogen lamps placed far away from the fibre arrays.

2 Design of the Optical System

As we have said, we use an optical fibre illumination system, because such a system allows us to place the light sources far away from the heads of the fibre. With this configuration, our third requirement can be met, because, the heads of the fibre can be placed at the showcase of the plate, and the light sources can be replaced without open the showcase. Other advantage of a fibre-based illumination system is the possibility that we have to obtain structured light beams with these systems. In this way, the vertical uniformity of the illuminance can be matched. To do this, we have employed four fibre optical arrays in order to create two light lines placed at the right and left of the plate. A picture of the lightlines employed can be seen in figure 2.

Figure 2: Image of the optical fibre lightlines employed in the illumination system.

As we have said in the introduction, it is necessary to illuminate the plate at grazing incidence, but if we placed the lightlines on the same plane of the plate, the plate relieves may produce undesired shadows over the plate surface. To avoid this, we have placed the lightlines up to the plane of the plate (see figure 3). We have determined empirically the optimum distance between the plate surface and the plane of the lightlines.

We have said that the usage of a linear fibre optical array produces a vertical uniformity in the illuminance distribution over the plate surface. We have calculated numerically the illuminance distribution over the plate surface, which is produced by 1) an array of optical fibres and 2) a set of halogen lamps. The results are shown in figures 4 and 5. There are considerable differences between the usage of an array of optical fibres and the usage of a set of halogen lamps. The illuminance distribution produced by an optical fibre array presents the vertical uniformity desired while the distribution produced by a set of halogen lamps does not present this vertical uniformity.

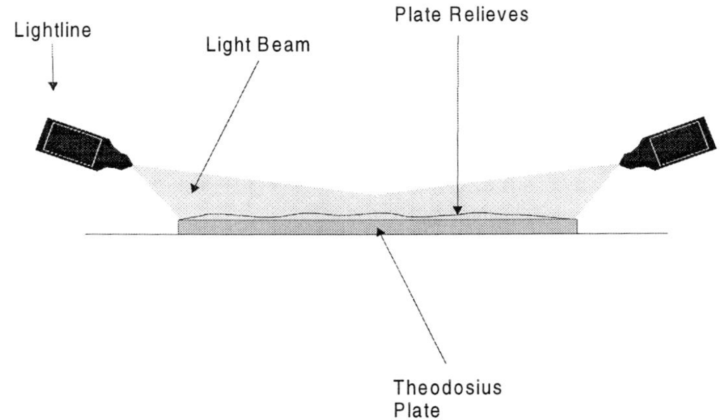

Figure 3. Lateral view of the Theodosius' Plate illumination system

This vertical luminance produced by the fibre array, joint to other advantages has leaded us to use this kind of illumination system.

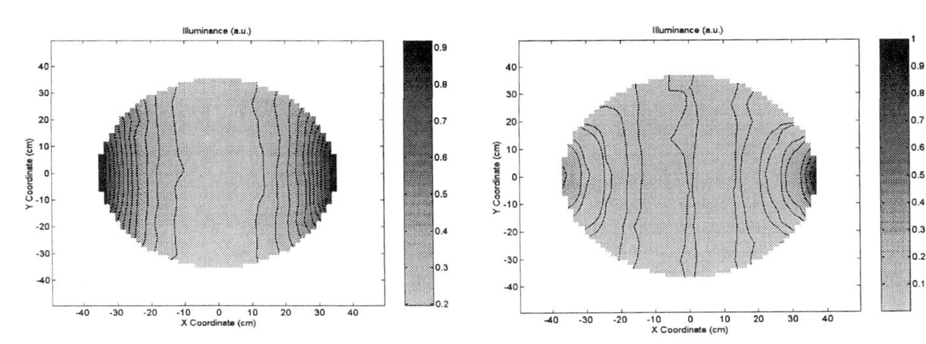

Figure 4: Calculated illuminance distribution of an optical fibre array over the Plate surface.

Figure 5: Calculated illuminance distribution of a set of 6 halogen lamps over the Plate surface

To avoid the non-desired effects of the infrared radiation we have employed four anticaloric filters and we have used glass fibres in order to avoid thermal degradation of the fibres. These anticaloric filters have been putted in front the halogens lamps before the optical fibre bundle, in this way no IR radiation arrives to the surface of the Theodosius' Plate.
Finally, we must note that it is possible to change the illuminance over the surface of the plate by means of a rheostat, so different lighting levels can be obtained allowing a more comfortable contemplation of this archaeological treasury.

3 Conclusions

We have designed a special illumination system for the Theodosius' Plate based in a linear optical fibre array. This system allows an optimum observation of this piece. At the same time, the maintenance operations of the lighting system designed do not interfere with the piece showcase and the IR radiation over the plate surface has been removed. In these conditions, the conservation state of the Theodosius' Plate remains unchanged.

References

[1] *The New Encyclopaedia Britannica*, 15th edition, Ed. Chicago University, vol. **11**, pp 689-691, (1990).
[2] A.H.M. Jones, *The Later Roman Empire*, 2 vol, pp 284-602, (1964).
[3] W.R. McCluney, *Introduction to Radiometry and Photometry*, Artech House, Boston, (1994).
[4] *Product Bulletin 4001* (Fostec Inc., 4950-C, Eisenhower Avenue, Alexandria, Va., 1995).

Research and Development of Raman Spectroscopy with Optical Fibre. Application to Pigments Identification

F.J. Sierra, J.M.Yúfera, S.Ruiz-Moreno, M.J.Soneira and C.Sandalinas
E.T.S.E.T. Barcelona, Universitat Politècnica de Catalunya (UPC)
c/Sor Eulalia de Anzizú, s/n, Mod.D5, Campus Nord, 08034 Barcelona
e-mail: yufera@tsc.upc.es, Ph: (34)-93-401 64 42

Abstract. A Raman spectroscopy equipment with fibre optics allows the analysis of great dimensions artworks. In this paper the performances of said equipment are explained and an application example, the analysis of an altarpiece from the 16th Century, is presented.

1 Introduction

For its non-destructive character and high specificity, the identification of pictorial materials with Raman spectroscopy, finds an immediate application in fields as conservation, restoration, dating and cataloguing of artworks, [1].

This communication has as general aims to describe the investigation, implementation and application of a Raman spectroscopy system that is able to offer the best performances and quality/price rate obtainable with the present technology. For such end, we suggest, as a way of design, to incorporate optical fibre technology. One can obtain, in this way, a triple objective: good spectral quality, cost reduction and maximum distance between the analysed artwork and the laboratory environment.

2 Instrumentation for Raman spectroscopy

In figure 1, the set-up of the used instrumentation can be seen. In outline, its behaviour is as follows. Includes a monochromatic source (He-Ne laser, output power of 17 mW at 632.8 nm) whose output is guided through the excitation optical fibre. This light spots, by means of an optical head, the non identified material and is then scattered and collected for being guided within the collection fibre, to the monochromator. Here it is spectrally splitted in order that the CCD can detect the spectrum and send the information to the computer, which, furthermore, controls the equipment.

The scope of the optical head is to focus the exciting light on the sample and send the Raman signal to the collecting fibre. Consists of different components: a narrow band interferential filter centred at the laser wavelength in order to eliminate the Raman spectrum of the fibre itself; the notch filters to reject the Rayleigh line at the laser wavelength and allowing only the pass of the Raman signal (the cut-off frequency can be reduced to 30 cm^{-1} using two sequential notch filters); and, finally, a dichroic mirror which allows the bi-directional light way of exciting/collection, so optimising the system efficiency.

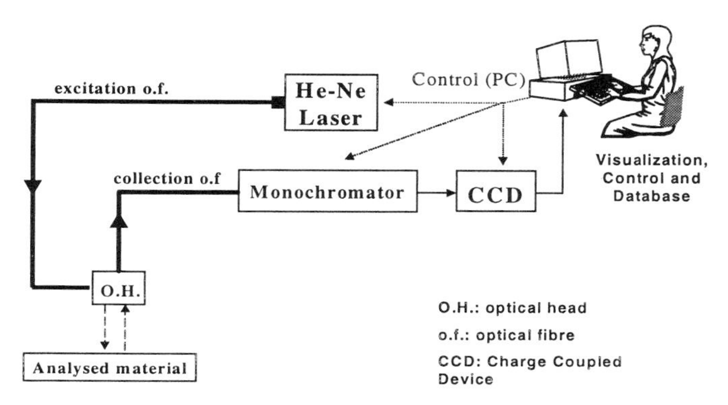

Fig. 1. Raman spectroscopy system with fibre optics.

The monochromator operation is as follows: the light enters the device through a slit, is collimated and spectrally splitted (with a grating) so that a spectrum band can be focused on the output slit. Due to the notch filters located inside the optical head, it is only necessary a simple monochromator (using just one grating), now that without them it would be necessary to use a triple monochromator to reject the Rayleigh line. In addition to put up its price, this would increase the size of the overall system. Two gratings are available: one with 1800 grooves/mm and the other with 600 grooves/mm.

In this Raman spectroscopy system, a CCD detector of high sensitivity measures the intensity of the scattered light after being spatially separated inside the monochromator. This is a device of high quantum efficiency and sensitivity, refrigerated by means of Peltier effect (-65°C), making negligible the dark current (<0.001 e/pixel/s).

The computer control is essential. It controls the monochromator, the detector and the laser. It allows to choose the wavelength range outgoing the monochromator, the CCD acquisition time and carries out the acquisition of the final information. It plays an important role on the dot of show and process the spectra, owing that, by means of signal processing techniques, the signal to noise rate can be optimised, [2]. The database of Raman spectra from standard pigments is stored in said computer.

3 Application example: an altarpiece

Attributed to the sculptor Manuel Álvarez (Castromocho, Palencia around 1517-around 1587), the "San Antolín y San Bernabé" altarpiece (see figure 2) could come from the "Antiguo Hospital de San Antolín y San Bernabé" from Palencia, and is dated, related to other documented artwork, between around 1560 and 1565.

Manuel Álvarez is one of the most interesting and prolific artist from the second half of the sixteenth century, because of its way of assimilating Berruguete and Juan de Juni's legates and fusing them with new trends.

The analysed artwork is a small religious altarpiece that consist of a *predela* and a body with its corresponding attic. In it, the architecture prevails over the sculpture, in the line of the artworks of that period due to its mannerist elegance.

Fig. 2. "San Antolín y San Bernabé" altarpiece.

The altarpiece body, of double dedication, presents the figures of the saints: Saint Antolín, Palencia's patron, and Saint Bernabe as healer.

The altarpiece presents a rich original polychrome, with motifs typically mannerist, with copious golds, *estofados* and fine artist's end brush works. In the *estofado* technique gilded areas are covered with opaque colours or translucent glazes in which pattern are then incised (*sgraffito*), so that the underlying gold is revealed. The oil painted flesh_colours has been totally conserved with some losses mainly at the socle and repolychrome in the figures of the saints.

The polychrome in its entirety presents a unity of style and technique, excluding the small sculpture in the San Antolín's attic which, because of its technical and stylistic characteristics, is not considered part of the whole.

The Raman spectroscopy equipment with optical fibre has been used in the identification of the materials composing the "San Antolín y San Bernabé" altarpiece. Figure 3 shows some of the Raman spectra of the identified and standard materials used for the comparison. Now, the identified pigments will be commented.

White pigments: lead white was obtained in the flesh colours, mixed with vermilion, and also mixed with lapislazuli in the small San Antolín sculpture.

Blue pigments: lapislazuli and a mixture of azurite and malachite were obtained in the polychrome. It has to be pointed out that our standard spectra used for the identification of the azurite and malachite were obtained from mineral samples and, because of this, its spectra were contaminated due to the natural impurities that were there. In any case, the number of Raman bands regarding both azurite and malachite is large enough as to identify these materials on the analysed altarpiece. The malachite found could be a transformation of the azurite due to the slow hydratation processes.

Red pigments: just vermilion was identified, as a red pigment, in the flesh colours polychrome and the chasuble of the little San Antolín sculpture.

Furthermore, gypsum (calcium sulphate) was identified in the preparation, and massicot in different areas, we think that used as secative. We have found iron oxides and gypsum (Venice red) composing the red bole layer between the preparation and golden layers.

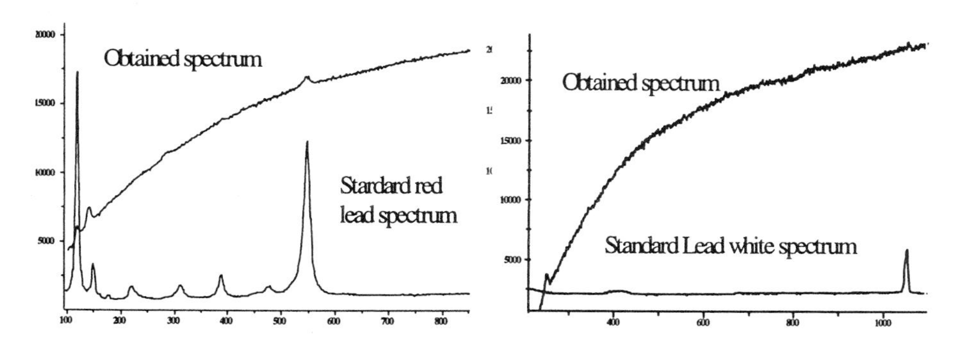

Fig. 3. Example of Raman spectra from pigments identified in the altarpiece.

4 Conclusions

Until now the conventional Raman spectroscopy has been associated to expensive equipments and devices, of big dimensions and heavies. The fundamental aim of this research is to overcome said limitations. The optical fibre allows to transform the Raman spectroscopy in a versatile technique with which every area that wants to be analysed can be reached even in artworks of big dimensions.

The foreseeable results in the research carried out can be summarised in:

- To carry out the study of artworks of big dimensions thanks to obtain a flexible Raman system, if it is necessary, when the artwork cannot be moved.

- To make evident the lowering of the price that offers this system in front of a conventional one, maintaining the required performances for a good analysis.
- Optimise the laboratory environment-analysed artwork distance.
- Identified any kind of pictorial material, organic or inorganic.

References

1 G.Turrell and J.Corset, *Raman Microscopy: Developments and Applications*, Academic Press Limited, 1996.
2 J.M.Yúfera et al., *Optical Technologies in the Humanities*, Vol.IV of Series of the International Society on Optics Within Life Sciences, Springer-Verlag, Heidelberg, 1997.

The Potential Uses of Lasers and Layer Manufacture in Conservation

P.S. Fowles
The Conservation Centre, National Museums and Galleries on Merseyside, Whitechapel, Liverpool L1 6HZ, U.K.
Tel. +44 (0)151 478 4904 Fax. +44 (0)151 478 4990
email: sculpture@nmgmc1.demon.co.uk

Abstract. The recording and replication of artworks is well established and methods such as photography and casting are common place in many museums. Conservation practice demands that we should be able to achieve a high degree of accuracy without any damaging contact occurring to the original object. Previous attempts to record and replicate artworks have not always been very successful in fulfilling this requirement. The aim of this paper is to highlight some of the possibilities presented by the latest laser-based recording and manufacturing techniques which require no physical contact with the original artwork.

1 Introduction

There has long been a need for the recording and replication of artworks for reasons including replacement of artworks at risk, public display and academic study. The replication of three dimensional (3D) works of art has not changed in concept for many years. Traditional techniques rely either on the application of a mould to the surface of artwork or on copying by eye by a sculptor or craftsman. Both techniques are however fraught with potential problems which must be addressed in the light of current conservation practice [1].

The use of moulds [2,3] requires considerable contact with the surface of the original artwork. In many cases this can not be considered safe practice since the surface of many artworks are not only fragile but may contain important detail or remnants which provide an art historical context for the piece. It is therefore of the utmost importance that the surface of artworks should only suffer contact when no other option remains. The use of moulds is also limited by two other factors. Firstly, the range of materials available which can be used to cast replicas is limited and may possibly not be sympathetic to the original. Secondly, when a mould is used there will inevitably be shrinkage in the casting material leading to a slight but distinct size difference between the original and copy.

The copying of artworks by eye does not have the problems of contact associated with moulding. It is however not an accurate technique but depends on the skill of the sculptor and is therefore a re-interpretation rather than a replica.

What is required then is a method of non-contact replication that provides an accurate but flexible method for the replication of 3D works of art. The advent of digital measurement systems and modern manufacturing techniques provides a potential solution to the problems of non-contact replication

2 Measurement in 3D

For many years the method of recording 3D objects has been dominated by photography. The principal problem has always been one of trying to adequately represent such objects using a 2D method. Now, however, the acquisition of 3D data is possible using 3D digital cameras and laser scanners [4]. These devices create 'clouds' of data points which map out the surface of an object and can provide a record of a 3D surface down to an accuracy of fractions of a millimetre and better. The recorded data is stored on computer and may be viewed by means of software which can translate this quantifiable information into a visual representation. At the heart of the problem of non-contact replication is the ability to translate objects in the virtual space of the computer into actual 3D. This is a well known problem in a modern industrial context, for example in the automotive and aircraft industry, where many complex 3D objects exist as computer models and physical prototypes must be made prior to production. There are many automated devices which can translate 3D computer models into a tangible form. The most common is computer numerical control (CNC) machining which enables multiple axes cutting using a wide variety of materials. In many cases CNC machining provides a excellent solution to the problems of non-contact replication [5]. Replicas can be machined very accurately using scanned data in material which is sympathetic to the original. There are however cases where CNC machining cannot easily provide a solution, such as when there are very complex forms and undercuts which occur in many sculptural works. A different kind of industrial process could provide an answer to the requirements of non-contact replication, this process is known as 'layer manufacture' or sometimes as 'rapid prototyping'.

3 Layer Manufacture

The development of layer manufacturing techniques [6] grew out of a need to reduce time scales for the production of physical prototypes of computer models. This is still a relatively new area of technology which emerged in the mid 1980s. The computer model to be made must have closed surfaces which unambiguously define an enclosed volume, that is the model must be complete without any 'holes'. The model data is converted into a polygonal format and then sectioned into a number of adjacent cross-sections or layers (akin to slicing a loaf of bread). The choice of orientation of the layers is important since the model will display some terracing due to the finite layer thickness. The cross-sections are then systematically created through the solidification of either liquids or powders and combined into a 3D form. Alternatively the layers are cut from thin sheets of material and then glued together into a 3D form. The range of materials is wide and is also rapidly increasing, and includes plastics, wax, sand, resin, paper. As an example, two of the more common techniques are discussed below.

An example of layer manufacture by solidification is stereolithography (SLA) which was the first of the layer manufacture techniques to be developed. The computer data is used to control a ultraviolet (UV) laser beam which draws consecutively each layer of the model on the surface of a tank of UV sensitive resin which is cured where the laser strikes. The layer depth of the computer model is arranged to be the same as the penetration depth of the laser into the resin. The model is built on a platform within the resin tank. At the beginning of the process, the platform is positioned just below the surface of the resin. After each layer is drawn, the platform descends to allow new liquid resin to cover the cured layer and the next layer is built on top. The model is thus built from the base upward as the platform descends.

A second technique is laminated object manufacture (LOM) which operates by forming an object from successive layers of a heat bonded sheet material, usually paper. The computer data is used to control an infrared carbon dioxide laser which cuts around the perimeter only of each layer in the material. The laser is calibrated so that it only cuts through one sheet of material at a time. Once each layer is complete a new sheet of material is heat bonded on top and the laser the next outline. In this way, a solid version of the 3D model is built from the base upward.

4 Conclusions

Layer manufacture can be used to produce very accurate replicas of 3D art objects. The question remains as to whether they provide a useful alternative to other techniques. Once we have the layer manufacture model, it is easy for moulds can be taken and more replicas produced if required. This is often necessary since many objects created with layer manufacture are not particularly robust. The disadvantages of moulds in terms of limiting the range of materials of production obviously still applies, however the fact that we can precisely manipulate the size and scale of the model produced, allows for compensation to be made for shrinkage which occurs in the moulding process. The volume of the object built is also limited by the dimensions of the layer manufacture machinery, nevertheless it is a relatively simple operation to build the object in sections which can then be assembled.

At first it appears that layer manufacture has little to offer over traditional CNC techniques, further analysis reveals that this is not the case. The process is very accurate, the algorithms used are simple, there are no tool changes, no complex surfaces to follow and no pockets to machine. Most significantly, there is no need for complex tool paths to avoid collisions or problems with access to complex detail or undercuts. This feature alone makes possible the replication of some objects which would not otherwise practical. The strength of objects made using layer manufacture is improving, as is the range of material available with techniques using ceramics and metals currently in development. The use of layer manufacture in non-contact replication provides a powerful addition to the range of techniques already available.

References

[1] J.Larson,: New approaches to the conservation of external stone sculpture: the twelfth century frieze at Lincoln Cathedral, Proceedings of the 7th International Congress on Deterioration and Conservation of Stone (1992) 1167-1175, Eds: J.Rodrigues, F.Henriques, F.Jeremias, ISBN: 972-49-1483-6

[2] T.Bryce, D.Caldwell: Scottish mediaeval sculpture. The making of reproductions and their uses, Museums Journal 81 (1981) 67-70

[3] R.Payton, A survey of two traditional moulding techniques for stone inscriptions, Retrieval of objects from archaeological sites: Ed R.Payton (1992) 147-156, Archetype publications, ISBN: 1 87 3132 30 1

[4] P.Boulanger, M.Rioux, J.Taylor, F.Livingstone: Automatic replication and recording of museum artefacts, 12th International Symposium on the Conservation and Restoration of Cultural Property (1988) 131-147, Tokyo National Research Institute of Cultural Properties

[5] P.S.Fowles and J.Larson, The Replication of an Egyptian Relief, Restauratorenblatter, Proceedings of LACONA II (Liverpool, 1997), Verlag Mayer, (in press)

[6] C.Kai, Three-dimensional rapid prototyping technologies and key development areas, Computing and Control Engineering Journal (1994) 200-206

Identification of Pigments by Raman Microscopy: Relevance to the Authentication or Otherwise of Egyptian Papyri

Lucia Burgio and Robin J. H. Clark*
Christopher Ingold Laboratories, University College London, 20 Gordon Street, London WC1H 0AJ, UK

Raman microscopy is now established to be an excellent technique for the scientific investigation of materials used on works of art, especially pigments[1]. In fact, it combines high sensitivity and spatial resolution with non-destructiveness; it is reasonably free from interference from surrounding materials and it can be performed *in situ*, excluding any sampling and consequently any damage to the object under examination. The analysis and identification of pigments on a work of art is desirable not only for conservation and restorative purposes but also for the purposes of dating and authentication, as the following case study makes clear.

Six Egyptian papyri were analysed by Raman microscopy. The papyri were stated to belong to different periods from the 13th to the 1st centuries B.C., the principal objective of the study being to authenticate the dates given. Five papyri were supposed to belong to the period of Ramses II, who was Pharaoh in the 13th century B.C. The sixth papyrus was supposed to be the contemporary portrait of the Egyptian Queen Cleopatra, who lived in the first century B.C.

As a result of the investigation by Raman microscopy the following pigments were detected: phthalocyanine blue ($Cu[C_{32}H_{16}N_8]$), phthalocyanine green ($Cu[C_{32}H_{15}ClN_8]$), chalk ($CaCO_3$), berberine ($[C_{20}H_{18}NO_4]Cl$), ultramarine blue ($Na_8[Al_6Si_6O_{24}]S_n$), anatase (a crystalline form of titanium dioxide, TiO_2), Prussian blue (iron(III) hexacyanoferrate(II), $Fe_4[Fe(CN)_6]_3.14\text{-}16H_2O$), and at least two different organic lakes not yet identified.

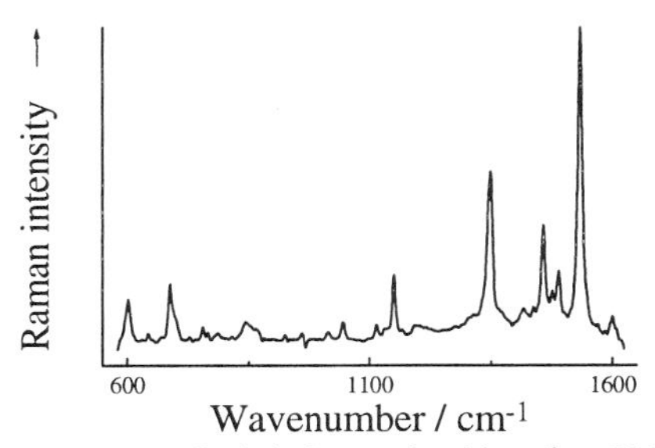

Figure 1: Raman spectrum of phthalocyanine blue (λ_0=514.5 nm, 1 mW)

While chalk, berberine and ultramarine blue are naturally occurring pigments, all the others detected on the Egyptian papyri are of synthetic origin. Prussian blue was first synthesised in 1704[2], while phthalo-cyanine blue and phthalocyanine green became available only from the early 1930's[2]. The mineral anatase occurs in nature, but it is very rare[3], and there is no evidence that it was ever used as a pigment in antiquity since the mineral always contains some inclusions which alter its colour from white to virtually black. It was only early in this century that two industrial processes were discovered[3] to produce pure, crystalline, forms of titanium dioxide (as rutile or anatase), suitable as paint pigments. Therefore the presence of white anatase as a pigment on works of art is indicative of a 20th century product.

The presence of ultramarine blue on some of the papyri does not necessarily prove or exclude their genuineness. Before the process to manufacture ultramarine blue was discovered in the 1820's, the blue pigment (lazurite) was obtained by grinding the semi-precious stone lapis lazuli, the presence of which on works of art can be traced back to the third millennium B.C[2]. However, the use of lapis lazuli as a pigment on ancient Egyptian artefacts has never been established[4], even if nothing excludes the possibility that it was actually employed (the Egyptians had a convenient blue pigment, a calcium-copper silicate called "blue frit" or "Egyptian blue"). However, the appearance under the microscope of the pigment on the papyri indicates that the pigment is of synthetic origin: the particles have a uniform appearance, they have smooth, round edges, and their size is very small (1 - 4 μm diameter), whereas natural lazurite would be characterised by irregular shapes with a wide range of particle sizes and by sharp edges due to the mechanical grinding of the stone, lapis lazuli. Moreover, the absence of impurities which always accompany the natural pigment, i.e. quartz, silica, pyrites and calcite, is also significant.

The Raman study has been complemented by the use of polarising optical microscopy and scanning electron microscopy coupled with EDX analysis. The use of the SEM has permitted the composition of the metals which were used for gilding the papyri to be determined (copper or an alloy of copper and zinc), information that was not possible to obtain by Raman microscopy.

As a consequence of this study, it is possible to conclude that most of the pigments applied onto the papyri are modern, albeit the papyri themselves could not be dated. The results obtained in this case study therefore confirm that Raman microscopy is one of the finest techniques available for curators and scientists for the non-destructive, *in situ* identification of pigments on art objects and thus for the dating and authentication of works of art.

References

1. R.J.H. Clark, Chem. Soc. Rev. 24, 187 (1995).
2. K. Wehlte, The Materials and Techniques of Painting, with a Supplement on Colour Theory, Van Nostrand Reinhold, New York, 1975.
3. M. Laver, in E. West Fitzhugh ed., Artists' Pigments, a Handbook of Their History and Characteristics, vol. 3, Oxford University Press, 1997, p. 295.
4. A. Lucas and J.R. Harris, Ancient Egyptian Materials and Industries, Arnold, London, 1962, p. 343.

Progress in the Use of Excimer Lasers to Clean Easel Paintings

A E Hill, A Athanassiou, T Fourrier, J Anderson, and C Whitehead

Department of Physics, University of Salford, The Crescent, Salford M5 4WT

Abstract. A programme is being conducted to determine the potential of using pulsed ultraviolet light from excimer lasers to clean discoloured varnish and surface detritus from the surface of old easel paintings.

The ablation characteristics of artificially aged and naturally aged varnishes have been examined. It has been found that careful control and monitoring of the cleaning process must be employed to ensure that a residual varnish layer remains to preclude damage to the underlying pigments. A number of appropriate techniques, which can be used alone or in combination, have been investigated. These include ultraviolet fluorescence observations and thermal mapping.

The effects of the pulsed excimer laser light on pigments have also been investigated. Adverse effects such as discolouration have been seen to occur in certain cases when the pigments were subject to direct laser radiation. X-ray diffraction experiments have investigated the change in the lattice structure of the pigments after UV irradiation and heating.

1 Varnish Ablation

The cleaning of easel paintings requires the removal of the oxidised and polluted varnish layers from the painting surface. However, it is essential to terminate the procedure while still leaving a thin varnish layer on top of the pigments. This plays a protective role for the pigments underneath, so that the laser light does not penetrate to them. Therefore the thickness of the varnish layer left should be larger than the penetration depth of the laser light.

The penetration depth of artificially aged varnishes was measured at 308 nm using a power meter, linked to a computer unit, mounted behind various fresh and matured varnish samples. The following values of absorption depth (α^{-1}), derived from Beer-Lambert's law [1] were determined.

Varnish	Absorption depth	Varnish	Absorption depth
Damar	$11.1 \pm 0.4\mu m$	Mastic	$14.0 \pm 1.0\mu m$
Copal Oil	$6.0 \pm 0.2\mu m$	Ketone	$7.3 \pm 0.2\mu m$

These values are typical for these materials but varnish absorption is strongly dependent on age and composition.

2 Discolouration of Pigments

If the protective layer of varnish is removed the laser light will penetrate to the pigments. We have reported [2] that this can cause discolouration. In order to investigate the nature of discolouration of pigments caused by the UV light, numerous experiments were conducted on pigments deposited on glass slide substrates. During these experiments two main observation were made:

- there exists a threshold fluence ($100\,\mathrm{mJ/cm^2}$) under which neither discolouration nor ablation take place.
- as little as one shot above this threshold suffices to discolour the pigments.

To gain an insight into the discolouration phenomenon, X-ray diffraction [3] was used to look for potential structural changes in the pigments.

2.1 X-ray Diffraction (XRD)

X-ray Diffraction experiments were performed on different pigment samples to identify any structural changes after discolouration. A Siemens/Bruker D5000 Diffractometer was used at the CuK^{α} wavelength. The angle of the X-rays' output tube was fixed at 1°and the detector was scanned at angles between 10°and 60°.

The results shown here are for the Iron Oxide pigments Yellow Ochre and Raw Sienna with linseed oil binder discoloured by UV laser light (248 nm) and for the same pigments in powder form discoloured in an oven at 350° C.

2.2 X-ray Diffraction Analysis of Iron Oxide Pigments

Yellow Ochre and Raw Sienna are iron oxides based on the natural occurring mineral Goethite. Goethite is a hydrated Iron Oxide ($Fe_2O_3.H_2O$) of a combined form FeOOH. The XRD patterns of these pigments after heat treatment showed that the pigment particles' composition converts from FeOOH to Fe_2O_3, the mineral Hematite. In this process, molecules of entrapped water are released from the pigment particles.

The clearly generated peaks at $2\theta \approx 35°$in the XRD spectra of the laser discoloured pigments can either be Hematite or Magnetite (Fe_3O_4), since their characteristic peaks at this particular 2θ angle are very close. Looking at the black colour of the sample, it can be assumed that it is likely to be Magnetite. It is reported [4] that heating of Fe_2O_3 can give Fe_3O_4.

2.3 Colour Change in the Minerals Goethite, Hematite and Magnetite Crystal Field Transitions

The transition element in these oxides is Iron and it is characterised by having an unfilled 3d orbital. The energy levels of the d-orbitals split when the atom

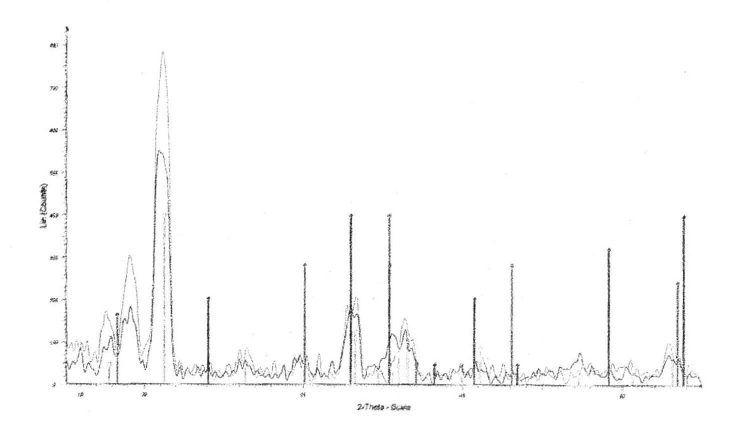

is located in the crystal field. When the Fe cation is placed in a coordination site in the mineral, a non-uniform crystal field created by the co-ordinating anions will interact with the d orbitals. The geometry of the coordinating anions will determine the splitting of the energy levels. The difference in energy of the orbital states, created after the split due to the crystal field, is similar to the energy of the visible light [5]. Therefore, electrons may be excited from the lower energy levels to the higher ones by visible light absorption.

The cations in Goethite and Hematite are in the same oxidation state (Fe^{3+}) and therefore the number of 3d electrons is the same. However, their crystal structure is different and hence the crystal field is different. Therefore the amount of the split energy levels varies and the same anion produces different absorptions. Goethite is characterised by the hydroxyl (OH)- group or H_2O molecules. The presence of the hydroxyl group causes bond strengths to be weaker than those in Hematite. The valence state of Iron in Magnetite is different (Fe^{4+}) so the number of electrons at the 3d orbital and hence the orbital energy states are different.

3 Control and Monitoring Techniques

For the use of laser light to be an efficient technique in the cleaning of easel paintings, the procedure must be carefully controlled to ensure that the light does not penetrate to the pigments. Two of the control techniques investigated so far are fluorescence and thermal mapping. A combination of both techniques may be used in the final control system.

3.1 Laser Induced Fluorescence (LIF)

The fluence of the laser used for the excitation of the sample is very low and does not need to exceed 1 mJ/cm^2. The emitted fluorescence is collected by a fused-silica optical fibre and analysed by a spectrograph. An intensified photodiode array detector is used for the detection of the fluorescence spectrum which is then recorded on an optical multi-channel analyser (OMA). The fluorescence signal was recorded from areas of paintings that had previously been treated with laser light. All the experiments performed so far showed that, when fluorescence is induced by laser light at 248 nm, the results are not pigment dependent. In this case the penetration depth of the light is smaller than the varnish thickness.

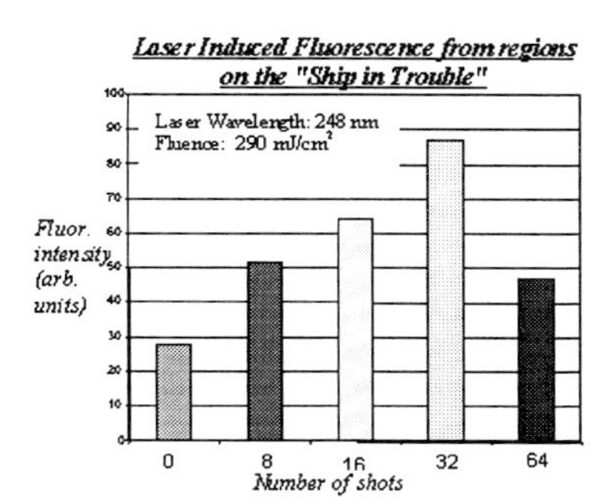

The results are always the same: the signal increases after the removal of the first polluted/oxidised varnish layers, reaches a maximum point and then starts to decrease when the thickness of the remaining varnish is less than the penetration depth of the laser light. It becomes a minimum when no varnish is left and therefore the pigments are exposed. On the diagrams it is clear that the signal has a maximum point. This is the point where the cleaning process should terminate.

3.2 Thermal Wave Detection

This is a non-contact method of measuring the absolute thickness of the varnish over the pigments. The technique is based on thermal wave interferometry [6]. A thermal wave is generated on the varnish surface by a frequency modulated solid state laser (870 nm) directed onto the target via an optical fibre. The end of the fibre is situated 20-30 mm from the target surface. A pyro-electric detector, linked to a phase sensitive detection unit, determines the beat frequency of the original and detected signals.

The thermal wave, generated by the laser, is reflected by the varnish-pigment interface where a thermal mismatch exists. The phase difference between the modulation waveform and the detected waveform (thermal wave phase shift), is indicative of the varnish thickness.

4 Summary

The laser cleaning of easel paintings implies the ablation of varnish from the painting surface. At the end of the procedure a layer of varnish should be left

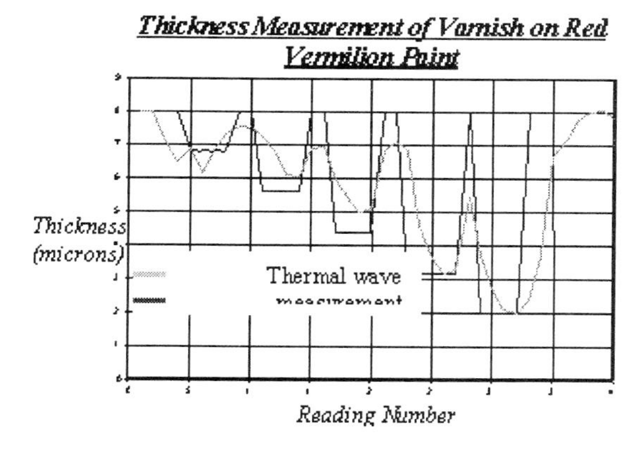

to protect the pigments. Its thickness should be larger than the absorption depth of the laser light and therefore an efficient control system is essential. A control system, which uses the painting's fluorescence under UV light, relies on the fact that clean varnish fluoresces more than the surface dirt or the pigments. Consequently the maximum fluorescence signal occurs at the point when the procedure should terminate.

The thermal interferometry technique may be used to determine the varnish thickness. A knowledge of the penetration depth of 308 nm laser light in many varnishes used in paintings is then required before a decision can be made whether it is safe to proceed with the cleaning.

The discolouration of certain pigments was examined by comparing X-ray diffraction spectra of pure and discoloured pigments. Heating pigments containing Goethite (FeOOH) at 350° C, causes this mineral to convert to Hematite (Fe_2O_3). However, under UV laser light, Goethite may turn to Hematite or Magnetite (Fe_3O_4).

5 Acknowledgement

The authors wish to acknowledge the financial support of British Nuclear Fuels plc for the work performed in this project.

References

1. Oxford Dictionary of Physics, Oxford University Press, 1996
2. Proceedings of LACONA II Conference, Liverpool, April 1997 (To be published)
3. Warren B. E., X-Ray Diffraction, Dover Publications, INC., New York, 1990
4. Addison W. E., Structural Principles in Inorganic Compounds, Longmans, 1961
5. Burns R., Mineralogical Applications of Crystal Field Theory, Cambridge University Press, Second Edition,1993
6. Almond D.P. and Patel P.M., Photothermal Science and Techniques, Chapman and Hall, 1996

Discrimination of Photomechanical Effects in the Laser Cleaning of Artworks by Means of Holographic Interferometry

V. Tornari, V. Zafiropulos, N. A. Vainos, D. Fantidou, and C. Fotakis.
Foundation for Research and Technology-Hellas (FORTH), Institute of Electronic Structure and Laser (IESL), P.O. Box 1527, Heraklion 71 110, Greece.
E-mail: vivitor@iesl.forth.gr

Abstract. A holographic interferometric system incorporating image processing and fringe analysis has been employed to study laser-induced photomechanical effects on artworks. The observed fluctuation of the fringe density, the discontinuity and the repeatability of interference patterns are used to characterize a number of diverse photomechanical influences. The information extracted from holographic interferograms enables comparative studies leading to the prediction of artworks' long term structural behavior, which depends on structural composition and nature. An evaluation procedure is proposed and first results are presented here. This method provides a quantitative assessment of materials susceptibility to selective laser-based layer-removal, through the study of the structural relaxation processes, and is also proved to be suitable in evaluation, indexing and classification operations.

1 Introduction

Holographic interferometry may provide useful tools for the quantitative analysis and study of effects associated with the laser cleaning procedures. This is of great theoretical and practical importance in the conservation on artworks. Pulsed laser light is usually used for the selective ablative removal of surface layers of an artwork [1], but is seen to be responsible for a potential damage of the target-the artwork-or for the generation of unpredictable physical defects in the long term [2].

Potential laser damage in various materials naturally depends upon the power density of the beam, the pulse duration, the pulse repetition rate, the thickness and nature of the target and a number of materials parameters is thus of great importance [3, 4]. In laser cleaning applications the light intensity used is usually low (<0.6 J/cm^2) and, therefore, the direct damage induced by intense radiation may not be of great interest. Experimental evidence indicates, however, that the laser cleaning action produces long-term effects in the artwork structure, depending on its nature. This is certainly of main importance as it determines its future status.

The effects induced in the structure by laser radiation may be distinguished as photothermal, photochemical and photomechanical and are usually treated separately. The energy deposited into the target is converted to mechanical waves, which in turn propagate in the material until fully absorbed causing irreversible damage with long-term results. The current study is based on the analysis of the structural deformations manifested and observed as displacements of the target's surface by means of holographic interferometry.

The recording and study of holographic interferograms leads to the observation of defects induced by the generated mechanical shocks and the methods proposed here allow for the discrimination of such photomechanical effects on individual artworks. By measuring short-term deformation produced by the pulsed irradiation, usually applied in laser cleaning, its physical influence can be deduced and long-term effects may be predicted. Indeed, for simple materials, such procedures are already producing reliable results [5, 6]. For multilayer composites, however, which may be of unknown origin, the situation is extremely complex. The study of the influence of this released shock energy will, thus, provide further insights and is of main importance. Such experiments for studying the dynamic response of the target after laser treatment have been performed and first results are presented here. The study of the holographic interferometric fringes shows, that laser-induced shocks yield large variations on the amount and direction of deformation of the artworks, for a long time period after the termination of the pulsed exposure. This is contrasted to single layer specimens in which a monotonic relaxation is usually observed.

2 The Experimental Methods and Results.

A continuous wave He-Ne laser emitting 30mW at 632nm was employed as the coherent illumination source for holographic interferometry. The off-axis optical arrangement used to record the holograms incorporates especially equipped micrometer-adjustable film holder and magnetic bases for maintaining a well-stabilised set-up for long-term use. A KrF excimer laser (Lambda Physik, Compex 100 model) emitting ultraviolet laser pulses of 30 ns duration and 300 mJ energy at 248 nm, was also used for laser cleaning. The excimer laser beam was weakly focused onto the panel using a cylindrical lens with focal length f=300 mm. Multi-layered square panels simulating an original painting panel were also used as the test specimens. The panels were steadily fixed at their bottom edge. A superficial evenly applied varnish layer, widely used in painting conservation, was applied to the specimens to provide a diffuse optical surface well suited for holography. The procedure was to, first, acquire the primary holographic interferogram of the panel and then ablate the top part of the varnish layer, simultaneously initiating a timed sequence of holographic interferogram recordings. A selection of typical interferograms exhibiting characteristic changes of the complicated fringe pattern is depicted in Figure 1. The vertical size of the reconstructed area in this case is 12 cm.

A multilayered symmetric panel of dimensions 10cm×10cm×1.5cm was used for the systematic study. The laser fluence used for the particular test panel was set at 0.36J/cm^2 with a pulse repetition rate of 10 pulses per step.

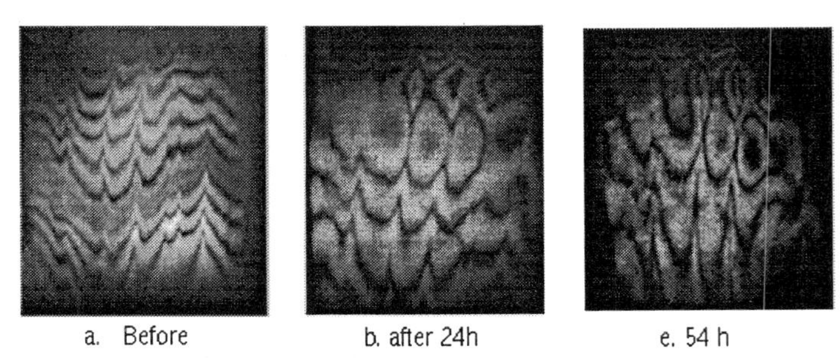

Fig. 1. Holographic interferometric testing for photomechanical effects.

Several interferograms were produced first at regular one-hour (1h) time-intervals for a 24-hour period after the ablation of the varnish layer. A strong and fast relaxation was observed and the procedure was extended further to a 500h period. All recordings were made in an environmentally stable laboratory environment held at a temperature of 20-21°C and a 50-52% RH. The double-exposure interferograms were reconstructed and captured via a CCD camera and stored with the aid of a frame grabber for further processing and analysis. By using an especially designed algorithm [7] each interferogram yields phase distribution plots and, thus, even minor changes of the interference field of the panel could be suitably detected and studied comparatively. This relatively long-term observations aim to reveal the relaxation process of the panel after the varnish ablation and interferograms were recorded for as long as fringe density and diffraction efficiency values were fluctuating (~500h). All experimental parameters including geometry, exposure and chemical development were kept identical in all recordings.

A photodetector-pinhole assembly connected to a digital power meter was used

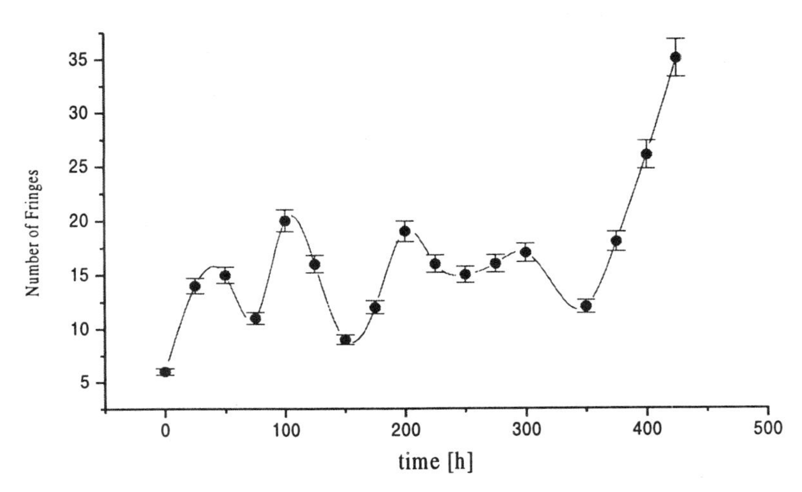

Fig. 2. Fluctuation of the number of fringes with time.

to measure directly the reconstructed interferometric pattern along a significant vertical contour over the entire length of the panel (i.e. ~10cm). In Figure 2, the Number of Fringes (a parameter proportional to the surface displacement) is plotted as a function of time, which is measured from the first holographic exposure.

The value of the number of fringes fluctuates intensely for many hours and increases, indicating an inhomogeneous stress distribution throughout the volume of the test panel. The deformation of the panel follows a long-period oscillation mode with some abruptly changing values for the first 150 hours later. In the interval of 150 to 300 hours after irradiation, the values are becoming more stable to increase in a threefold manner during the next 100 hours. It is noted here that such behaviour is seen to be characteristic of the multilayer structure. Testing of simple single-layer materials reveals a simpler monotonic relaxation and such differences are currently under study. Existing theoretical models of elasticity used to predict the behaviour of the panel were not sufficient to provide a

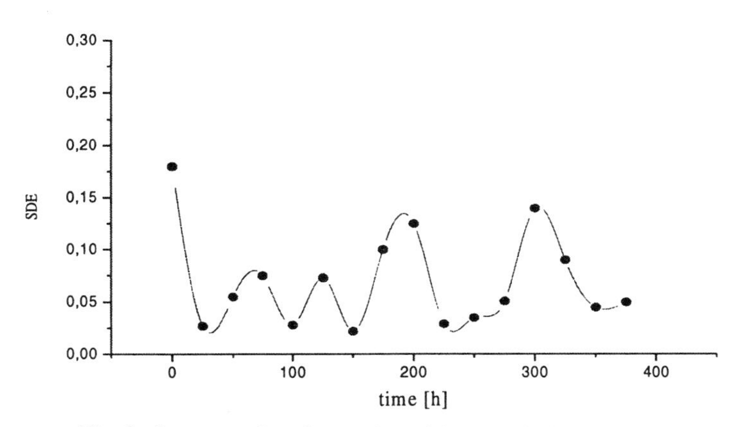

Fig. 3. Corresponding fluctuation of the spot diffraction efficiency (SDE).

correlation with the experimental data.

A simpler, less tedious approach has also been developed based on diffraction efficiency measurements, which is seen to produce quite systematic and correlated results. The method requires recording of well diffusing objects in the Fresnel region. In this case, a reliable reconstruction of the interferogram is obtained, though with a much-reduced resolution, by reconstructing only a small part of the hologram at a specified position. Consequently, in a realistic case of a non-uniform fringe distribution, this spot-diffraction-efficiency (SDE) is a measure of the number, size and contrast of the fringes, as well as the hologram quality. Systematic fluctuations of the spot diffraction efficiency, at a well-specified position, have also been observed, as shown in Figure 3. The diffraction efficiency of the holographic interferograms falls extremely fast when very dense interference patterns are observed due to the reduced interferometric stability attained during hologram recording, as implied by the fast relaxation due to stress accumulation. It should be noted that, although the SDE fluctuates with the visibility of the interferograms, measurements were made even in adverse cases in which counting and digitising of fringes were not possible. The method is currently under study, as it may provide simpler practical means for a wider application.

3 Conclusions

The laser induced photomechanical shock waves may yield destructive effects in an artwork undergoing pulsed laser cleaning. The presented preliminary results confirm that this resultant deformation follow a long-term relaxation procedure which can be monitored and quantified precisely by holographic interferometry. The feasibility of studying the long-term influence on real artworks is shown here. Exact modeling may not be reliable due to the diversity of materials and the limited knowledge of their nature. Nevertheless, the experimental results obtained for a period immediately after the laser cleaning process may ensure the prevention of further destructive structural effects. These latter would certainly be amplified in a real situation by the intrinsic localized deterioration of the artwork, which is existing prior to laser processing. Further work in progress is involving the on-line monitoring of the propagation of defects, which may lead to the prediction of the behavior of the object under treatment and the determination of the parameter space appropriate for the processing and the preservation of each individual item.

Acknowledgements. This work was partially supported in the framework of the ongoing EU projects LASERART (SMT-CT96-2062) and TMR Access to Large Scale Facilities (ERBCHGECT920007), Ultraviolet Laser Facility at FORTH/IESL.

References

1 S. Georgiou, V. Zafiropulos, V. Tornari, and C. Fotakis, Laser Phys., 8, 307(1998)
2 V. Tornari, D. Fantidou, V. Zafiropulos, N. A. Vainos, and C. Fotakis, Proc. SPIE, 3411, 420 (1998)
3 B. Steverding and H. P. Dudel, J. Appl. Phys., 47, 1940 (1976).
4 D. Zweig, V. Venugopalan, and T. F. Deutsch, J. Appl. Phys., 74, 4181 (1993)
5 R. Hunter and T. M. Morton, Experimental mechanics, p. 153, April issue (1975)
6 S. Anghel, I. Iova, S. Leval, I. Iorga-Siman, Opt. Eng. 35, 1396, (1996)
7 W. Osten, F. Elandaloussi, U. Mieth, in *Akademie Verlag Series in Optical Metrology*, vol.3, pp.98, (Academie Verlag; Bremen, Germany, 1998)

Cleaning of Ceramics Using Lasers of Different Wavelength

Theodosia Stratoudaki[1], Alexandra Manousaki[1], Vassilis Zafiropulos[1]*, Nathalie Huet[2], Stephanie Pétremont[3], Armand Vinçotte[2]
[1]FO.R.T.H. - Laser and Applications Division, P.O. Box 1527, 71110 Heraklion, Greece
[2]Laboratoire Arc'Antique 26, rue de la Haute Forêt 44300 NANTES - FRANCE
[3]ISITEM La Chantrerie CP 3023 44087 NANTES – FRANCE
*Corresponding Author: Tel: (+30) 81 391-485 E-mail: zafir@iesl.forth.gr

Abstract: Until now, cleaning of ceramic objects is done by solvents or sandblasting. At the present study, a laser based technique is presented. Three different wavelengths have been tested in order to clean ceramic objects. Laser Induced Breakdown Spectroscopy (LIBS), x-ray diffraction analysis (XRD), EDX and SEM were used for the validation of the cleaning results.

1 Introduction

Up to now, the use of lasers on cleaning ceramic materials has not been extensively studied. Traditional techniques for cleaning impose serious conservation dilemmas when used for the treatment of these objects. The present study was focused on glazed ceramics. Especially ceramics found outdoors, have been exposed to severe conditions such as pollution, weathering and mechanical stresses. Such conditions can lead to deterioration of the objects and growth of microorganisms. Laser cleaning has been successfully used in art conservation in cases such as marble, limestone, glass, removal of varnish from paintings and many others [1], [2]. These achievements gave the idea of the present work, so as to consider the lasers as a solution for the conservation of ceramics.

2 Experimental and Results

Two different types of late 19th century outdoor ceramics were examined. The first was a green-glaze ceramic (sample 1 – Fig. 1) and the second a blue-glaze one (sample 2 – Fig. 2).

They were both from Roubaix (France) and were used for the support of washstands exposed to exterior atmospheric conditions at the National College of Art and Textile Manufacturing (Roubaix-France). They were covered with a black layer of pollutant residues. The removal of this extremely hard layer using chemical techniques is ineffective and on the other hand, sandblasting is damaging because of the impact of abrasives on glaze (behavior of glass).

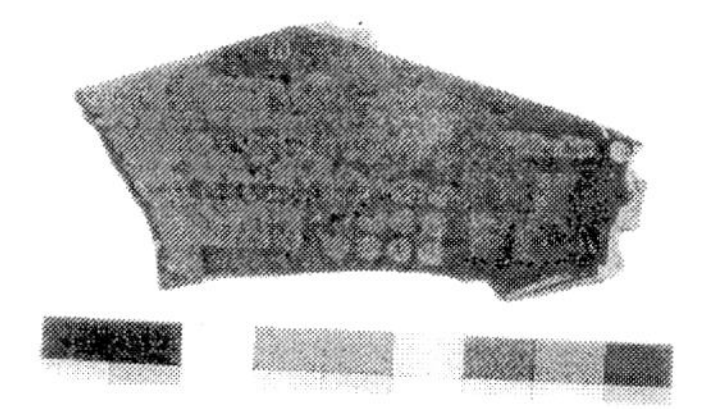

Sample 1: Green glaze ceramic

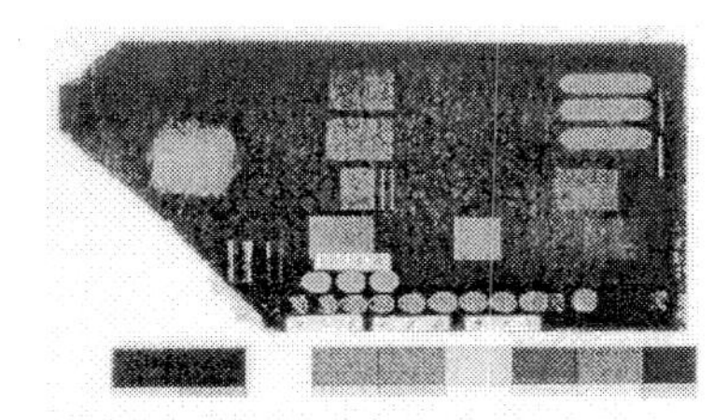

Sample 2: Blue glaze ceramic

Different types of nanosecond pulsed lasers were tested in order to obtain the best results. The lasers used were a B.M. Industries Nd:YAG laser (Fundamental λ=1064 nm and 3rd harmonic λ=355 nm) and a LambdaPhysik KrF Excimer laser (λ= 248 nm). The results (quality of layer removal, possible surface alterations, composition of surface layers etc.) were assessed by spectral techniques such as Laser Induced Breakdown Spectroscopy (LIBS) [3] and more standard characterization methods, such as microscopic observation, SEM, EDX and x-ray diffraction (XRD).

2.1 General Results

For sample 1, satisfactory results were obtained using KrF excimer laser at a fluence of 0.8 J/cm^2 and 1.6 J/cm^2, after a scanning of 5 pulses per spot in both cases. Nevertheless, the cleaned surface turned a bit yellow. The discoloration of the sample was more evident at the higher fluence. Microscopic observation showed that already at 0.8 J/cm^2 the surface was clean. For sample 2 satisfactory results were obtained at fluence values of 0.7 J/cm^2 (after a scanning of 30 pulses per spot) and 1.6 J/cm^2 (after a scanning of 10 pulses per spot). Again a yellowish coloration of the sample was more evident at the higher fluence.

Using the fundamental of a Nd:YAG laser on samples 1&2 and scanning the surface with 10 pulses per spot and a fluence of 1 J/cm^2, the removal of the black encrustation was also satisfactory. Like in the case of KrF excimer laser though, a slight yellowish alteration of color was observed. At fluence values higher than 1.1 J/cm^2 traces of heat alteration of the glaze appeared.

Finally, when using the third harmonic of a Nd:YAG laser very good results were obtained at fluence values of 0.6 J/cm^2 and 0.9 J/cm^2 and a scanning of 10 pulses per spot in both cases. Color alteration was not observed. At this wavelength (355 nm) the difference between the fluence threshold for the removal of the encrustation and the ablation threshold for the glazed substrate becomes maximum (latter much higher than former). Therefore the self-limiting process criterion [2] is satisfied and further analysis shows no macroscopic or microscopic alteration of the glazed surface.

XRD analysis has been performed on both samples. The glaze was found to contain mainly quartz (SiO_2), cristobalite (SiO_2) and mullite ($3Al_2O_3{\cdot}2SiO_2$). The black layer of dirt that was covering both samples was found to contain mainly iron oxide (Fe_2O_3) and carbon (C). After cleaning, LIBS analysis (see Figs. 3&4)

was performed *in situ* [4], for both samples. The elements Ti, Fe and Si appeared in the spectrum of the dirt layer. On the contrary, the clean glaze was dominated by the presence of Al and some Ti, but there were no more traces of Fe. Ca is present in both cases. These results were also confirmed by EDX analysis.

The above elements indicate that the black layer of dirt is owing to the industrial environment of the region Nord-Pas-de-Calais, where Roubaix is situated. The iron oxide and the traces of titanium are possibly originated from the iron industry of this region.

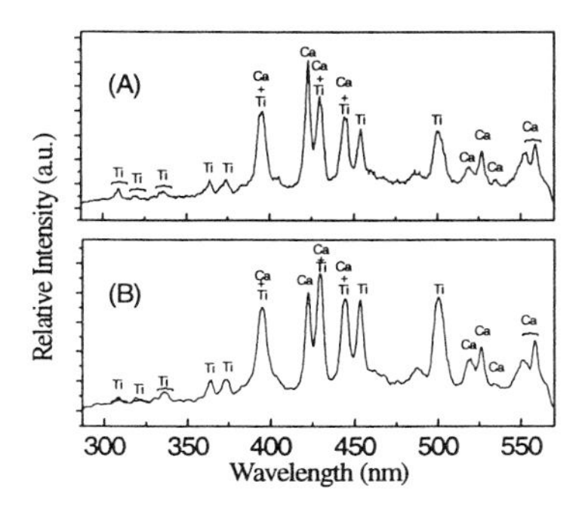

Fig. 3 LIBS spectra on sample 1 (A) Dirt layer, after 5 pulses (B) Clean, unaltered area, after 1 pulse

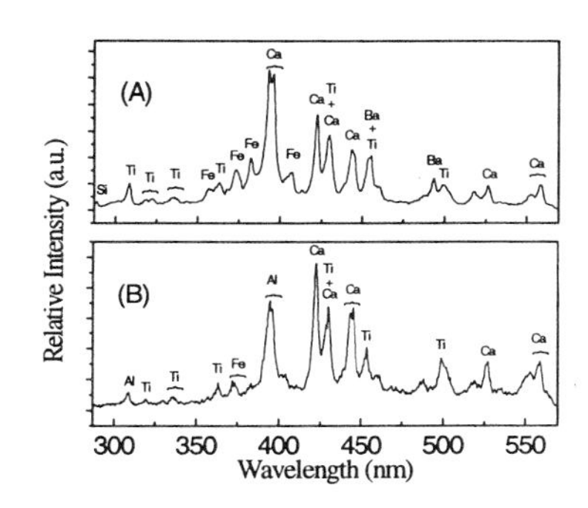

Fig. 4 LIBS spectra on sample 2 (A) Dirt layer, after 4 pulses (B) Glaze, after 10 pulses

2.2 Results from Scanning Electron Microscopy (SEM)

Using SEM the black layer of dirt was visualized, the quality of the cleaning procedure was validated and the alterations caused by the laser irradiation were revealed.

Fig. 5 shows the dirt layer of an area of sample 2 before laser treatment. Figures 6&7 show areas of the same sample after cleaning with the KrF excimer laser (248nm) with a fluence of 0.65 J/cm^2 and 1 J/cm^2 respectively. Fig. 8 shows an already clean area (not treated with laser). After comparison of these pictures, it is evident that the fluence of 0.65 J/cm^2 is barely sufficient to clean the glaze (just above the ablation threshold), since a dirt layer is still apparent. The higher fluence has better results but some dirt is still left. Comparison of Figures 7 and 8 indicates the possible formation of new cracks in the glaze, likely due to the mechanical effect (shock wave) of the laser assisted cleaning procedure.

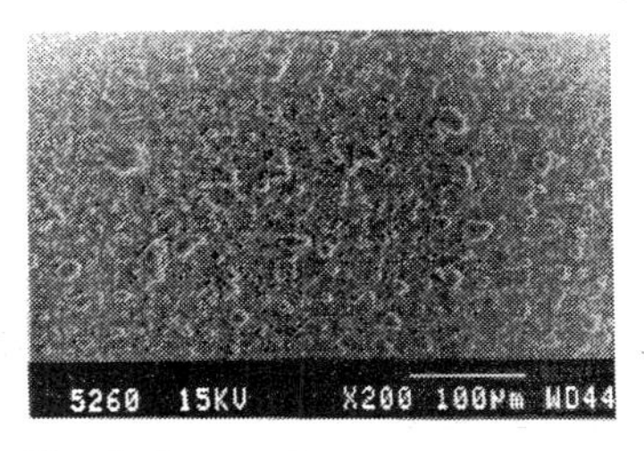

Fig. 5. Sample 2: Dirt layer before cleaning (x200)

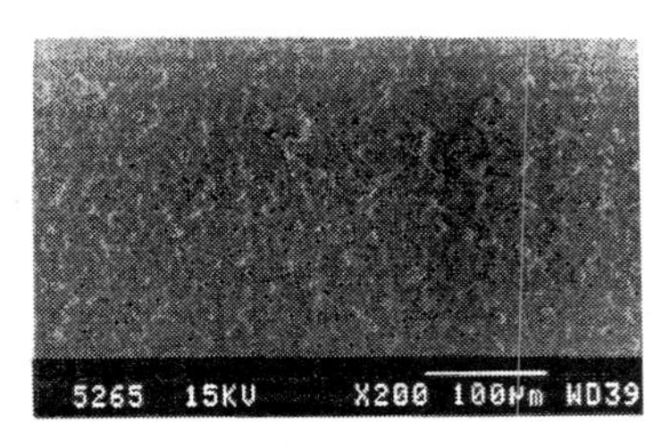

Fig 6. Sample 2: 248nm, F=0.65J/cm^2 (x200)

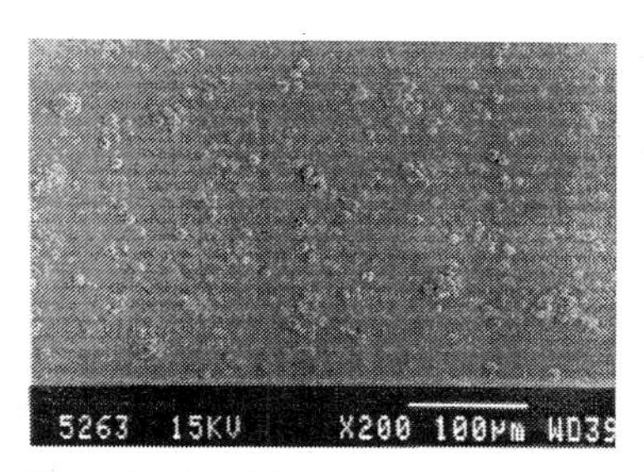

Fig 7. Sample 2: 248nm, F=1 J/cm^2 (×200)

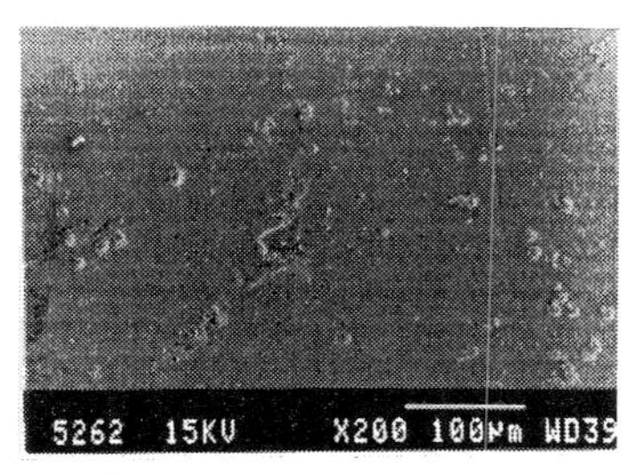

Fig 8. Sample 2: clean, untreated glaze (×200)

3 Conclusions

The overall results show that the effectiveness of laser assisted removal of encrustation from glazed ceramics depends on the choice of laser as well as the energy density used. On glazed outdoor ceramics, it was impossible to completely remove the last layer of dirt composed of the smaller particles with the excimer laser (248nm), even at high energy fluence. In order to clean these objects, the fundamental wavelength of a Nd:YAG laser - whose efficiency in the removal of dark encrustation from stone surface is high [5] - had drawbacks. Most of the time, a yellowing of the surface is observed and attempts to remove this coloration by increasing the laser energy caused damages on the glaze. For example fusion, cracking or explosion, which are owing to the severe thermal effect and the mechanical waves [6] generating high local stresses in the glaze and at the interface of the glaze-underlying surface. The most promising wavelength for this kind of objects is the 3rd harmonic of a Nd:YAG laser. Using this wavelength on glazed ceramics, no surface damage or unwanted coloration was observed and the removal of dirt seems to be total. Even at this wavelength, the laser beam must be free of hot spots in order to obtain the optimal results.

References

1. Proceedings of LACONA I, 4-5 October 1995, Heraklion, Crete, Greece."Comparison of three cleaning methods - Microsandblasting, Chemical pads and Q-Switched YAG laser - on a portal of the Cathedral Notre-Dame in Paris, France", V. Vegres-Belmin, LACONA I (1995).
2. "Laser Cleaning in Conservation - An Introduction", M. Cooper, *ed. Butterworth-Heinemann,* (1998)
3. "Laser Induced Plasmas and Applications", L.J. Radziemski and D.A. Cremers, *eds Marcel Dekker, New York* (1989).
4. "Laser-Induced breakdown spectroscopy as a diagnostic tecnique for the laser cleaning of marble", P.V. Maravelaki, V. Zafiropulos, V. Kilikoglou, M. Kalaitzaki, C. Fotakis, *Spectrochimica Acta,* Part B, **52** (1997), pp. 41-53
5. Same as reference [2], pp. 57-78
6. "Shock wave generated during laser ablation", C. Stauter, P. Gerard, J. Fontaine, *Proceedings of SPIE* (1998) Vol. 3343

Non-Destructive Analysis of Two Post-Byzantine Icons by Use of the Multi Spectral Imaging System (MU.S.I.S 2007)

O. Theodoropoulou, G. Tsairis
Conservation Laboratory, 5th Ephorate Of Byzantine Antiquities, Office Of Kalamata, Greek Ministry Of Culture

1 Introduction

The non-destructive analysis and documentation of artworks by use of visible, ultraviolet and infrared radiation offers analytical information both about the state of preservation and the construction technique while it proves to be essential to the selection of the appropriate methods of conservation. This poster depicts the information and the preliminary conclusions which resulted from the study of two post-Byzantine icons using the Multi-Spectral Imaging System MUSIS 2007 (Figure 1). The MUSIS 2007 offers sensitive imaging in a range from 320nm to 1550nm. The capability to have several imaging modes in a wide spectral range as well as the comparative evaluation between them gives us very important analytical information before and during the process of conservation. In this poster the preliminary results concerning the drawing, the paint layers, the overpaintings and the cleaning are reported by the comparative evaluation of all imaging modes.

2 Discussion

Identification of the materials regarding both icons has not been completed yet, since such a procedure demands comparative evaluation and study of standards composed of original reference samples.

Co-operation, on research terms, has been developing between Conservation Laboratory and Institute of Electronic Structure & Laser FO.R.T.H. aiming at:

- the collection and access of data
- the creation of original reference samples
- the evolution of software

Use of Multi-Spectral Imaging System MUSIS 2007 supports the study of the materials and the construction techniques, facilitates comparison between different artworks and also helps both the dating and the assay of the integrity of them. Furthermore, by means of Multi-Spectral Imaging System MUSIS 2007 we can get a better report as to the state of preservation of artworks while in parallel, we are able to choose the appropriate conservation method.

3 Results

The study concerned two post-Byzantine icons belonging to the orthodox church of St. Nicolas situated at Alagonia in the region of Messinia, Peloponnes. The portable icons of St.Virgin (94cmx47cmx3cm) and of Jesus Christ (94cmx50cmx3cm) quote the date AXYZ (1697 A.C.). The technique used here is egg tempera, canvas and gesso on wood. Figures 2, 4 show the icons before conservation while Figures 3, 5 show the icons after partial cleaning of the oxidized varnish layer and after the cleaning sample of the overpaintings.

The infrared reflection of the icons revealed the image of the paint layers which were under the oxidized varnish layer and the overpaintings. Figure 6 marks a detail from the blue tunic of Jesus Christ in visible spectrum. This very detail in near infrared disclosed dark crosses (Figure 7). With the help of the infrared imaging mode the initial paint layer with the dark crosses and its damaged areas could be discerned all over the area of the blue tunic of Jesus Christ. By the removal of the blue overpaint layer in a small area (Figures 8, 9) the initial blue paint layer with the crosses came out. Regarding Figure 10 it is evident that the presence of the crosses in the frame was a basic feature of the initial layer.

A detail from the blue tunic of St.Virgin, before cleaning(Figure 14), in infrared radiation gives information about the drawing (Figure 15). The figure and the shape of the drawing lines let us assume that brush and pigment were used. The recording of the ultraviolet reflection (Figure 12) in areas where the paint layers present great transparence showed that the drawing was traced on the preparation.

Infrared, false-colour infrared and visible fluorescence imaging mode can be used in order to distinguish pigments of different chemical composition that appear similar in the visible. In Figure 8, in the large frame the blue overpaint layer appears similar to the initial blue paint layer in the small frame. The same paint layers in infrared mode differ in brightness while in false-color infrared imaging mode they appear with different false-colour (Figure 11). In visible fluorescence imaging mode both paint layers present different percentage of fluorescence (Figure 13). Fluorescence imaging clearly reveals varnish layers appearing green-blue.

Figure 16 shows a detail of the portable icon of St. Virgin before conservation. After partially cleaning the varnish and removing the green overpaint layer in a small area, a yellow inscription over a green background came out (Figure 17). These two green layers appear with different false-colour in false-colour infrared imaging mode (Figure 19).

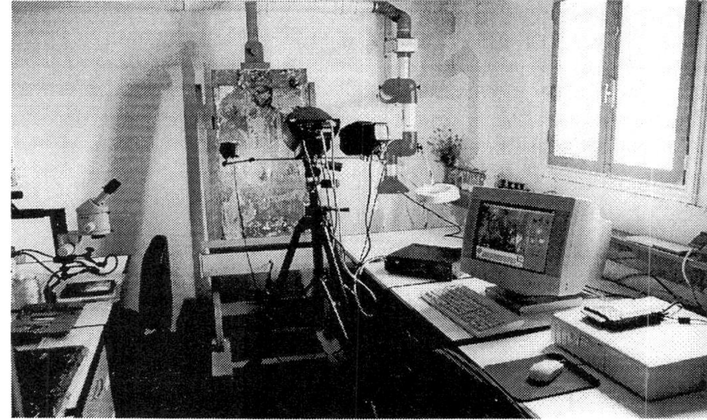

Fig 1. The Conservation Laboratory with the Multispectral Imaging System MUSIS 2007

Fig 2. Before Conservation

Fig 3. After Conservation

Fig 4. Before Conservation

Fig 5. After Conservation

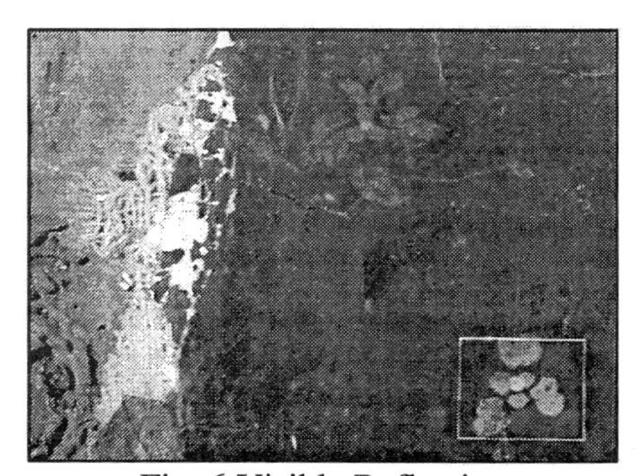

Fig. 6 Visible Reflection

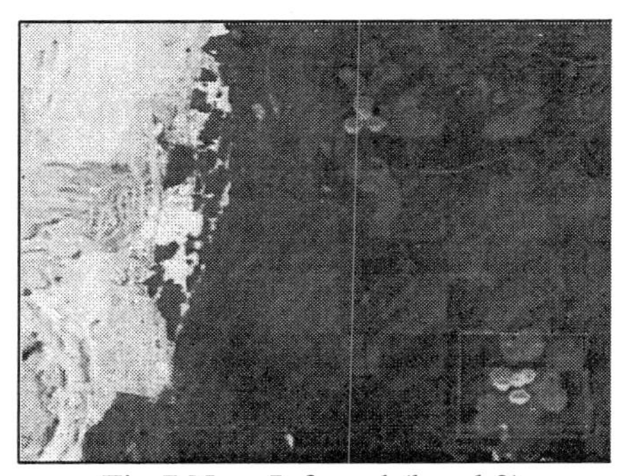

Fig 7 Near Infrared (band 2)

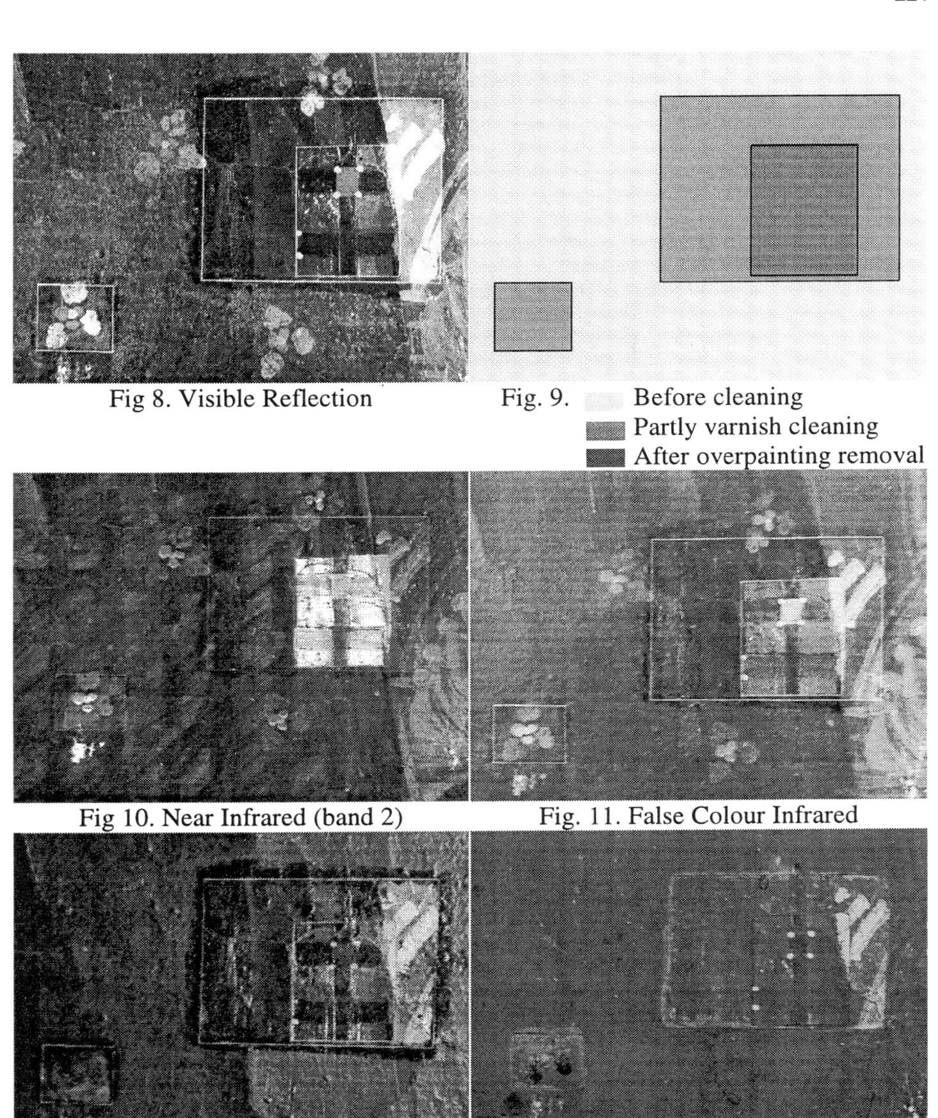

Fig 8. Visible Reflection

Fig. 9.

Fig 10. Near Infrared (band 2)

Fig. 11. False Colour Infrared

Fig 12. Ultra Violet

Fig. 13. Visible Fluorescence

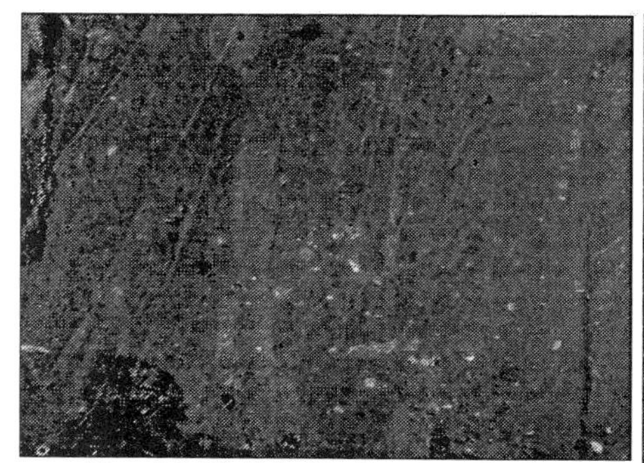

Fig. 14 Visible Reflection (before conservation)

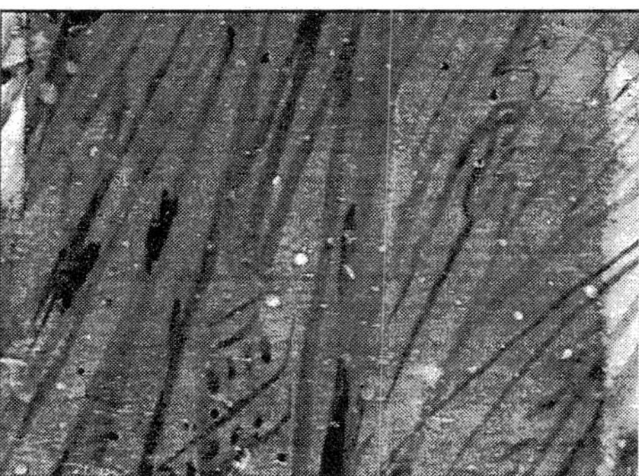

Fig. 15 Near Infrared (Band 2)

Fig. 16. Visible Reflection (before conservation)

Fig. 17. Visible Reflection (After partial cleaning)

Fig. 18. Near Infrared (band 2)

Fig. 19. False Colour Infrared

Lasers in Medicine

TMLR: Potential Mechanisms of Action and Design Parameters for Excimer-Based Clinical Systems

R.E.N. Shehada, T. Papaioannou And W.S. Grundfest
Laser Research and Technology Development, Cedars Sinai Medical Center, 650 South Sant Vincente Blvd, Los Angeles, CA 90048

1. Introduction

The idea of transmyocardial revascularization (TMR) was to provide blood flow to ischemic myocardium by creating a reptilian heart, in which blood from the left ventricle perfuses the myocardium by a series of channels. The clinical advantages of the technique are evidenced by the decreased intensity and frequency of angina, reduction in the need for antianginal medications, and improved work capacity [1-5]. The original laser-based technique (TMLR) was proposed by Mirhoseini *et al.* [6-8] who applied a specially designed CO_2 laser to drill channels from the epicardial to the endocardial surface. The epicardial bleeding stops after few minutes of drilling and it is believed that the channels remained patent to perfuse the myocardium directly from the ventricle. However, many investigators [9-14] have found that the majority of the channels close relatively quickly and believed that other mechanisms must be responsible for the observed clinical improvement and reduction of anginal symptoms. Current research [9-14] suggests that the original hypothesis of the reptilian heart may not be correct, and that TMLR may provide its benefits, if any, through other mechanisms. Various investigators have proposed the following hypotheses as mechanisms of action of TMLR, however, the observed clinical improvement may be resulting from a combination of some or all.

- *Hypothesis 1*: Laser-drilled channels reendothelialize, remain patent, and permit blood flow from the ventricle directly into the myocardium [6-8].
- *Hypothesis 2*: Laser-drilled channels close, but as clot lysis occurs, interconnections are formed between various capillary beds (sinusoids). Release of growth factors, due to cellular injury and the presence of platelets and thrombus in a confined space, induce angiogenesis, which increases vascular interconnections [13-17].
- *Hypothesis 3*: Laser-drilled channels change the conduction characteristics of ischemic myocardium and improve myocardial contractility and ventricular function [18].
- *Hypothesis 4*: Laser-drilled channels close, produce small scars and improve the compliance of the ventricle wall, allowing the heart to operate more efficiently [19,20].

- *Hypothesis 5*: Laser-drilled channels and surrounding damage zones denervate the myocardium thereby reducing the anginal symptoms [21].
- *Hypothesis 6*: The observed clinical improvement may bear no relationship to the laser-drilled channels but may be a response to other factors, including optimization of and adherence to medication regimens, improved diet, appropriate rest and exercise, and placebo effects.

The initial choice of the pulsed CO_2 laser for TMLR was based on the desire to create straight regular channels, since this was believed to be essential for maintaining patency. In addition, the ability of the CO_2 laser to penetrate the myocardium in a single, short (<100ms) pulse would minimize the thermal damage to the tissue surrounding the channel. Unfortunately, the CO_2 laser has several inherent drawbacks for clinical use. First, the laser requires an articulated arm to deliver its energy, which precludes percutaneous transcatheter applications. Second, the ventricle should be filled with blood upon firing the laser such that the blood would act as a "beam stop" and prevent perforation of the posterior wall. The requirement of a full ventricle places constraints on clinical application during surgery. Third, the laser can cut through the chordae tendineae, causing unsuspected mitral valve incompetence or drill unwanted channels that cannot be seen by the operator. The inability to precisely regulate the deposition of laser energy is a significant limitation of CO_2-based TMLR.

These limitations of the CO_2 laser-based TMLR had instigated our investigation of alternative energy sources that could be precisely controlled, delivered fiberoptically and produce minimal thermal damage. Any effort to rationalize choice of laser source for TMLR must be based on an understanding of how TMLR works. However, since the mechanism of action of TMLR is currently unknown, the alternative is to choose a laser based on a *hypothetical* mechanism of action and optimize the ablation process to produce the desired effects. The inability to define the physiologic basis for TMLR makes optimization difficult. For each of the above mechanisms, parameters can be optimized to produce the specific tissue effects. Past experience with lasers in cardiovascular applications suggests that the less thermal and shock-wave injury, the better. Thermal injury leads to intense collagen deposition and scarring. If producing small scars in the myocardial wall is the mechanism of action, then lasers are not required, since this can easily be accomplished using heated needles or RF (radio frequency) current. The concept of denervating the myocardium and thus reducing anginal symptoms may provide clinical relief of angina but it may also produce undesirable consequences, such as over exertion leading to muscle damage.

This investigation examines the various laser choices for TMLR with emphasis on optimizing the channel drilling parameters using the XeCl excimer laser and delivery systems. A series of *in-vivo* experiments were performed to identify the optimal delivery systems and laser settings that produce the most uniform tissue ablation with the least thermal and mechanical damage.

2 Materials and Methods

2.1 Lasers and Delivery Systems

Three types of lasers with their associated delivery systems were tested for the TMLR procedure:

A XeCl excimer laser (Model CVX 300, Spectranetics, Colorado Springs, Colorado) was used. The laser generates pulses of ultraviolet (UV) radiation at a wavelength of 308 nm and duration of 145 ns. Laser energy is delivered to the myocardium via UV-grade multifiber bundles (Spectranetics, Colorado Springs, Colorado) with diameters of 1 and 1.4 mm. A computer controlled electro-mechanical mechanism (custom designed) is used to drive the optical-fiber bundle into the myocardium. This mechanism advances the optical-fiber bundle out of the tip of a hand-held probe at a constant preselected speed. To determine the optimal channel drilling settings, we varied the pulse repetition rate and the catheter advancement speed between 10-80 Hz and 0.1-15 mm/s, respectively. A constant fluence of 35 mJ/mm^2 was used throughout the experiments.

A holmium:YAG laser (Model 1-2-3, Schwartz Electro-Optics Inc., Orlando, Florida) was used. The laser generated pulses of near-infrared radiation at a wavelength of 2.014 μm and duration of 1.5 μs. Laser pulses were delivered to the myocardium via low OH optical fibers ranging in core diameter from 400 to 1000 μm. The pulse repetition rate and the energy were varied from 2 to 10 Hz and 0.2 to 2 J/pulse, respectively.

A CO_2 laser (PLC Medical Systems Inc., Franklin, Massachusetts) was used. The laser generates pulses of mid-infrared radiation at a wavelength of 10.6 μm and duration of 50 ms. The laser was operated at its clinical TMLR settings of 45 J/pulse. This laser radiation could not be delivered through fiberoptics at the intensities and pulse duration used in this study. Hence, an articulated arm consisting of a series of lenses and mirrors is used to guide the laser light to the tissue. A focusing zinc-selenide lens was used to provide a long beam waist. The laser system is synchronized to the ECG such that it is triggered at full ventricle.

2.2 Animal Model and Experimental Protocol

Twenty seven farmer pigs weighing between 30-50 kg were used to model myocardium ablation in TMLR. The animals were pretreated with bretylium 7-10 mg/kg, lidocaine 2-3 mg/kg, and propanolol 0.025-0.03 mg/kg. The animals are intubated and mechanically ventilated at a constant respiratory rate of 10 bpm with a tidal volume of 450 ml. An intramuscular injection of a ketamine (20 mg/kg), acepromazine (0.5 mg/kg) and atropine (0.05 mg/kg) cocktail was used for sedation. Anesthesia was induced by injecting thiopental into an ear vein and maintained by using isoflurane (1-2%). The pigs were placed in the supine position. The heart was exposed by either of two surgical approaches: a midline sternotomy or an intercostal incision between the third and fourth ribs. The pericardial sac was carefully opened to expose the anterior lateral wall of the heart. In both surgical approaches, the heart was kept in its anatomical position to avoid any disturbance to the normal blood flow and was moistened periodically with a saline soaked gauze to prevent dryness.

The animals are sacrificed after 1 h of the channel drilling procedure. The heart was then excised, and the section of myocardium surrounding each channel was dissected. Each section was placed in its own container for fixation in a 10% formalin solution for more than 48 hours prior to embedding. A reference map of the heart and a grid of the drilled channels were made to assure that the drilling settings for each channel were clearly defined and properly marked on each specimen bottle. The specimens were embedded in paraffin, stained with hematoxylin and eosin, and sectioned for light microscopy. All pigs received humane care in compliance with the recommendations of the Institute of Laboratory Animal Resources, National Institutes of Health [22].

2.3 Histological Analysis

The slides were initially viewed under transmitted light with an Olympus microscope (BH-2 multiport at 12.5x or 40x objective lenses). A color video camera (Hitachi KP-C553 CCD, 722x492 pixels) was used to image each slide twice under both normal and polarized lighting. The polarized light was generated by using a sub-stage polarizer and a first order red filter. Each of the images was digitized with frame grabber (Matrox Illuminator Pro CCIR-601) housed in a 486-based personal computer. A magnification reference grid (0.5x0.5 mm) was digitally superimposed on each digitized image so as to relate the image structures to the real-world dimensions. A digitizing pen tablet (Wacom SD420 E 12x12 digitizing tablet) is used to outline the boundaries of the channel on the image acquired with polarized light. This outline also represents the inner boundary of the thermally damaged region surrounding the channel. The outer boundary of this region was then outlined by marking the edge of decreased brightness or loss of birefringence. Increase in muscle temperature to above 60 °C leads to a decrease in its birefringence and hence reduce its brightness when viewed with polarized light [23,24]. The assessment of changes in birefringence can be an indicator of the extent of thermal injury [23,24]. The outlined 720x486 image is converted into a 648x486 VGA format and stored as a TARGA (Truevision Inc., Santa Clara, CA) file for image processing. A specialized image processing software (Matrox Inspector Version 1.7, Matrox Electronic Systems Ltd., Dorval, Quebec, Canada) was used to segment the TMLR channel and its surrounding thermal damage regions. The software is initially calibrated for two-dimensional measurement using the 0.5 mm scale superimposed on the image. Precision validation showed the ability to resolve 0.003 mm on the 0.5 mm grid system. The areas of the segmented channel and thermal damage regions were then measured in mm^2. A total of 560 tissue specimens were imaged and analyzed: 515 from excimer, 37 from holmium:YAG and eight from CO_2 irradiated myocardial sites.

3 Results

Histologic analysis revealed wide variations in the channel shapes, sizes and surrounding thermal damage for all the three lasers, even at the same energy setting for a given laser. The histology of a typical channel drilled in porcine myocardium using a CO_2, holmium:YAG and XeCl excimer is shown in Fig. 1.1 a, b and c, respectively. The channel and its surrounding region of thermal damage are digitally outlined.

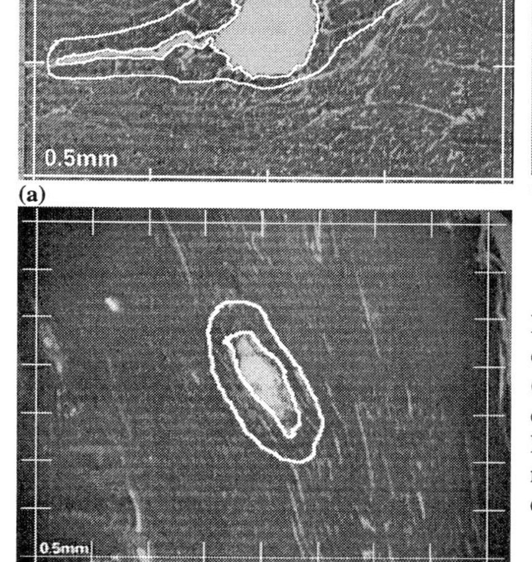

(a)

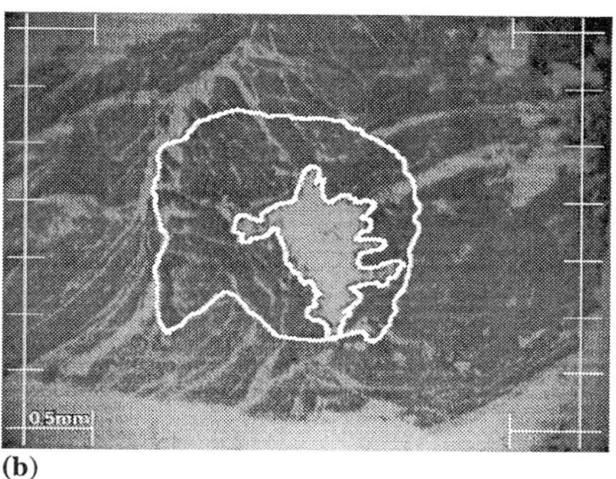

(b)

(c)

Fig. 1. Histology of a typical channel drilled in porcine myocardium using **(a)** CO_2, **(b)** holmium:YAG, and **(c)** XeCl excimer laser, as imaged under polarized light. The channel and its surrounding region of apparent thermal damage are outlined

The channels drilled using the CO_2 laser showed a wide variation in shape, size and their associated thermal damage zones. On visual inspection, most of the channels were oval in shape and surrounded by a small ring of blanched tissue (thermal denaturation). Of particular note is the presence of extended zones of thermally denatured tissue, which follow small capillaries away from the center of the crater. In some cases, the thermal damage zones extended about 3 mm from the crater.

The channels drilled using the holmium:YAG laser are irregular in shape with ragged lumens. This is probably due to tissue tearing and thermal coagulation. The size of the channels ranged from 0.2 mm in the short axis to 0.9 mm in the long axis. The zones of thermal injury range from 0.4 mm on the short axis to 0.3 mm on the long axis. Pronounced tissue tearing, as evidenced by fissures of 1-

4 mm in length, were prominent particularly at higher energies. At lower energies channel sizes were dramatically reduced (0.2-0.4 mm in diameter).

The channels drilled using the XeCl excimer laser demonstrated sensitivity to the lasing parameters. The relationship between the mean cross-sectional area of the channel and the catheter advancement speed for two catheter diameters (1.0 and 1.4 mm) is shown in Fig. 1.2a at a pulse repetition rate of 60 Hz. As shown, the mean area of the channel increases with the increase of the catheter diameter. However, the magnitude of its increase is dependent on the catheter advancement speed. At faster catheter advancement speeds, the channel area is less sensitive to changes in catheter diameter and vice versa. Similarly, the mean area of thermal damage tends to drop as the catheter diameter decreases as shown in Fig. 1.2b. A 40% increase in fiber diameter (i.e. from 1 to 1.4 mm) leads to an approximately 100% increase in the area of thermal damage at the catheter advancement speeds used in this study.

The relationship between the mean area of the channel and the catheter advancement speed at different pulse repetition rates is shown in Fig. 1.3a for a catheter diameter of 1.4 mm. The graphs indicate that the area of the channel tend to drop almost exponentially with increasing catheter advancement speed. The decay rate of this exponential relationship appears to decrease with increasing pulse repetition rate. In other words, the area of the channel is less sensitive to changes in the catheter advancement speed at higher pulse repetition rates and vice versa. At a given advancement speed, the area of the channel tend to increase with the increase of the pulse repetition rate.

Similarly, the size of thermal damage drops almost exponentially with increasing advancement speed and is also affected by the pulse repetition rate as shown in Fig. 1.3b. The area of thermal damage at any given advancement speed tend to increase with the increase of the pulse repetition rate. In general, faster advancement speeds produce smaller channels with even less thermal damage and slower advancement speeds produce large channels but with substantially greater thermal damage.

The channel area increase with the pulse repetition rate as shown in Fig. 1.4a. The magnitude of this increase is dependent on the catheter advancement speed.

At a given pulse repetition rate, slower catheter advancement speeds tend to produce larger channel areas and vice versa. However, the channel area was insensitive to the doubling of the catheter advancement speed at a pulse repetition rate of 10 Hz.

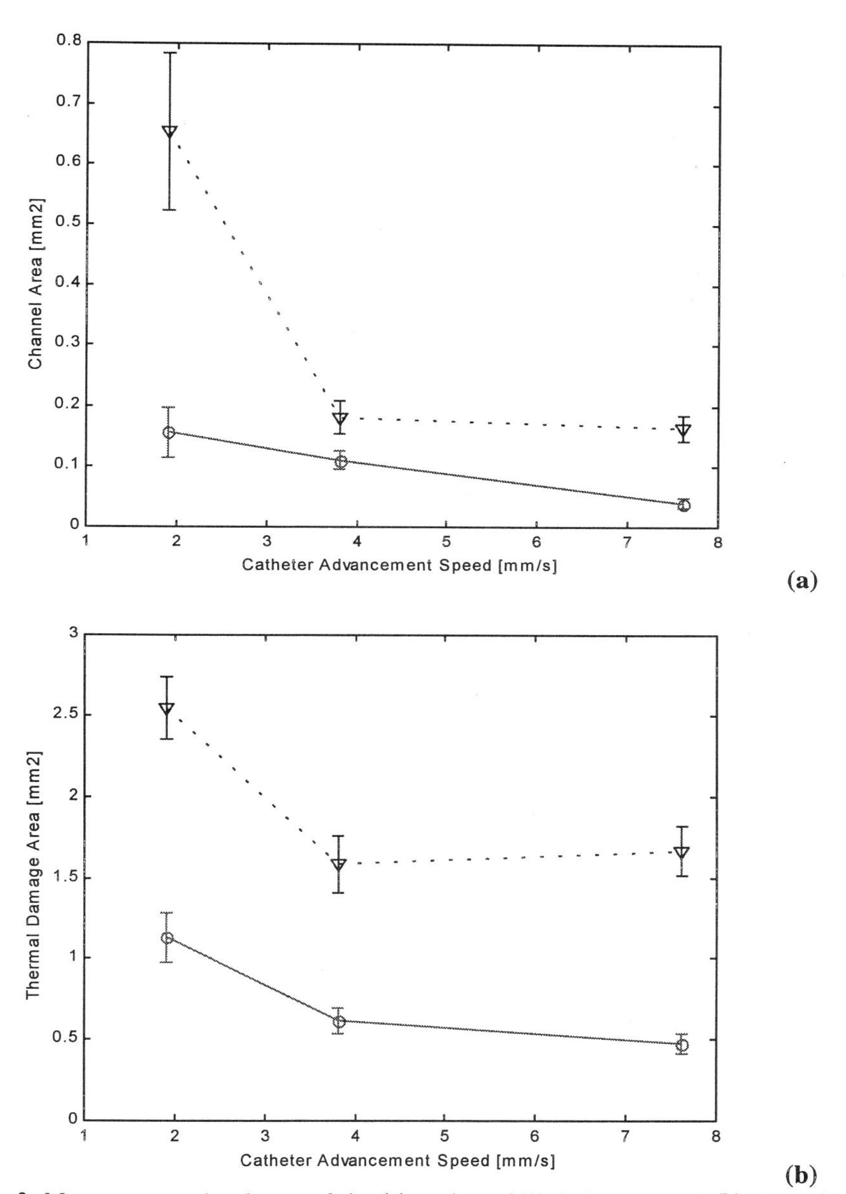

Fig. 2. Mean cross-sectional area of the **(a)** excimer-drilled channels and **(b)** thermal damage versus advancement speed for a catheter diameter of 1 (o) and 1.4 (∇) mm. Fluence was set at 35 mJ/mm^2 at a pulse repetition rate of 60 Hz. Error bars represent ±1 standard error of the mean (SEM)

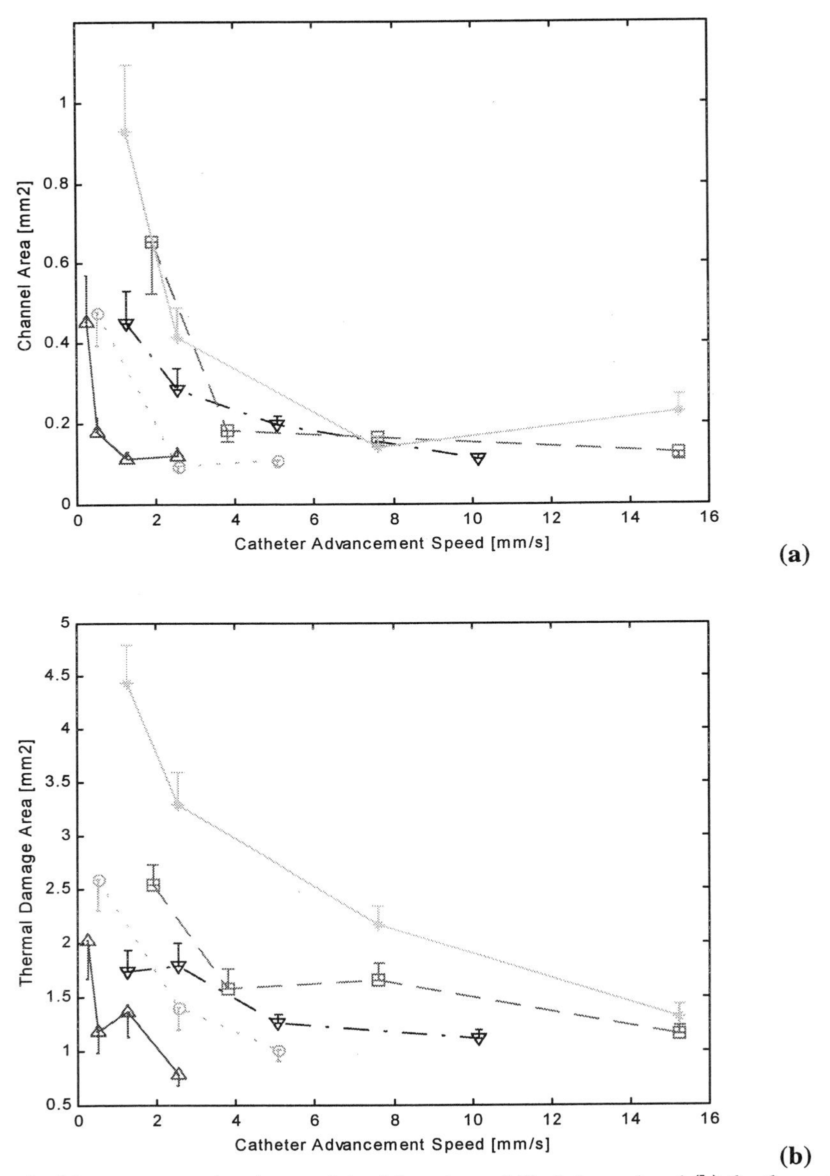

Fig. 3. Mean cross-sectional area of the **(a)** excimer-drilled channel and **(b)** the thermal damage versus the catheter advancement speed at pulse repetition rates of 10 (Δ), 20 (o), 40 (∇), 60 (□) and 80 Hz (✳). The catheter diameter and the fluence were 1.4 mm and 35 mJ/mm^2, respectively. The error bars represent one standard error of the mean (SEM).

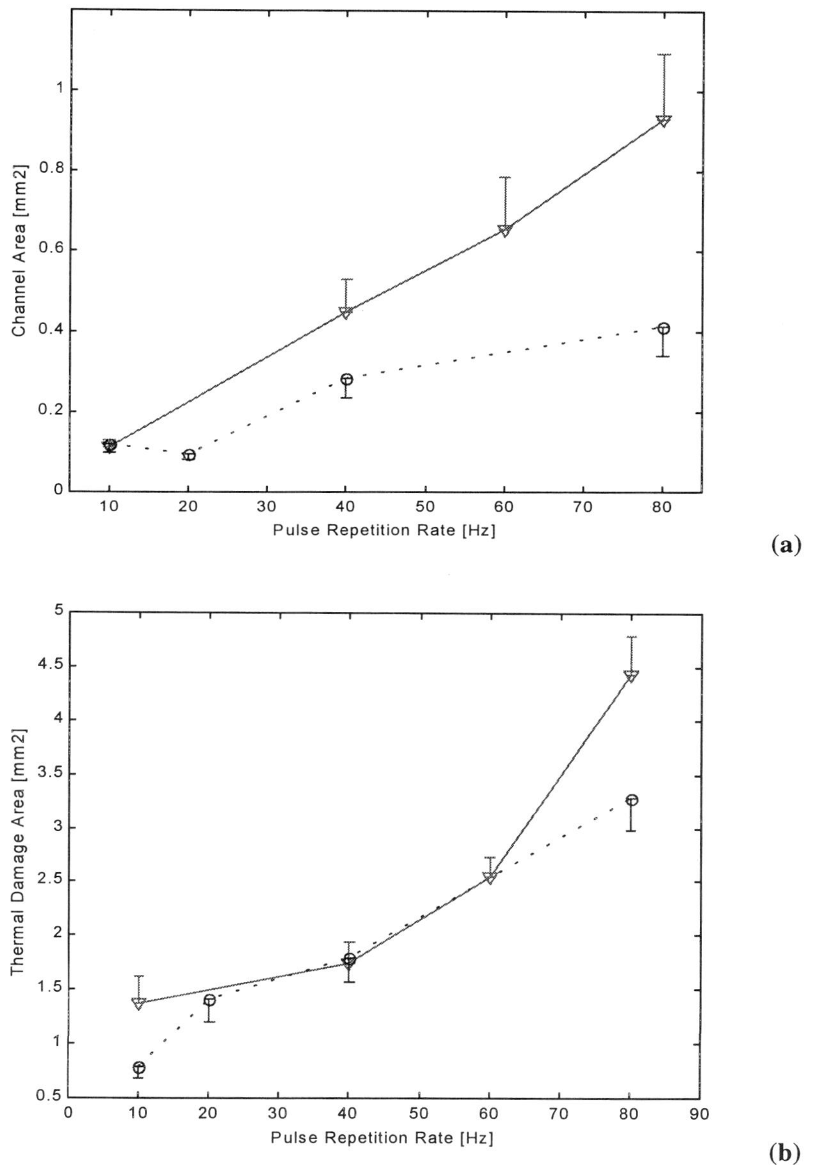

Fig. 4. Mean cross-sectional area of the (**a**) excimer-drilled channels and (**b**) their associated thermal damage zone versus the pulse repetition rate at a fluence of 35 mJ/mm^2, fiber diameter of 1.4 mm and fiber advancement speed of 1.27 (∇) and 2.54 mm/s (o)

The dependency of the thermal damage area on the pulse repetition rate, at different advancement speeds is shown in Fig. 1.4b. At a constant advancement speed, the extent of thermal damage increases with rising pulse repetition rates and vice versa. It is also noticeable that within the pulse repetition range of 20 to 30 Hz, both the 1.27 to 2.54 mm/s advancement speeds appear to cause the same level of thermal damage.

4 Discussion

We have examined three different lasers and their applicability for drilling channels in the myocardium for TMLR. All three lasers could drill channels, but each has its own advantages and limitations. The histological results presented above could provide some guidance in the process of TMLR optimization.

The CO_2 laser system used in our experiments could be only operated at its preset optimal clinical settings. Its laser energy cannot be delivered through catheters, which limits its use in percutaneous TMLR applications. In addition, data from the literature and our own investigations have revealed that the CO_2 laser produces considerable thermal damage around the drilled channels. The finger-like projection of thermal damage shown in Fig. 1.1a could be the result of steam being pushed into local capillaries or dissected myocardium during the ablation process. Steam escaping from the channel and driven out through local capillaries can cause substantial thermal injury along its path. Theoretically, these capillaries are supposed to interconnect with the newly formed channel to improve the blood perfusion. This type of thermal injury is undesirable because of its considerable spread and unpredictability. Although channel drilling is much faster with the CO_2 than with the holmium:YAG or the excimer lasers, histological evidence of unpredictable thermal damage and the lack of precise energy deposition constitute major drawbacks of this modality.

The holmium:YAG laser energy can be delivered fiberoptically which makes it suitable for percutaneous [25] and thoracoscopic [26] TMLR. This laser can easily drill channels into the myocardium but would also inflict significant thermal and mechanical injury to the surrounding tissue as indicated by the substantial fissures and loss of birefringence in Fig. 1.1b. Each pulse creates a crater 2-5 mm in depth with thermal damage zones of 1 mm or greater beyond the channel lumen. Tissue fissuring (tearing) could, theoretically, open new channels to adjacent blood vessels, but the thermal coagulation (burn) which accompanies these fissures is a strong stimulus for collagen deposition and scar formation. A recent comparative analysis [27] between CO_2 and holmium:YAG lasers in TMLR has indicated that the initial myocardial effects of both lasers are similar and differ predominantly in the amount of acute thermoacoustic injury. However, their appearances were indistinguishable by six weeks following organization and neovascularization of the channel region. Parameter optimization for the holmium:YAG laser is difficult, since there is a substantial trade-off between channel size and thermal injury. For example, the energies required for ablating a

0.8 mm diameter channel can cause substantial thermal damage and pronounced tissue fissuring.

The excimer laser has already been used in clinical applications, such as laser angioplasty, because of its unique ablation characteristics. In comparison to the CO_2 and holmium:YAG lasers, it seems to produce the least thermal and mechanical damage to the tissue around the site of ablation. The excimer laser can operate over a broad range of energies and repetition rates. In addition, its energy can be readily delivered via optical fibers. This flexibility requires the evaluation of a variety of lasing parameters to determine those producing the optimal myocardial ablation for TMLR. Of particular interest, were the histological effects of the catheter advancement speed and the pulse repetition rate. Our results indicate that the catheter advancement speed might be a crucial channel-drilling parameter. An increase in the advancement speed leads to a drop in the size of both the channel and thermal damage for all repetition rates. For example, doubling the advancement speed from 1.27 to 2.54 mm/s leads to more than 50% reduction in the channel size for repetition rates ≥ 20 Hz as shown in Fig.1.4a. This suggests that at higher advancement speeds, the drilling process might be primarily due to the mechanical tearing rather than ablation. The latter explanation is supported by histological evidence of increased tissue tearing and reduced thermal damage at the higher speeds. Advancement speeds that are slower than the laser ablation speed (= ablation rate x repetition rate) would increase the thermal damage mainly as a result of deposition of subablative laser fluences. On the other hand, advancement speeds that are faster than the laser ablation speed would increase tissue tearing by the force of the advancing catheter. It is therefore reasonable to assume that the optimal catheter advancement speed should be comparable or slightly faster than the laser ablation speed. Since the laser advancement speed is a function of the repetition rate, the value for optimal advancement speed would depend on the choice of repetition rate. For example, at an ablation rate of about 40 μm/pulse and a pulse repetition rate of 40 Hz, the ablation speed is about 1.6 mm/s, and hence, in this case the optimal advancement speed should be around or slightly higher than the latter value. This way the catheter tip will follow the ablation front and will produce the largest possible channel with minimal mechanical tearing. An advancement rate that is slightly higher than the ablation speed may also prove advantageous in displacing the blood from the ablation site, thereby minimizing its interference with the ablation process. An automated catheter advancement mechanism was developed and used to provide control on the ablation process and ensure operation at the optimal advancement speed. A fluence of 35 mJ/mm^2 was used throughout the current *in-vivo* investigation. This value has been previously determined by a set of in-vitro experiments in our laboratory, to be the optimal fluence for myocardial ablation. Finally, since larger channels are desirable, it is evident from Fig. 1.2a that the larger diameter (1.4 mm) catheter should be used.

5 Conclusions

The histologic measurements from our *in-vivo* experiments indicates that XeCl excimer laser provides the surgeon with a precise method of creating TMR channels in the myocardium with minimal and largely predictive tissue damage. In comparison to the CO_2 and holmium:YAG lasers, the excimer laser appears to cause the least thermo-mechanical damage to the tissue surrounding the TMR channel. The cross-sectional area of an excimer-drilled channel and the associated thermal damage increase with the decrease of the catheter advancement speed for all pulse repetition rates, and vice versa. The optimal catheter advancement speed should be comparable or slightly faster than the ablation speed. Within the parameters tested, larger catheters (1.4 mm), pulse repetition rates between 35 to 40 Hz, and advancement speeds of about 1.3 mm/s (~0.05 "/s) seem to provide reasonably optimized parameters for excimer-based TMLR. While the mechanism of action of TMLR remains unknown, the 308 nm laser radiation offers a channel drilling technique that is compatible with either open-chest or percutaneous applications. The flexibility of the delivery system and the precise control of energy deposition afforded by ablation in the ultraviolet region (308 nm) could permit development of a safe percutaneous myocardial revascularization (PMR) technique.

References

[1] Horvath K.A., Cohn L.H., Cooley D.A., Crew J.R., Frazier O.H., Griffith BP. Kadipasaoglu K., Lansing A., Mannting F., March R., Mirhoseini M.R., Smith C. (1997) Transmyocardial laser revascularization: results of a multicenter trial with transmyocardial laser revascularization used as sole therapy for end-stage coronary artery disease. J Thoracic & Cardiovascular Surgery 113:645-53

[2] Krabatsch T., Tambeur L., Lieback E., Shaper F., Hetzer R. (1998) Transmyocardial laser revascularization in the treatment of end-stage coronary artery disease. Annals of Thoracic & Cardiovascular Surgery 4:64-71

[3] Horvath K.A. (1997) Clinical studies of TMR with the CO_2 laser. Journal of Clinical Laser Medicine & Surgery 15:281-5

[4] Milano A., Pratali S., Tartarini G., Mariotti R., De Carlo M., Paterni G., Boni G., Bortolotti U. (1998) Early results of transmyocardial revascularization with a holmium laser. Annals of Thoracic Surgery 65:700-4

[5] Diegeler A., Schneider J., Lauer B., Mohr F.W., Kluge R. (1998) Transmyocardial laser revascularization using the Holium-YAG laser for treatment of end stage coronary artery disease. European Journal of Cardio-Thoracic Surgery 13:392-7

[6] Mirhoseini M., Muckerheide M., Cayton M.M. (1982) Transventricular revascularization by laser. Lasers in Surgery & Medicine 2:187-98

[7] Mirhoseini M., Cayton M.M., Shelgikar S., Fisher J.C. (1986) Laser myocardial revascularization. Lasers in Surgery & Medicine 6:459-61

[8] Mirhoseini M., Shelgikar S., Cayton M.M. (1988) New concepts in revascularization of the myocardium, Annals of Thoracic Surgery 45:415-20

[9] Krabatsch T., Schaper F., Leder C., Tulsner J., Thalmann U., Hetzer R. (1996) Histological findings after transmyocardial laser revascularization. Journal of Cardiac Surgery 11:326-31

[10] Gassler N., Wintzer H.O., Stubbe H.M., Wullbrand A., Helmchen U. (1997) Transmyocardial laser revascularization. Histological features in human nonresponder myocardium. Circulation 95:371-5

[11] Whittaker P., Kloner R.A., Przyklenk K. (1993) Laser-mediated transmural myocardial channels do not salvage acutely ischemic myocardium. Journal of the American College of Cardiology 22:302-9

[12] Kohmoto T., Fisher P.E., Gu A., Zhu S.M., Yano O.J., Spotnitz H.M., Smith C.R., Burkhoff D. (1996) Does blood flow through holmium:YAG transmyocardial laser channels? Annals of Thoracic Surgery 61:861-8

[13] Fleischer K.J., Goldschmidt-Clermont P.J., Fonger J.D., Hutchins G.M., Hruban R.H. Baumgartner W.A. (1996) One-month histologic response of transmyocardial laser channels with molecular intervention. Annals of Thoracic Surgery 62:1051-8

[14] Brilla C.G., Rybinski L., Gehrke D., Rupp H. (1997) Transmyocardial laser revascularization--an innovative pathophysiologic concept. Herz 22(4):183-9

[15] Spanier T., Smith C.R., Burkhoff D. (1997) Angiogenesis: a possible mechanism underlying the clinical benefits of transmyocardial laser revascularization. Journal of Clinical Laser Medicine & Surgery 15:269-73

[16] Pelletier M.P., Giaid A., Sivaraman S., Dorfman J., Li C.M., Philip A., Chiu R.C. (1998) Angiogenesis and growth factor expression in a model of transmyocardial revascularization. Annals of Thoracic Surgery 66:12-8

[17] Mack C.A., Patel S.R., Rosengart T.K. (1997) Myocardial angiogenesis as a possible mechanism for TMLR efficacy. Journal of Clinical Laser Medicine & Surgery 15:275-9

[18] Personal communications with Dr. Gerhard Muller, Institut fur Medizinische/Technische Physik und Lasermedizin des Universitatsklinikum Benjamin Franklin, Berlin, Germany.

[19] Horvath K.A., Greene R., Belkind N., Kane B., McPherson D.D., Fullerton D.A. (1998) Left ventricular functional improvement after transmyocardial laser revascularization. Annals of Thoracic Surgery 66:721-5

[20] Kadipasaoglu K.A., Pehlivanoglu S., Conger J.L., Sasaki E., de Villalobos D.H., Cloy M., Piluiko V., Clubb F.J. Jr., Cooley D.A., Frazier O.H. (1997) Long- and short-term effects of transmyocardial laser revascularization in acute myocardial ischemia. Lasers in Surgery & Medicine 20:6-14

[21] Mueller X.M., Tevaearai H.H., Genton C.Y., Bettex D., von Segesser L.K. (1998) Transmyocardial laser revascularisation in acutely ischaemic myocardium. European Journal of Cardio-Thoracic Surgery 13:170-5

[22] Committee on Care and Use of Laboratory Animals, Institute of Laboratory Animal Resources (1985) Guide for the care and use of laboratory animals. National Institutes of Health, DHHS publication no. (NIH) 86-23

[23] Whittaker P. (1997) Detection and assessment of laser-mediated injury in transmyocardial revascularization. Journal of Clinical Laser Medicine & Surgery 15:261-7

[24] Jansen E.D., Frenz M., Kadipasaoglu K.A., Pfefer T.J., Altermatt H.J., Motamedi M., Welch AJ. (1997) Laser-tissue interaction during transmyocardial laser revascularization. Annals of Thoracic Surgery 63:640-7

[25] Kim C.B., Oesterle S,N. (1997) Percutaneous transmyocardial revascularization. Journal of Clinical Laser Medicine & Surgery 15:293-8

[26] Milano A., Pietrabissa A., Bortolotti U. (1997) Transmyocardial laser revascularization using a thoracoscopic approach. American Journal of Cardiology 80:538-9

[27] Fisher P.E., Khomoto T., DeRosa C.M., Spotnitz H.M., Smith C.R., Burkhoff D. (1997) Histologic analysis of transmyocardial channels: comparison of CO_2 and holmium:YAG lasers. Annals of Thoracic Surgery 64:466-72

Lasers in Modern Cataract-Surgery

E. Alzner and G. Grabner
Eyeclinic, County Hospital Salzburg
Head: Univ.-Prof. Dr. Günther Grabner

A cataract may be defined as any opacity of the crystalline lens that interferes with vision. Cataract is the most common, visually disabling eye disease in the world, 45% of blindness is caused by cataracts. Age related changes in the lens is the most prevalent type of cataract in the civilised world. Due to this, cataract surgery is probably the most frequent surgery procedure of all. The principle of cataract surgery is to remove the opacified lens out of the eye.

In adult cataracts this is mostly done as a small incision procedure by emulsifiing the hard nucleus of the lens with ultrasonic power – this technique is named phacoemulsification. The phacoemulsification-technique was introduced to the clinic in 1967 by Charles Kelman [9] and after several modifications has now become a rather safe and effective procedure with a low complication rate. To achieve a good optical correction the opacified lens is replaced by a foldable intraocular lens.

But there are also some disadvantages of the phacoemulsification technique. The extremly high frequency of the longitudinal oscillations of the tip of the phacoinstruments (20-40 kHz) lead to thermal effects, which in unfavourable circumstances, cause corneal and/or scleral burns [2,4,11,22]. The high energy brought to the eye during the phacoemulsification can also lead to the loss of endothelial cells [5,21] of the cornea resulting in corneal clouding and secondarily, reduced visual acuity.

What is the advantage of lasers in cataract surgery: there could be

- smaller incisions for smaller lenses (e.g. injectable lenses in the future) and shorter recovery times
- fewer complications as corneal edema, corneal burns, capsular rupture
- a shorter learning curve because of an easier procedure

Attempts to utilise lasers with different wave lengths (short ultraviolet as well as near infrared wavelengths) in cataract surgery were started in the mid-eighties [12,13,16,19]. Based on a survey of the recent literature it can be stated that the use of both Excimer and Picosecond lasers was largely discontinued due to technical problems, high equipment costs and considerable risks associated with primary and secondary laser beam radiation [8,10,17]. Also the delivery of the beam by a fiberoptic was a criterium.

The extremly rapid development in laser technology has led to clinical testing of two types of solid-state laser systems for cataract extraction, the erbium-YAG (2936 nm) [20] and neodymium-YAG (1064 nm) [1,6,7].

Today, several companies in Europe and the United States are constructing, testing and even selling (under investigative circumstances) these new laser-systems for cataract surgery.

A considerable difference exists between these two laser systems.

While the beam of the erbium-YAG laser is targeted to the material of the lens itself, the neodymium-YAG laser hits a titanium target inside of the tip of the laser instrument, the nucleus of the lens is cracked by the resulting plasma formation and the shockwave. So the effect of neodymium-YAG laser on the lens material is indirect .

1. Nd-YAG-Laser in Cataract Surgery

Our own experience concerns the Nd-YAG-lasersystem, introduced by Jack M Dodick in New York. This method is termed Dodick–Laser–Lens–Lysis. For our clinical study a pulsed laser with a duration of 14 ns, an energy of 10 mJ and frequency between 1 to 5 Herz was used. The laser beam is carried to the tip of the handpiece by a 400 mμ quarz-fiber.

After the lysis of the lens, the material has to be removed out of the eye, therefore the laser handpiece is used as an aspiration probe. The end of the tip contains the titanium target with a diameter of 600 μm. The slightly divergent laser beam hits this target and produces the mentioned plasma and shockwave. To stabilise the anterior chamber of the eye, a second handpiece for infusion (or an infusion-pipe on the laser-handpiece) is needed. The diameter of the tip of the laser handpiece as well of the infusion handpiece is 1.2 mm.

The longitudinal oscillations with extremly high frequency occurring in the phacotip (20-40 kHz) lead to thermal effects, which in unfavourable circumstances, cause corneal and/or scleral burns. Although there is expectation that this method will not lead to heating at the "point of action" or entry incision used by the tip, we studied this in our lab.

For the temperature measurements we used custom built Neodym-YAG-laser phacolysis probes with a tip diameter of 1.2 mm, the laser frequency was set at a constant 10 Hertz based on initial clinical experiences.

As a control-group we used the "conventional" ultrasonic phacoemulsification handpiece of the Sonocat ™ produced by Örtli, Swizerland, with a frequency of ultrasound of 30 kHz.

The temperatures were measured with the digital thermometer GTH 175/MO produced by Giesinger, Germany.

Measurements were carried out in a special designed measuring device both under air and under balanced salt solution (4 ccm) and in the anterior chamber of donor eyes (unsuitable for corneal transplantation) from the Salzburg eye-bank with the laser phacolysis and the ultrasonic phaco handpiece. The "point of action" at the laser device is the region of the titanium target, the temperature was measured near the end of the tip. At the ultrasonic-phacotip the temperature was taken from the corresponding region.

While the heat at the phacotip originates from the friction of high-frequency oscillation, results a possible rise of temperature at the lasertip from the optical breakdown at the titanium target.

In our experiments we found, that there exists a significant difference in temperature generated in proxinity to the tip between the conventional ultrasonic phaco handpiece and the laser phacolysis handpiece.

In the dry artificial anterior chamber a heat difference of the laser phacolysis tip of up to 3.4° degrees centigrade (mean 3.13°C, SD 0.38) was measured in contrast to the increase of up to 55.3°C (mean 42.03° C, SD 22.9) with the ultrasonic phacolysis probe.

Although the difference of temperatures measured in donor eyes when comparing the laser and ultrasonic device is moderate, there is still a maximum rise in heat of 10.9°C (mean 7.06°C SD 2.51) with the ultrasonic phaco needle wereas the phacolysis needle produces only an increase of 1.2°C (mean 0.41°C, SD 0.41).

In another set of experiments we studied the resistance of the posterior capsule of the lens against both the ultrasonic device and the Neodym-YAG-laser device.

The rupture of the posterior capsule is one of the compliacatons of phacoemulsification, every new system should reduce this risc.

By allowing the laser tip (10 Hz, 10mJ) as well as the ultrasonic tip (70% energy-setting) to approach the posterior capsule of human crystalline lenses in an artificial anterior chamber setting we found that the breakage of the posterior capsule occurs earlier with the conventional phacoemulsification instrument than with the Nd-YAG-instrument.

Approved by the ethics commitee of the county of Salzburg we studied the clinical use of the Dodick Laser Lens Lysis on 25 patients. The mean age is 74.8 years, 19 of the patients were female, 6 male.

The inclusion criteria were:

- age 65 years or above,
- visual acuity beyond 0.4 – this is eqivalent to 20/50 on the Snellen chart
- axial length between 22.0 and 25.0 mm
- no pathology on the second eye except cataract, pseudophakia or mild maculopathy
- all patients had to sign informed consent

The following data were recorded:

- visual acuity and intraocular pressure
- slitlamp and fundus examination
- pachymetry and
- endothelial cell count

The examinations were done preoperatively and 1-2 days, 2 months, 6 months postoperatively

We found that there were no risks exceeding the conventional phacoemulsification technique in our selected cases, one was complicated by posterior capsule opening during the inspiration/aspiration phase of the cortex removal, additionally one patient sustained a high endothelial cell loss, we attributed this to the hard nucleus and the resulting large amount of irrigation fluid used, coupled with excessive laser pulses in the early stage of our learning curve.

The best corrected visual acuity on postop day 1-2 was 0.64 in mean, the best corrected visual acuity 6 to 8 weeks postoperatively is 0.93 (normal visual acuity of a healthy eye is 1.0) – this is equal to the results we get with the conventional ultrasonic procedure.

For endothelial cell analysis we excluded the above mentioned case from our data, we sustained a 3.90% endothelial cell loss in our study, copmared to the US-device this seems to be a rather good result [5,15,21].

There are still many open questions concernig the Dodick Laser Lens Lysis, currently we study the titanium ablation of the target (together with P. Wilhartitz of the Plansee Group for Metallurgical Products, Reutte, Austria) and the quality and amount of the secondary radiation of the plasma (together with W. Kautek and J. Bonse form the Lab for thin film technolgy, Federal Institute for Materials Research, Berlin, Germany).

2. Er-YAG-Laser in Cataractsurgery

On the other hand 3 companies with similar erbium-YAG-Laser-instruments are currently approaching the market – Eyesys with "Premiere", Wavelight with "Adadio" and Aesculap-Meditec with the "Phacolase". In contradiction to the Neodymium-YAG-Laser by ARC-Laser (Germany), the erbium-YAG-Laser-systems are multipurpose systems not for the cataract-surgery only, they also have capabilities in the fields of glaucoma surgery, vitrectomy and in dermatologic skin-resurfacing.

The effect of the erbium-YAG laser is based on the high absorbtion of its wavelength (2936 nm) in water, utilising a combination of photoablative and photoacoustic phenomena.

Although the laser beam is totally absorbed in the first 1-2 μm of water in the anterior eyechamber, the effect of the laser reaches 1 to 2 mm into the tissue because of the acustic shockwave and the effect of cavitation.

Performing cataract surgery, the instrument can be used not only for the lysis of the nucleus, it also allows (with a special 45 deg. probe) to perform the anterior capsulotomy, this is the procedure of the opening of the anterior capsule of the lens with a regular margin as the first step of the lens-extraction. When capsulotomy is desired, photoablative properties are employed. The laser setting is just above the ablation treshold.

Also with the erbiuim YAG, there is the question about thermal effects in the anterior chamber and at the incision site. Investigations by Berger and coworkers in 1996 [3] have shown, that similar to the Nd-YAG there is no significant rise of temperature with the erbium-YAG Laser.

First investigations about the rupture of the lens capsule have been performed by Snyder and coworkers (18), they could show, that at high energy levels, the erbium-YAG-Laser ist comparable to and at lower energy levels it is safer than phacoemulsification in its ability to damage the lens capsule.

Clinical studies are currently being performed in several sites, the first results are just in press.

These studies show (H. Höh, Neubrandenburg, Germany; A. Francini, Florence, Italy; M. Zato, Madrid, Spain) , that – similar to the result of our study with the Nd-YAG Laser – the loss of endothelium cells is low (Höh 2.25%), breaks of the posterior capsule are rare.

Also with this instrument – like the Neodymium-YAG-Instrument - there are cases with hard nuclei, where it is useful to change to conventional phacoemulsification.

In conclusion – both lasers – the Neodym-YAG- as well as the erbium-YAG are capable of removing cataracts in an effective and safe manner, but they are only at the very beginning of clinical use. There is still the need for modifications of handpiece, lasersettings and surgical technique. If this happens, laser-surgery could potentially be the surgery of choice for cataracts in the near future.

References

1. E Alzner, J M Dodick, R Thyzel, G Grabner (1998) Experimentelle Laserphakolyse mit dem ARC-Puls-Nd-YAG-Laser, Spectrum Augenheilkd 12/1: 24-27
2. Benolken R M, Emery J M, Landis D J (1974) Temperature profile in the anterior chamber during phacoemulsification, Invest. Ophthalmol 1: 71-74
3. Berger J W, Talamo J H, LaMarche K J, Kim S H, Snyder R W, D'Amico D J, Marcellino G J-Cataract-Refract-Surg. (1996) Temperature measurements during phacoemulsification and erbium:YAG laser phacoablation in model systems, J Cataract Refract Surg 22: 372-8
4. Bissen-Miyajima H, Goto E, Shimazaki J, Tsubota K (1997) Thermal effect on corneal incision with differnt tips and sleeves, Presentation at ASCRS-Meeting, April 26-30, 1997, Boston, USA
5. Dick B, Kohnen T, Jacobi K W (1995) Endothelzellverlust nach Phacoemulsification und 3.5 vs. 5 mm Hornhauttunnelincision, Ophthalmologe 92: 476 - 483
6. Dodick J M (1991) Laser Phacolysis of the Human Cataractous Lens, Dev. Ophthalmol 22: 58-64
7. Dodick J M, Lally J M, Sperber L T D (1993) Lasers in cataract surgery, Current Opinion in Ophthalmology 4: 107-109
8. Forster W, Emmerich K H, Busse H, Scheid W, Weber J, Traut H (1991) Die Induktion chromosomaler Aberrationen in menschlichen Lymphocyten als Modell zur Prüfung der mutagenen Wirkung von Excimerlaserstrahlung in der Ophthalmologie, Fortschritte Ophthalmol 88: 377 – 379
9. Kelman C D (1967) Phacoemulsification and aspiration: a new technique of cataract removal. A preliminary report, Am J Ophthalmology 64: 23-35
10. Lubaschowski H, Kermani O (1992) 193 nm Excimer-laserphotoablation der Hornhaut. Spektum und Transmissionsverhalten von Sekundärstrahlung, Ophthalmologe 89, 134 - 138
11. Mackool R J (1994) Scleral and corneal burns during phacoemulsification with viscoelastic materials, J Cataract Refract Surg 20: 367-368

12. Müller-Stolzenburg N, Stange N, Kar H, Dörschel K, Bath P, Müller G (1989) Endocapsuläre Kataractoperation mit dem Eximerlaser bei 308 nm, Fortschr Ophthalmol 86: 561-565
13. Nanevicz T, Prince M R, Gawande A A, Puliafito C A (1986) Excimer laser ablation of the lens, Arch Ophthalmol 104: 1825-1829
14. Nishi O, Nishi K, Mano C, Ichihara M, Honda T (1997) Controlling the capsular shape in lens refilling, Arch-Ophthalmol. 115: 507-10
15. Ohrloff C, Zubcov A A (1997) Comparison of phacoemulsification and planned extracapsular extraction, Opthalmologica 211: 8-12
16. Puliafito C A, Steinert R F, Deutsch T F, Hillenkemp F, Dehm E, Adler C M (1985) Excimer laser ablation of the cornea and lens: experimental studies, Ophthalmology 92: 741-748
17. Seiler T, Bende T, Winkler K, Wollensack J (1988) Side effects in excimer corneal surgery. DNA damage as a result of 192 nm excimer laser, Graefes Arch 226: 273-276
18. Snyder R W, Noecker R J, Jones H (1994) In vitro comparison of phacoemulsification and the erbium YAG laser in lens capsule rupture, Investigative Ophthalmology 35: 1934
19. Vogel A, Capon M R, Asiyo-Vogel M N, Birngruber R (1994) Intraocular photodisruption with picosecond and nanosecond laser pulses: tissue effects in cornea, lens and retina, Invest Ophthalmol Vis Sci 35: 3032-3042
20. Wetzel W, Brinkmann R, Koop N, Schröer F, Birngruber R (1997) Laserphakoemulsifikation mit dem Er:YAG-Laser, in D. Vörösmarthey et al (Hrsg) 10. Kongreß der DGII 1996, Springer Berlin Heidelberg 356-359
21. Wirbelauer C, Anders N, Pham DT, Holschbach A, Wollensack J (1997) Frühpostoperativer Endothelzellverlust nach korneoskleralem Tunnelschnitt und Phakoemulsifikation bei Pseudoexfoliationssyndrom, Ophthalmologe 94: 332-336
22. Wirt H, Heisler J M, Domarus V D (1995) Phacoburns: Experimental Study for Evaluation of Risk Factors, Eur J Implant Ref Surg 7:275-278

Correlation of Thermal and Mechanical Effects of the Holmium Laser for Various Clinical Applications

M.C.M. Grimbergen, R.M.Verdaasdonk, and C.F.P. van Swol
Department Of Clinical Physics and Biomedical Engineering
University Hospital Utrecht, The Netherlands

Abstract. The Holmium laser has become established in orthopedic surgery and urology due to its unique combination of mechanical and thermal properties induced by explosive vapor bubbles. In a specialized setup, real-time high-speed and thermal images of dynamic vapor bubbles and thermal relaxation at a water/tissue interface were obtained simultaneously. The thermal effects in the tissue model were correlated to the characteristics of the bubbles dependent on pulse energy (0.2-4 J), pulse repetition frequency (5-40 Hz), distance and angle of fiber delivery system (diameter 365 μm) in respect to the tissue surface. Up to a fiber-to-tissue distance of 50% of the radius of the bubble, only a superficial tissue layer was heated. During bubble implosion, the tissue surface was attracted to the fiber, ripping of irregularities, and was effectively cooled by turbulence. In case of hard tissues, the bubble detached from the fiber imploding towards the hard surface. At closer distances (<50% of bubble radius), the tissue itself was vaporized resulting in mechanical damage and thermal relaxation into the tissue, especially above repetition rates of 5 Hz.

There is a strong correlation between the path length of the free beam within the bubble and the degree of mechanical and thermal damage in the tissue directly irradiated by this beam.

During clinical applications the surgeon should be aware of the size of the vapor bubble in relation to the distance and angle with the tissue for safe and optimal use of the mechanical and thermal properties of the Holmium laser

1 Introduction

Since the introduction of the holmium laser in the medical field, the number of applications has grown rapidly and include prostatectomy and lithotripsy (urology), cartilage defects (orthopedics), discectomy (neurosurgery) and dacryorhinocystotomy (ophthalmology). Above mentioned procedures comprise as well hard as soft tissue applications. Depending on the tissue properties, the surgeon needs a controlled extent of either thermal or mechanical tissue effects.

The aim of the study was to investigate the correlation between mechanical and thermal effects of the Holmium laser in relation to pulse energy, the pulse repetition rate and the angle of irradiation to provide a ‘rule of thumb’ for the surgeon for a safe and optimized treatment strategy

2 Holmium Laser Tissue Interaction

Holmium laser light is usually delivered to the tissue through a fiber under endoscopic guidance. Therefore, the fiber tip will be submerged in water or saline. The 2.1 μm wavelength of the holmium laser light is absorbed in the first 500 μm layer of (tissue)water in front of the laser beam. At the start of the typically 300 μs laser pulse, the water is instantly vaporized forming a rapidly expanding vapor bubble (Fig.1). The initial bubble creates an 'opening' through the liquid in front of the fiber. While the laser pulse continues, the beam vaporizes the front-end of the bubble increasing the size and changing its shape to a 'pear' (the 'Moses effect'[1]). At the end of the laser pulse the expansion continues due to mass inertia until the underpressure will finally cause implosion of the vapor bubble.[2] Although the penetration of the laser beam at the start of the laser pulse is only 500 μm, a 4 J laser pulse can penetrate up to 15 mm into the (tissue)water [3] A distinction can be made between direct interaction with the tissue due to direct exposure to the laser beam and indirect interaction due to the bubble.

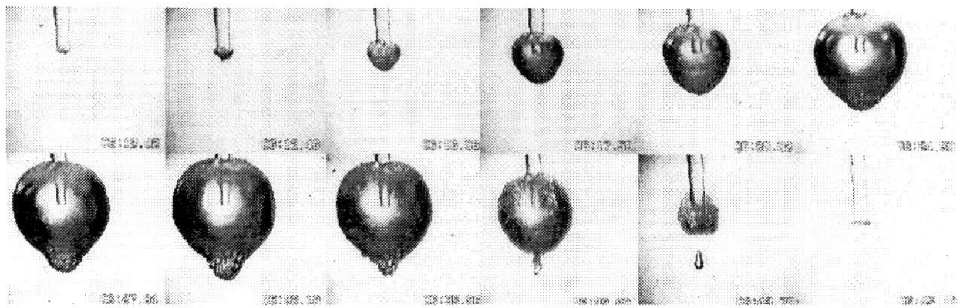

Fig. 1. Holmium laser light induced expanding and collapsing vapor bubble. Time frame of the sequence is about 1 ms

During bubble implosion the heat of condensation is dissipated in the surrounding liquid resulting in thermal effects in the liquid or tissue. The thermal energy is effectively dissipated due to turbulence induced by the bubble implosion. In tissue, however, the temperature increase can be substantially.

3 Material & Methods

3.1 Imaging techniques

Time delayed high-speed photography with a temporal resolution of 1μs was used to capture the development of the bubble during the vaporization process and the implosion. A 'start of pulse' signal from the laser was time-delayed with a preset time to trigger an arc flash lamp illuminating the vaporization process at the fiber tip. It is assumed that the vapor bubble formation is reproducible, so the sequential images of the ablation process were obtained from different individual bubbles at time-delays from 0 - 1000 μs.

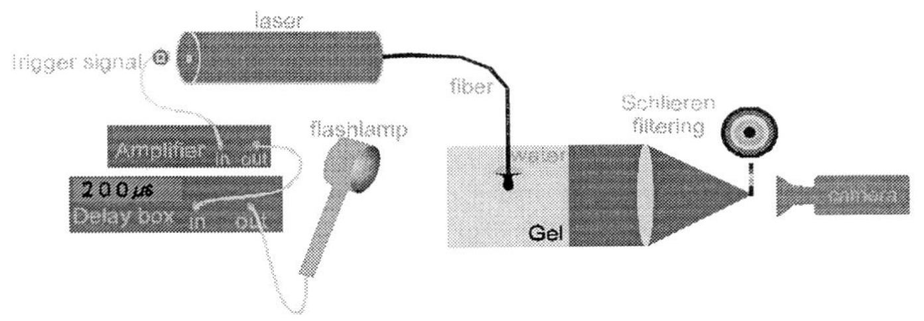

Fig. 2. Experimental setup combining Schlieren optics and fast photography

The thermal effects were visualized by means of Color Schlieren techniques based on an optical processor [4]. Using this method, very small changes in optical density, induced by temperature gradients inside an optically transparent medium, are color-coded resulting in color images. These images are obtained with resolutions in the millisecond region and are comparable with thermal-images produced by a thermo-camera. The feasibility of this technique was shown in previous studies of thermal effects of cw and pulsed lasers.

For the setups used in this study, the ablation effects of the irradiated tissue can only be visualized in transparent media. Therefore, a polyacrylamide gel (PAA) was used which resembles tissue such that it consists of merely water in a matrix of organic molecules and consequently thermal characteristics are comparable to biological tissue. For the highly absorbed wavelength of the Holmium laser, it was assumed that beam scattering could be neglected so the optical properties are also similar. The use of polyacrylamide gel ensures reproducibility of the samples and enables molding of the material in the required geometry for a Schlieren setup.

3.2 Laser and experimental settings

An 80W Holmium laser (Coherent Versa Pulse, Palo Alto, Ca) was used. The experimental settings used were; pulse energy varied from 0.5 to 2.0 J where the pulse frequency varied between 5 and 10 Hz. The experiments were performed under conditions and settings, simulating the clinical situation as observed during orthopedic surgery and urologic treatments. In a water environment, the distance and angle of a 365 μm bare fiber tip was varied in relation to the tissue surface.

4 Results

4.1 Bubble size and thermal zone

The experiments were aimed at determination of the relation between the thermal and mechanical effects. The size of bubble was directly related to the energy content of the laser pulse as shown in Fig. 3.

The left image shows a composition of a fast photography and a thermal image from the same `bubble formation'. The dark area represents the artificial tissue with water on top. Arrows refer to the dimensions obtained for bubble size, extent of the thermal and ablated zone. The right graph displays the extent of the thermal

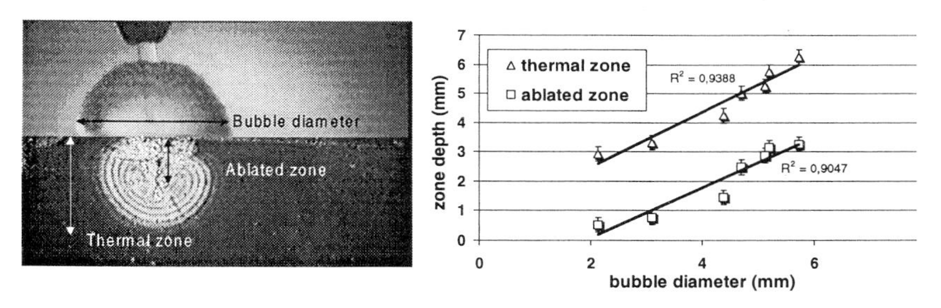

Fig. 3. Composition of a fast photography and thermal image from the same 'bubble formation'.

as well as the ablated zone in relation to bubble diameter, which is directly related to the pulse energy. The thermal zone was determined after 5 pulses and shows a good correlation with the bubble diameter.

4.2 Thermal effects in relation to irradiation angle

The tissue was irradiated at particular angles and distances. For various angles at close distance to the tissue, the dimensions of the thermal zone were determined as illustrated in Fig.4. The thermal area was characterized by the extent of thermal effect into the depth and in the direction of the fiber (or path of the beam). The results in relation to the angle of irradiation are presented in Fig 4 (right). The asymmetry of the thermal area in the tissue is closely related to the path of the beam irradiating the tissue and is especially pronounced at shallow irradiation angles.

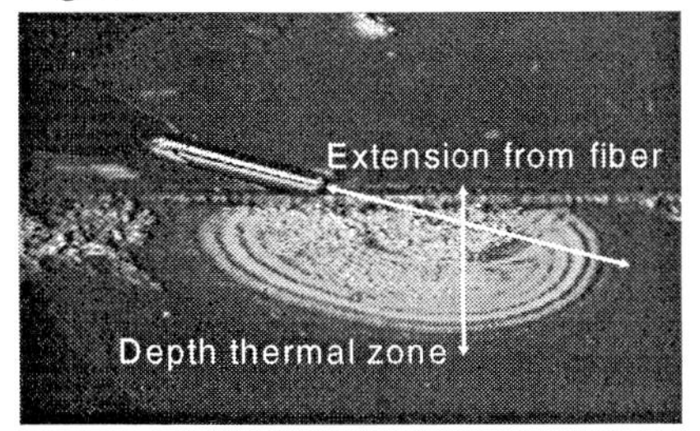

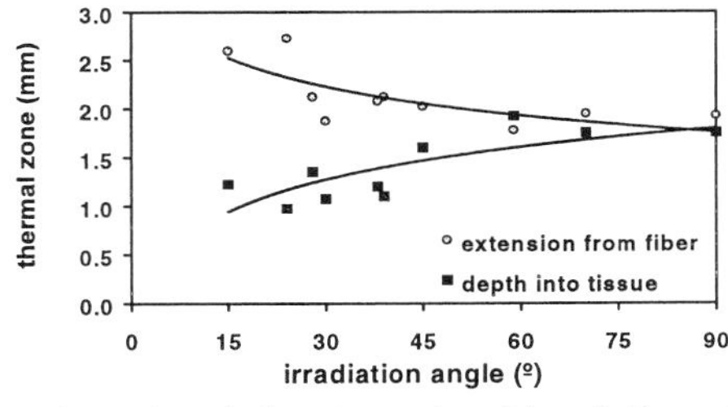

Fig. 4. Dimensions of thermal zone in the tissue in relation to angle of irradiation at close distance to the tissue

4.3 Bubble implosion

Depending on pulse energy, the bubble implodes between 500 to 1000 μs after the start of the laser pulse. During bubble implosion, a distinctive difference was observed between the mechanical effects near hard and soft tissue surfaces. In case of soft tissue, the tissue surface was attracted towards the fiber while in case of hard tissue the bubble detached from the fiber tip and the implosion was concentrated at the hard tissue surface. Soft tissue easily deforms under these pressures and will give in towards the implosion center. If the elastic properties are high enough the tissue will bounce back into its original shape after the implosion. However, soft tissues with a weaker structure will fractured into small pieces.

5 Discussion

There are many parameters that contribute to the resulting thermal and mechanical effects in tissue and that can be both beneficial and adverse at the same time. To provide the surgeon with a rule of thumb what tissue effects to expect for a specific parameter, the results of this study are summarized in the simple graphical representation of the bubble in Fig.5. The tissue effects are divided in mechanical and thermal.

5.1 Thermal tissue effects (Fig.5, right)

Direct irradiation area. The central area within the bubble represents the area directly vaporized by the laser beam. Although the first 500 μm layer will be vaporized at first instance, the vapor will create an opening to more distal layers during the laser pulse.

Condensation area. During and after implosion of the bubble, condensation heat will be dissipated into the liquid or tissue layer that was in direct contact will the bubble surface. This layer will at first be near 100°C with a steep temperature gradient and consequently, due to thermal conduction to surrounding layers, cool rapidly to ambient temperatures effecting tissue up to a few millimeters.

Heat transfer area. If multiple pulses are given heat is conducted to deeper layers. An equilibrium is reached depending on repetition rate and pulse energy. For 5 Hz, dimensions of this area are comparable with the original bubble size.

5.2 Mechanical effects (fig.5, left)

Vaporized tissue area The central area within the bubble represents the laser beam emitted from the fiber. Due to mechanical resistance of the tissue the mechanical effect extends to only about half the bubble diameter as shown Fig. 3

Vapor rupture area Alongside and in front of the vaporized tissue area, tissue is ruptured due to the explosive expansion of the vapor. Depending on the mechanical properties of the tissue, the vapor will find its way of lowest resistance Typically, this area reaches about twice the dimensions of the vaporized tissue area.

Implosion ruptured are. Soft tissue at the boundary of the bubble will be sucked towards the fiber tip during implosion and parts will detach from the underlying structure. During implosion near a hard surface, there is no inflow of liquid from the hard surface, therefore implosion is directed towards the hard surface

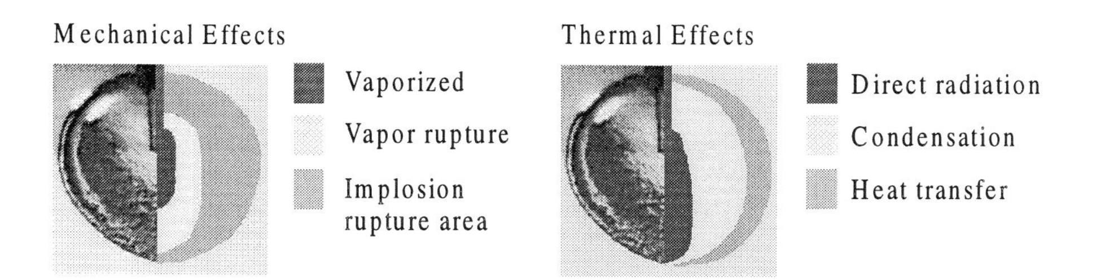

Fig. 5 Thermal and mechanical effective areas in relation to the original bubble shape.

6 Conclusion

There is a strong correlation between the path length of the free beam within the bubble and the degree of mechanical and thermal damage in the tissue directly irradiated by this beam

Knowing the dimensions of the vapor bubble, the surgeon is able to control the mechanical and thermal effects in tissue in relation to the distance and angle of the fiber with the tissue for safe and optimal application of the Holmium laser

References

1. Leeuwen AGJM v, Veen MJ vd, Verdaasdonk RM, Borst C: Non-contact tissue ablation by holmium:YSGG laser pulses in blood. *Las.Surg Med* 1991;11:26-34.
2. Jansen ED, Asshauer T, Frenz M, Motamedi M, Delacretaz G, Welch AJ: Effect of pulse duration on bubble formation and laser-induced pressure waves during holmium laser ablation. *Las.Surg.Med* 1996;18:278-293.
3. Leeuwen AGJM van, Veen MJ vd, Verdaasdonk RM, Borst C: Tissue ablation by holmium:YSGG laser pulses through saline and blood, in Jacques SL (ed): *Laser-tissue interaction II.* Bellingham, SPIE Vol 1427, 1991, pp 214-219
4. Verdaasdonk RM: Imaging laser induced thermal fields and effects, in Jacques SL (ed): *Laser-Tissue interaction VI.* Bellingham, SPIE Vol 2391, 1995, pp 165-175.

Intra-operative Fluorescence Imaging of ALA-induced PpIX using the Double Ratio Imaging Technique in Malignant Brain Tumours

Armand J. L. Jongen[1], Ricardo E. Feller[2], John G. Wolbers[2], and Henricus J. C. M. Sterenborg[3]

[1] Laser Centre, Academic Medical Centre, Meibergdreef 9, 1105 AZ, Amsterdam, The Netherlands
[2] Department of Neurosurgery, VU Hospital, Amsterdam, The Netherlands
[3] Department of Clinical Physics, Daniel den Hoed Cancer Centre, University Hospital Rotterdam, Rotterdam, The Netherlands

Abstract. Contrast in fluorescence signals measured in vivo from tumour-selective photosensitizers, is blurred by spatial variations in optical properties of the tissue, its intrinsic fluorescence and geometrical parameters of the fluorescence measurement setup. A method using two wavelength excitation and two wavelength regions detection, the Double Ratio technique, corrects the measured signals for the influence of variations in absorption, scattering and setup geometry, leaving a result depending only on the photosensitizer concentration. A fluorescence imaging system based on this technique, was used intra-operatively during removal of malignant brain tumours from patients that received ALA pre-operatively. Comparison of preliminary imaging results with fibre spectroscopic measurements and histological results, show that the Double Ratio fluorescence imaging system can reveal differences in ALA-induced Protoporphyrin IX concentrations, where normal fluorescence imaging fails.

1 Introduction

In the framework of the treatment of malignant brain tumours, the aid of an independent measurement method, that can distinguish between normal and malignant brain tissue during operation, is of great importance. A method that is capable of doing this will enhance the surgical reduction of tumour load and successively enhance the results of adjuvant therapies (e.g. radiotherapy and/or chemotherapy).

Fluorescence imaging of a tumour localizer concentration difference to distinguish between normal and malignant tissue is a potential method for this improvement. A clear contrast in measured fluorescence signals from malignant and normal tissue, corresponding with differences in tumour localizer concentrations, is of great importance for the clinical use of this method. Absorption and scattering of light in tissue however, cause complications in the detection of generated fluorescence light.

The Double Ratio (DR) method, explained in the next section, corrects for these complications [1]. A specially developed fluorescence imaging system based on this method, was used intra-operatively to measure concentration differences of Aminolevulinic Acid (ALA) induced Protoporphyrin IX (PpIX) after surgical removal of the tumour bulk in patients with malignant brain tumours.

2 Double Ratio Fluorescence Technique

The DR method uses two excitation wavelengths and two detection wavelength regions (see Fig. 1). The excitation wavelengths are chosen to have a different quantum efficiency for photosensitizer fluorescence. This is equal to choosing 405 and 435 nm when using PpIX. Detection wavelength regions are chosen around a peak of the photosensitizer fluorescence (670 nm for PpIX) and around a reference wavelength where there is mostly intrinsic tissue fluorescence (550 nm). For each excitation wavelength the ratio between the two

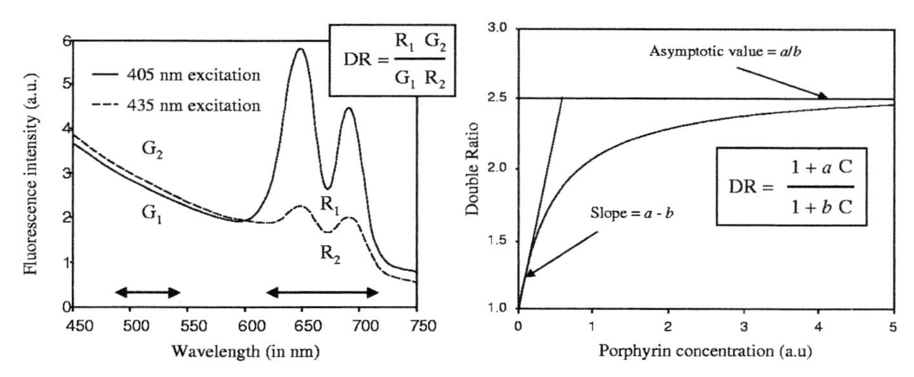

Fig. 1. Schematic representation of the DR method. On the left-hand side the emmision spectra for PpIX excited at 405 and 435 nm are given. The difference in quantum efficiency for fluorescence is clearly seen. The arrows indicate the wavelength region where fluorescence is measured. They are chosen around a peak of the PpIX fluorescence and around a reference wavelength where there is mostly intrinsic tissue fluorescence. The dependance of the DR value on the PpIX concentration is shown in the figure on the right-hand side. A value of one corresponds with zero porphyrin concentration

measured emission intensities is calculated. Hereby variations in excitation light intensity and in setup geometry are corrected for [2]. The DR value is obtained by taking the ratio of the two single excitation wavelength ratios. As the influence of absorption and scattering on the fluorescence is equal for both single excitation wavelength ratios, the DR value is independent of these parameters. In theory the DR method gives a result which, in a nonlinear way, is dependent on the concentration of the photosensitizer. Moreover, it is independent of variations in optical properties of tissue and geometrical parameters in the setup.

The nonlinear relation between the DR-value and the concentration of photosensitizer is given by

$$DR = \frac{1 + aC}{1 + bC} \tag{1}$$

where C is the concentration photosensitizer, and a, b are parameters that are unknown, but do not depend on the optical properties of the tissue nor on the geometry of the measuring setup [1]. On the right-hand side of Fig. 1, the relation in (1) is plotted. From this plot we can see that the DR value is highly sensitive for low porphyrin concentrations, making it an ideal technique to use in diagnostic applications.

3 Preliminary results

A specially developed fluorescence imaging system based on this technique was used intra-operatively during removal of malignant brain tumours of five patients. Three hours before the induction of anesthesia, the patients were

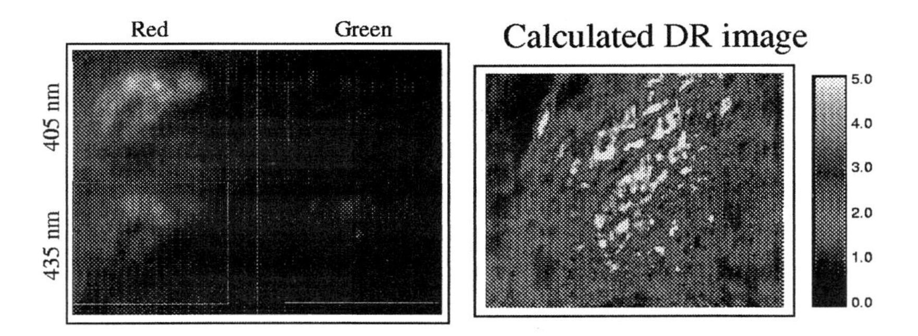

Fig. 2. Example of a measured fluorescence image (left) using the DR fluorescence imaging setup, which projects four different fluorescence images on the CCD-chip of an image intensified camera (Philips IP800) simultaneously. The right image represents the calculated DR image. An increased DR-value is represented by a lighter colour as indicated by the colorbar

administered 10 mg 5–ALA per kg body weight orally, inducing an increased PpIX concentration in malignant tissue. After initial debulking of the tumour by the surgeon, the DR-fluorescence imaging system was used to detect spots with an increased PpIX concentration. Care was taken to use sufficiently low excitation powers (80-120 μWcm^{-2}) in order to prevent the induction of unintended photodynamic damage. Biopsies were taken at the places with increased DR-value, indicating an increased PpIX concentration, and neuropathologically examined. An example of a measured fluorescence image and the calculated DR-image is shown in Fig. 2.

In the red fluorescence image, excited at 405 nm, a bright spot can be seen. Because the DR-value at this spot is also increased (3.3±1.7) (see Fig. 3), indicating an increased PpIX concentration, a biopsy was taken and neuropathologically examined. It was proven to be malignant brain tissue throughout the complete biopsy, meaning there would still be malignant tissue left behind. After biopsy the same site was measured, revealing an increased DR-value of 4.7±1.2, confirming this result. Additional fibre spectroscopic measurements showed that there was indeed an increased PpIX concentration at the investigated site after biopsy (see Fig. 3), although the red fluorescence image clearly showed a decrease. As the tumour was located right on top of the motoric centre of the brain, the surgeon decided to refrain from further treatment.

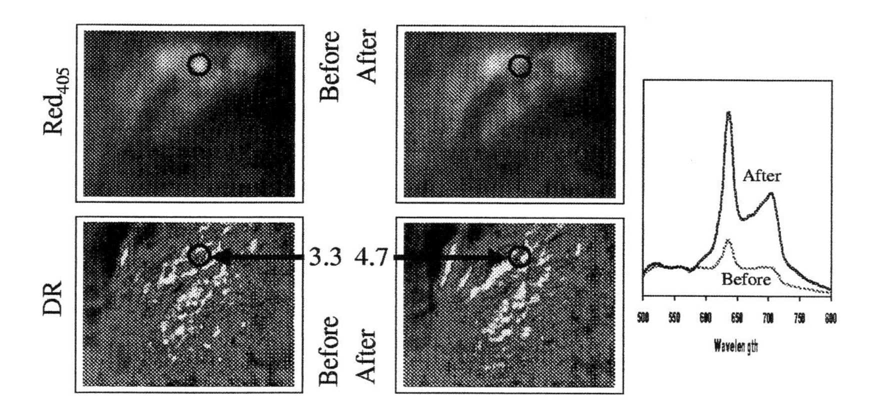

Fig. 3. See text for explanation (due to limitation of pages)

Acknowledgements

This research is supported by the Dutch Technology Foundation (STW), project number AGN 44.3414.

References

1. Sinaasappel M., Sterenborg, H. J. C. M. (1993) Quantification of the hematoporphyrin derivative by fluorescence measurements using dual-wavelength excitation and dual-wavelength detection. Applied Optics **32**, 541–548
2. Profio, A. E. (1984) Laser Excited Fluorescence of Hematoporphyrin Derivative for Diagnosis of Cancer. IEEE Journal of Quantum Electronics, Vol. QE-20, **12**, 1502–1507

Photon Interactions with Ceramic Bone Implants

St. Szarska, K.Sarnowska
Institute of Physics, Wroclaw University of Technology, 50-370 Wroclaw, Wyb. Wyspianskiego 27, Poland
E-mail: szarska@rainbow.if.pwr.wroc.pl; Fax: +48-71-229 696

Abstract: Bioglass(r) and HAP (hydroxyapatite) are important modern materials, which are applied in medicine to reduce disability and thus to improve the level of human life. These materials must first of all be compatible with human tissue. The influence of UV radiation and different simulated physiological solutions on the surface of these materials is presented in this paper. The applied measurement techniques were absorption spectra, SEM and optical stimulated exoelectron emission method. Before measurements the glass samples were submitted to the action of three sorts of simulated physiological solution. The results of these measurements are discussed in terms of point defects in the structure.

1 Introduction

Biomaterials play a very important role in tissue engineering, which is aimed to provide replacement for tissue and organs which have been damage or lost as a consequence of disease, aging or accident. A part of biomaterials is designed for replacing bone and inducing bone formation. Cumulative research in field of bone repair, particularly bioceramics, has introduced concepts, which relate the bioactivity of a material to its calcification properties [1]. A surface, which can induce the formation of an apatite layer in vivo, or is pre-coated with an apatite layer, will demonstrate good bone-bonding properties. Ceramics are made of hydroxyapatite (HAP) and bioactive glasses (BG) are successfully applied in surgical orthopedics and dentistry, mainly for filling of osteal defects and bone heightening. The nature and mechanism of osteogenesis - the process of bone tissue formation in HAP or BG in the biological environment resulting in physicochemical integration of ceramics with living bone - are not still clear. Existing model for this phenomenon is based on dissolution - precipitation reactions proceeding on surfaces of ceramics in vivo. In case of bioactive glasses exposed to body fluids, undergo corrosion with leaching of alkali ions resulting in the formation of a silica gel and a calcium phosphate layer will recrystallize into hydroxycarbonate apatite. In the process, the dissolution rate of the material depends on its phase structure [2]. Bone-bonding properties of bioactive glasses are based on the formation of this layer. Compared to synthetic HAP, the surface layer of these glasses is more similar, in terms of crystallinity, to the apatite of bone tissue and consequently, a greater bone bonding has been reported for BG than for HAP [3]. These materials are attractive because of its bioactive properties, but have extremely poor mechanical properties and may provide little strength reinforcement in particulate form. Indeed, these bioactive materials use some type of composition structure to achieve particular combination of properties. Modern implant alloys are chosen both for their mechanical properties and high degree of biocompatibility. Bioglass fibers have recently been promoted as possible reinforcement materials [4], HAP with particulate zirconia [5,6] or plasma spraying of HAP coating on mechanically superior materials of ceramics and metal [7]. The use of composite materials provides many new options and possibilities in implant design, but it also requires a better understanding of the objectives and limitations involved [8,9]. In most of the cases, the surface properties of the BG and HAP play an important role in its application, as their surfaces are in direct contact with the environment when in use. Therefore, to control or to manipulate the surface properties of them is of great importance.

Valuable information has been gained from the analysis of osteogenesis by in vitro method. The process of bone tissue formation in HAP or BG in body solution is highly influenced by a number of factors, which are not taken into account by existing models. These factors are the defective structure, (vacancies, and dislocations), the charge and its distribution, photon interactions that cause charge transfer. BG and HAP were tested in tris buffer solution or simulated body fluid for investigation of various aspects such as bioactivity, durability, quality assurance [10]. If it could to application these materials to new composite structures it must be study also another properties. Recently, an attention in present work has also been paid to the optical and electrical properties of these materials after the action of tree kind of body fluids.

2 Materials and Methods

The subject of investigations were two types of Bioglasses(produced by Jelenia Góra Optical Factory according to Hench [11] with the following, slightly differentiates composition (percentage by weight):

BG I (45S5) 45 SiO_2, 24,5 Na_2O, 24,5 CaO, 6 P_2O_5

BG II 46 SiO_2, 25,5 Na_2O, 24,5 CaO, 4 P_2O_5

The second investigated material was hydroxyapatite: calcium phosphate with this follow chemical formula: $Ca_{10}(PO_4)_6(OH)_2$ in which weight % composition: CaO -55,8; P_2O_5 - 42,45; H_2O - 1,8. The hydroxyapatite powder (HA BIOCER PR) with 0,06 mm grains mixed with three kinds of solutions. Obtained paste was solidification by UV light source for 4 hour.

The samples were submitted to treatment of the following simulated physical solutions: 0,9%NaCl (A) 0,9%NaCl buffered (B) phosphate buffer (C) SBF (Simulated Body Fluid)(D)[12] The SBF contains inorganic ions in concentrations close to those in blood plasma and is buffered with tris.

During measurements of transmission spectra we used spectrophotometer Specord UV VIS.

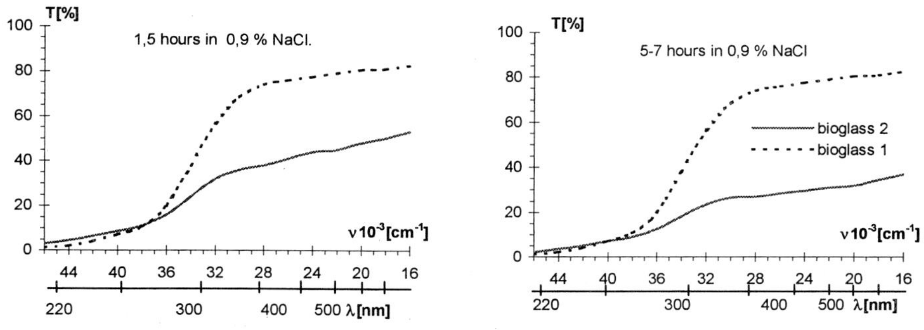

Fig .1a). Transmittance spectra for BGI and BGII; dissolution in 0,9%NaCl

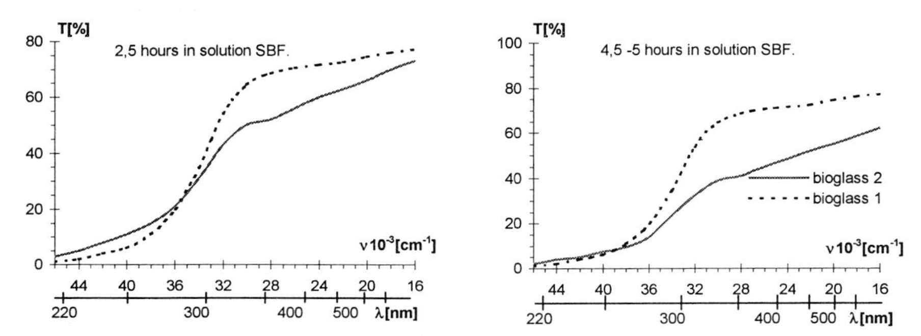

Fig .1b). Transmittance spectra for BGI and BGII; dissolution in SBF.

The OSEE current was registered by a secondary electron multiplier (10-18 A) in vacuum chamber with 10-4 Pa. Measurements of optical-stimulated kinetics were carried out with an optical stimulation using 6 interference filters giving the wavelengths ranging from 225 to 325 nm. As the parameters characterizing the OSEE decay curves was chosen I_0 intensity of slowly decaying component normalize to the same quantity of the quanta of stimulating light. This method has been described elsewhere [13]. The number of quanta, falling on the glass surface was constant for these measurements.

3 Results and Discussion

The investigation of influence of SBF on surface in BGI reported by many authors [12-15].The transmission spectra for BGI and BGII treated by different solution shows on Fig.1 On Fig.1a and b there could compare the influence on dissolution time of 0,9%NaCl and composition of glass on transmission spectra. For the Bioglass composition the shape of transmission isn't changing in this earlier stages of surface dissolution. In case BGII, there could observe the big difference of the value of transmittance (60%) before virgin and 1,5 hour dissolution sample. This difference increase with the time of dissolution. The influence of SBF dissolution on transmittance spectra is smaller than 0,9%NaCl,but the dependence on time

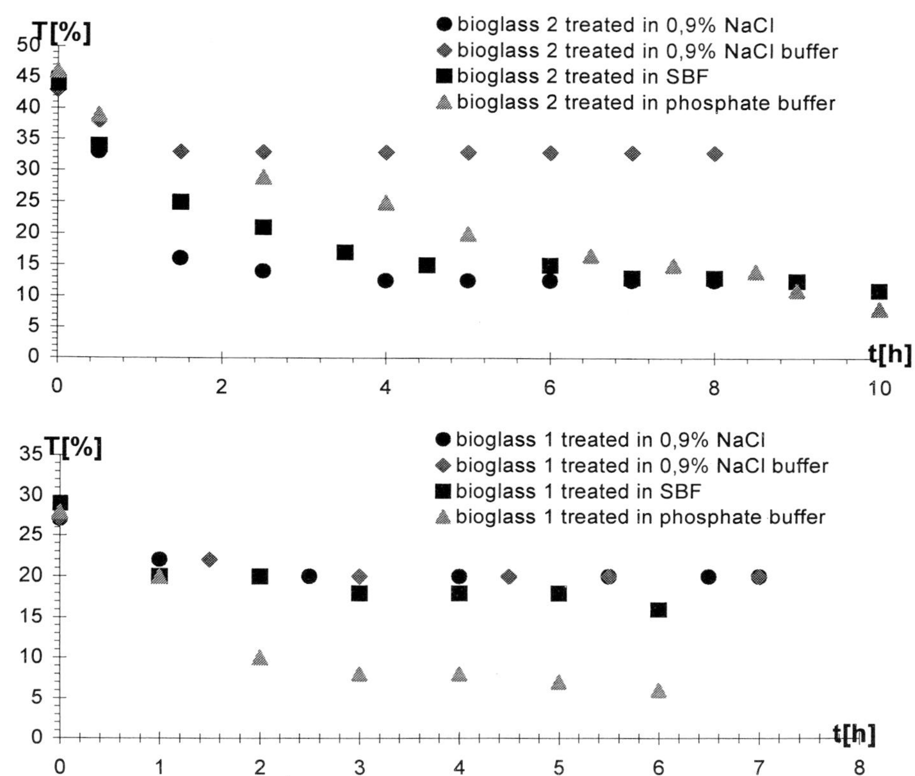

Fig 2 Time dependence of 275mm transmission change of BGI and BGII treated of different solutions.

dissolution for BGII is clear. The dissolution time dependence have been observed on Fig.2.

Fig.2a concerned the transmittances change for 275 nm wavelength on dissolution time for BGII and Fig.2b for BGI. Comparison shapes transmission curve for both glasses, it had been stated that the best influence on surface of BGI Bioglass had phosphate buffer, but for BGII 0,9%NaCl solution. Rapid decrease of transmittance value for the first hours of the solution interaction for both the glass composition indicate on most intensive changes in surface structure of these glasses in this time. Except the impurities, lattice defects, cracks, the analyses of experimental data in a number of glassy materials support that localized states for electronic and ionic excitations arise owing to the existence of some disorder in the perfect structure of the solid. From Truchin's [16,17] investigations for sodium silicate glasses suggests that the intrinsic absorption tail is formed by localized states and the mobility edge for the colour center formation is somewhere near 6 eV. He proposed model of L-centers a complex of the Na^+ - O^- - Si = type, capable of both becoming ionized and capturing an electron.From the former OSEE investigations for the Bioglasses results that the mean value of the emission intensity is reproducible, but the emission distribution along the sample exhibits distinct fluctuations. It points out for existence of the changing in time electric charge on the surface which source are different centers. The OSEE curves for these glasses treated by different body solutions was described earlier. The shape of these curves is typical decay curve of intensity, characteristic for many dielectrics.[18] In Fig.3 is presented a change in time of intensity of the optically stimulated exoelectron emission for HAP treated by SBF in different wavelength of stimulation light. These curves show an initial increase in emission and then a slow decay. The area under the curve presenting the changes of electron emission intensity in time is proportional to the surface charge gathered on the sample. The source of electrons are chemical interaction between the physiological solution and HAP, which leads to the local deformation, microfractures or breaking of the bonds which can cause emission of the charge. It has been proved that the SBF is related to the charges of particles transferring to the material. Electron transfer from their occupied valence band into the free states of the material surface and cause the decomposition of the HA.

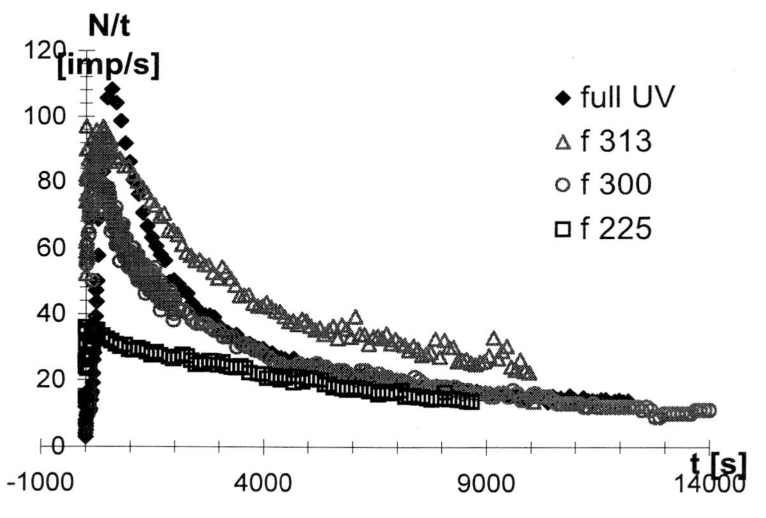

Fig 3. Dependence of OSEE intensity on time for HAP in wavelength of stimulated light function.

4 Conclusions

Investigations of electromagnetic radiation absorption confirm the SEM conclusions about influence of solutions on the surface. In this case the braking - down of the surface in result of action of the physiological solution reveals by decrease in transparency caused probably by the light scattering on the strongly defected surface of the Bioglass, especially BGII. Basing on results of these investigations we suppose that the changes of the structure of the surface layer occur the most intensively in the first hours of keeping the implant in the solution and depend in a high degree on the kind of this solution.As the investigations have been carried out for two kinds of the Bioglass differing with the chemical composition, we could show that bioactive properties depend on the chemical composition of glasses, because the same solution in another degree influences the surface of BGI than the surface of BGII. Process of gathering of surface charge is not uniformly distributed on whole surface it does not mean that in the case of the Bioglass it is not present.

References

[1] L. L. Hench Summary in future directions in: L. L. Hench, J. Wilson An introduction to bioceramics. (Singapore; World Scientific, 1993, 365)

[2] C. Rey, Biomaterials, 11,13(1991)

[3] K. J. J. Pajamaki, T. S. Lindholm, O. H. Anderson, J. Mater. Sci. Matre. Med. 6, 14 (1995)

[4] M. Marcolongo, P. Ducheyne, E. Schepers, J. Garino, Proc .5th World Biomaterials Congress 2, 444 (Toronto,1996)

[5] M. R. Towler, H. K. Varma, S. M. Best, W. Bonfield, Proc. 5th World Biomaterials Congress 1, 975 (Toronto, 1996)

[6] W. Bonfield, J. Biomed. Engng. 10, 522 (1988)

[7] P. Cheang, K. A. Kohr, Biomaterials, 17, 537 (1996)

[8] S. L. Evans, P. J. Gregson, Biomaterials 19, 1329 (1998)

[9] R. Labella, M.Braden, S. Deb Biomaterials 15, 1197, (1994)

[10] O. H. Andersson, K. Vahatalo, A. Yli-Urpo, R. P. Happonen, K. H. Karlsson Bioceramics 7, 67 (1994)

[11] L. L. Hench, R. J. Splinter, W. C. Allen, T. K. Greenlee J. Biomed. Mater. Res. 2, 117 (1971)

[12] T. Kokubo, J. Non-Cryst. Sol. 120, 138 (1990)

[13] St. Szarska in Biomaterials and Human Body, (eds. A. Ravaglioli, A. Krajewski) pp.396 (ElsevierP.C.,Amsterdam,1992)

[14] M. Ciecinska, L. Stoch, A. Duda Ceramics 49, 109 (1996)

[15] L. L. Hench, J. Wilson in: Silicon Biochemistry, pp. 231(Wiley, Chester, 1986)

[16] A. N. Trukhin, M. N. Tolstoi, L. B. Glebov, V. L. Savelev Phys. Stat. Sol. (b) 99, 155 (1980)

[17] A. N. Trukhin Ceramics 49, 7 (1996)

[18] St. Szarska Ceramics 49, 95 (1996)

Easy Human Visual MTF Measurements

Naoyoshi Nameda[1], Masato Masuya[1], Yuji Shimizu[2], and Eiichi Hatae[2]

[1] Faculty of Engineering, Kagoshima University
[2] Matsushita Electric Industrial Co., Ltd.

1 Introduction

This investigation concerns measurement of the human visual MTF properties using a computer display. At the beginning of the 21st century, 30 % of the Japanese population will be over 50 years old. One of the countermeasures for visual degradation is the improvement of displayed pictures. The human visual MTF properties are fundamental data for this improvement. In this paper, easy MTF measurement is discussed in comparison with precise MTF measurement. In the framework of this study, subjects were distinguished in the following way: people over fifty years old are referred to as "aged" and those around twenty years old are called "young".

2 MTF Measurements and Discussion

Fig. 1 shows the visual MTF properties obtained by both easy and precise MTF measurements.

In the precise measurements, the spatial frequency and contrast were varied for each trial. Many trials were implemented until the threshold contrast

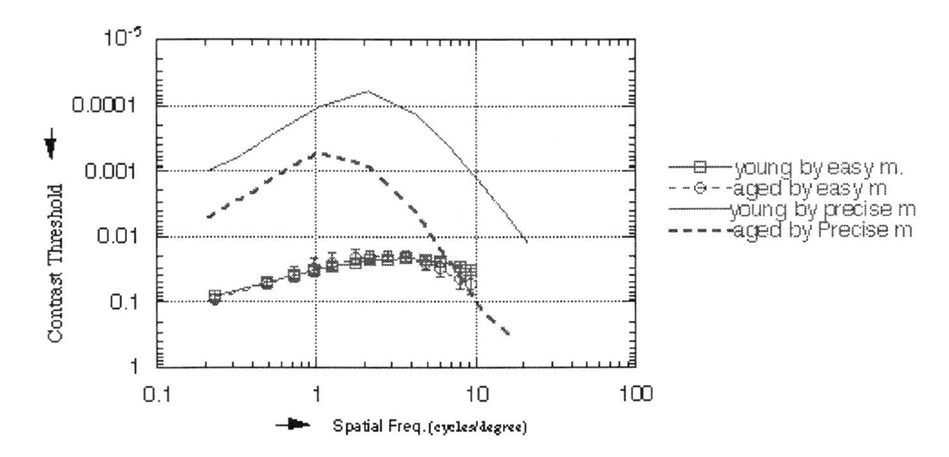

Fig. 1. Results of easy and precise measurement.

value was obtained [1]. In order to improve this defect, the easy MTF measurement was proposed that this measurement use only one pattern. This pattern has been known in the publications already [2,3]. This grating pattern consisted of horizontally varied spatial frequencies from 0.23 to 18.18 cycles/deg. and vertically varied contrast values from 0 to 1.0 as shown in Fig.2.

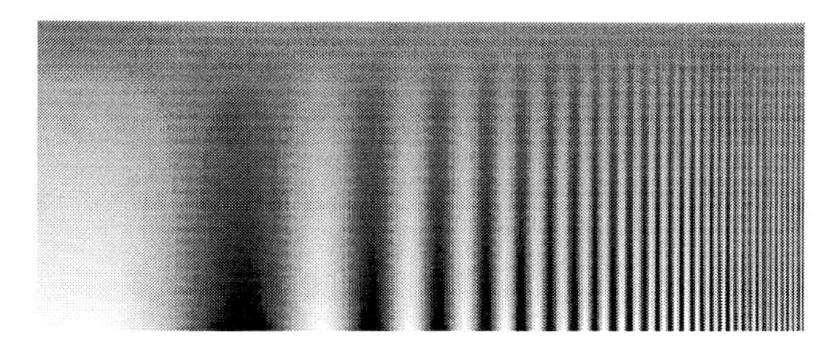

Fig. 2. MTF pattern for measurements using the easy method.

In this measurement, the subject drew a threshold line on the pattern, which was displayed on a computer screen, using a mouse. The data obtained are shown in Fig. 1. However, the sensitivity value is extremely low and the curvature is rather tranquil and shifted to higher spatial frequencies.

The reasons for the latter evidence seem to be that the subject observed several different spatial frequency gratings simultaneously in his vision in the reverse of the writing position. Taking this into consideration, the sensitivity was calculated by averaging the sensitivities of the precise properties contained within five degrees from the writing position (Fig. 3).

Next, the low sensitivity evidence was considered so that the threshold value became inaccurate because the lower contrast area was too narrow. Thus, the pattern was changed and the contrast was varied from 0 to 0.3 (Fig. 4).

The measurement results are shown in Fig. 3.

3 An Example of Application of the Visual MTF Properties

A complicated Chinese character printed on a gray background is difficult to read for many aged people. Readability can be improved by enhancing the character with an appropriate pattern filter. In this experiment, the enhancement filter utilized the difference between the MTF properties for "young" and the MTF properties for "aged". Fig. 5 shows the enhancement filter and the enhanced Chinese character.

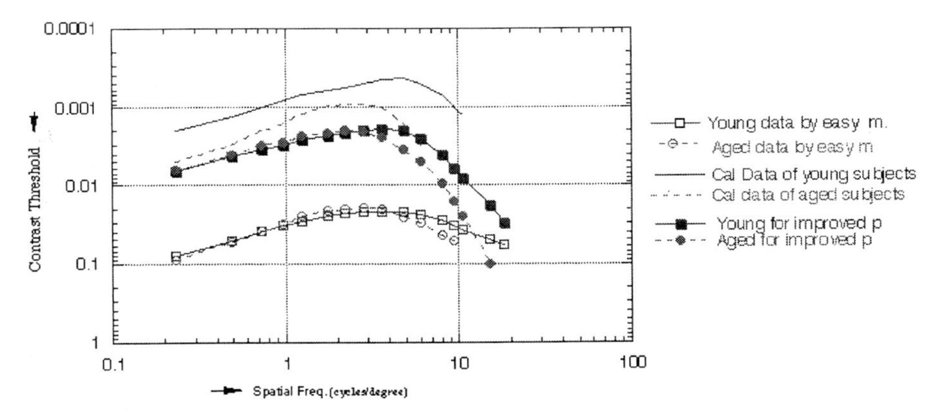

Fig. 3. Calculated data from the precise measurement data and measurement data for the improved patterns. They are compared with the easy measurements data.

Fig. 4. Improved MTF measurement pattern.

Moreover, the more desired speculation is with people of even greater age. The most troublesome eyesight degradation occurs in those over 70 or 80 years of age. Many such aged people suffer a variety of eye disease. The visual MTF properties for such people should present differences from the above mentioned aged properties, which were obtained from subjects with healthy vision [4,5] . For extremely aged people, the properties should be decreased around 1 cycle/deg. where the highest sensitivity exists. Thus, there are various visual properties. Therefore, many more measurements with subjects of advanced age must be done in the near future.

4 Conclusion

The easy visual MTF measurement is less accurate than the precise measurement. However, for classification of the properties to some standard, the easy visual MTF measurement is useful.

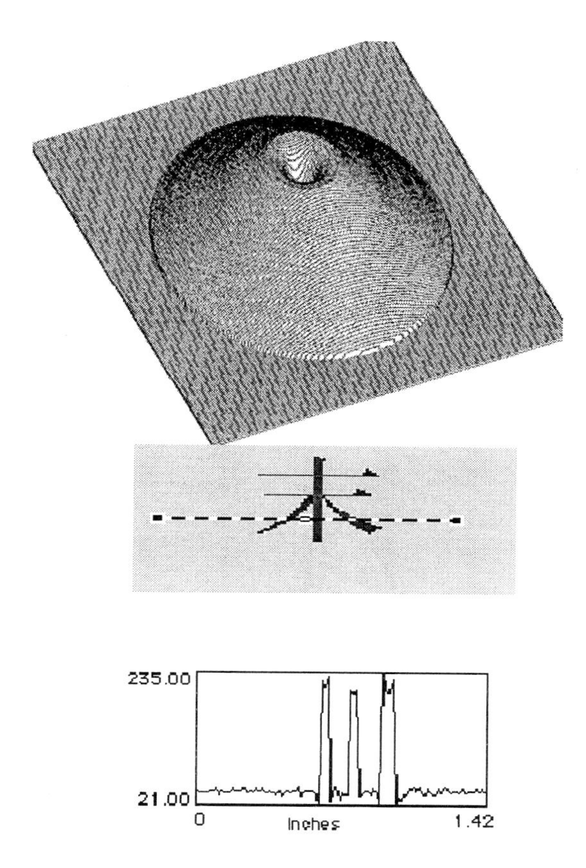

Fig. 5. Enhanced Chinese character: The upper figure is the enhancement filter; the middle figure is the Chinese character indicating cross section; and the lower figure is a cross–sectional figure of the density.

References

1. Ohyama et al ed: Sensation and Perception Handbook, Seisinn Shobou Co., Ltd. (1994) pp. 20 (in Japanese)
2. L.M.Proenza, J.M.Enoch and A.Jampolsky: Clinical applications visual psychophysics, Cambridge University Press, (1981)
3. Watanabe et al: Visual Science, Shashinn Kougyou Shuppannsha Co., Ltd., (1981) pp. 77 (in Japanese)
4. T. Kawara and H.Ohzu: Modulation transfer function of human visual system, Oyobuturi, Vol. 46, (1977), pp. 128-138 (in Japanese)
5. N.Nameda, T.Kawara and H.Ohzu: Human visual spatio–temporal frequency performance as a function of age, Optometry and Vision Science, Vol. 66, (1989) pp. 760-765

Measurement of the Anterior Segment of the Eye Using an Improved Slit Lamp

L.H.Liu and H.Ohzu
Waseda University, 3-4-1,Okubo, Sinjuku-ku, Tokyo, 169-0072, Japan
Tel: +81-3-5286-3225. Fax: +81-3-3200-2567
E-mail: 69615294@mse.waseda.ac.jp

Abstract An improved slit lamp that is composed of two light sources, one of which is slit beam and the other is parallel thin weak laser beams has been proposed and used in measuring the anterior segment of an extracted eye. The position and curvatures of the anterior and posterior surfaces of the cornea and lens are measured by analyzing the Scheimpflug photographs of the anterior segment eye as slit beam incident. The distribution of refractive index within the crystalline lens can be estimated from analyzing the Scheimpflug photographs of bent paths of laser beams passing crystalline lens being taken as thin weak laser beams incident. It is shown that this improved slit lamp may become a useful optical instrument for one who want to get the detail parameters of the anterior segment eye *in vivo*.

1 Introduction

Scheimpflug photography [1] provides an image of a sagittal section of the eye. For measuring the curvature of the posterior surface of cornea and the anterior-posterior surface of lens, it is necessary for both the Scheimpflug distortion and the refractive distortion on Scheimpflug photographs of the eye to be modified, because of the existence of several different refracting surfaces in the eye.

Pablo Lapuerta *et al* [2] constructed four-surface schematic eye of Macaque monkey by analyzing the Scheimpflug photography and checked it by keratometry and ultrasound. However, they did not consider the index distribution of the crystalline lens and they described the eye as a series of four spherical surfaces. In this study, an improved slit lamp that is composed of both slit beam and thin laser beams is proposed. Scheimpflug images of multi thin laser beams passing the anterior segment eye are used in investigating the gradient refractive index distribution of the crystalline lens. Scheimpflug photography of the eye at the direction of camera optical axis is to be used in modifying the refractive distortion on the Scheimpflug photography of the eye at the direction of visual axis.

2 Experiment

The species studied is pig. Eyes are obtained from Tokyo Shibaura slaughterhouses. The eyes are transported to the laboratory and measurements commence 70-80 min after enucleating.

Case1.When the slit beam is switched on:

Full focused pictures of the cross section of the anterior segment of the eye can be taken by using Scheimpflug principle. The curvature radius of the anterior surface of cornea at the direction of visual axis can be calculated by modifying the Scheimpflug distortion only. But the posterior surface of the cornea on Scheimpflug image taken at the direction of visual axis is included two kind of distortion, one is caused by Scheimpflug angle, the other is caused by the refraction of the anterior surface at the direction of the camera optical axis. Because of irregularity of cornea's shape, to modify this distortion, it is necessary to measure the curvature radius of the anterior surface at the direction of the camera optical axis.

Case2.when the laser is switched on.

The multi-thin weak laser beam incident on the eye at the direction of visual axis in a sagittal plane. Photography of the bent paths of laser beams passing lens is taken by using Scheimpflug principle. These paths is not the true one because of the distortion of the refractive surfaces and Scheimpflug angle. In order to modify the distortion caused by refractive surfaces, The curvatures of cornea and lens at the camera direction measured in (case 1) are used. The index distribution of crystalline lens is able to be estimated by comparing the modified true paths of laser beam with some ray tracing models of the eye[4].

3 Results

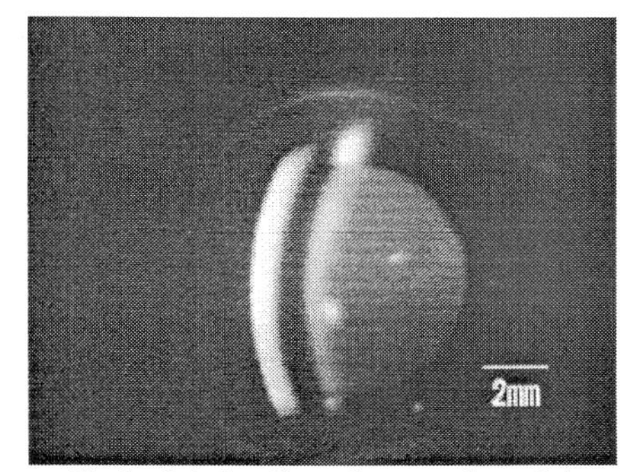

Fig.1. Scheimpflug photography of the anterior segment eye when slit beam incident on the eye at the direction of visual axis

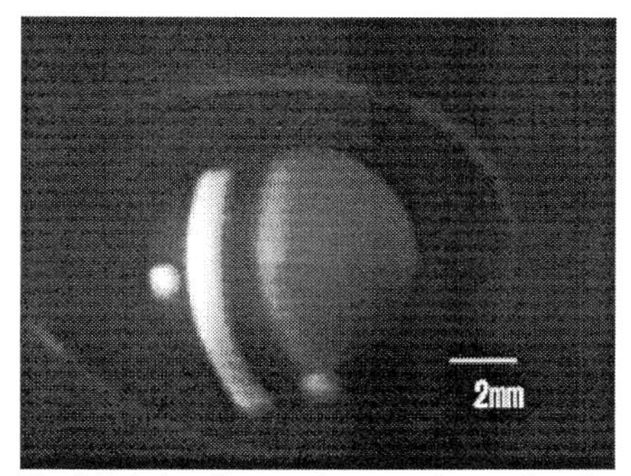

Fig.2.Scheimpflug photography of the anterior segment eye when slit beam incident on the eye at the direction of camera optical axis.

4 Discussion

The refractive index distribution of crystalline lens can be estimated either by calculating the bent paths of laser beams passing crystalline lens directly[3] or by comparing with the ray tracing results of the multi-layer lens models[4]. In this research, the second method is used. It may be less accurate than the first method because of the difficulty in measuring the posterior surface of lens.

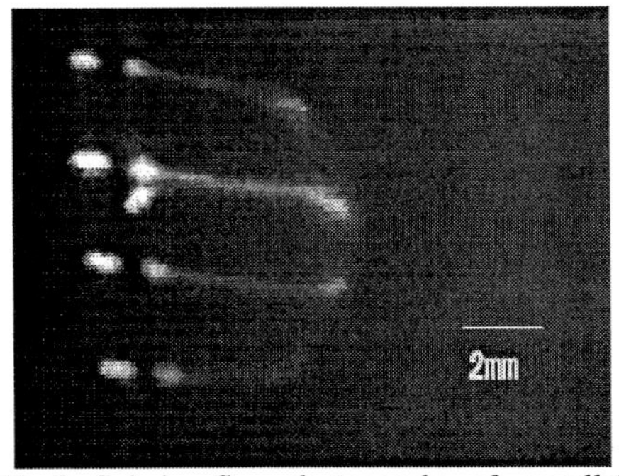

Fig.3. Scheimpflug photography of parallel laser beams passing the anterior segment eye

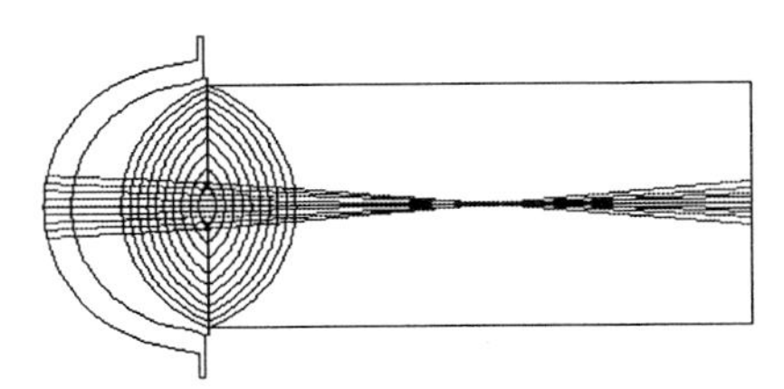

Fig.4. Model of the anterior segment eye.

Using the anterior surface measured at the direction of camera optical axis modifies the refractive distortion of posterior surface of cornea at the direction of visual axis. The refractive distortion of the posterior surface of cornea at the camera optical axis is modified by using the anterior surface measured at the direction of the visual axis. The shape of the anterior surface of the crystalline lens at the direction of visual axis can be calculated by using the anterior and posterior surface measured at the direction of the camera optical axis by ray tracing. The results measured by the method stated above may be more accurate than the results measured by Pablo Lapuerta *et al* [2], because of the existence of irregularity of cornea's shape.

It is concluded that both the shape of the anterior segment of the eye and the index distribution within crystalline lens can be measured by using the improved slit lamp. It may also be possible for this slit lamp to be used in measuring the anterior segment of the living human eye.

References

1. Scheimpflug, T.: Der Photoperspektograph und seine anwendung. (Photogr. Korr. 43: 516-531,1906)

2. Pablo Lapuerta and Stanley J.Schein, *A four-surface schematic eye of macaque monkey obtained by an optical method.* (Vision Res. Vol.35, No.16, pp.2245-2254,1995)

3. Gleb Beliakov and D.Y.C.Chan, *Analysis of inhomogeneous optical systems by the use of ray tracing..* (Applied Optics, pp.5106, 1 August 1998)

4. L.L.Liu, H.Ohzu, *Index distribution in a crystalline lens of the eye.* (Spie Vol.2778 p.1031. Proceeding of ICO 1996)

Optics in Photomedicine and Photobiology

STM of Light-Sensitive Biological Systems

P.B. Lukins and T. Oates

Department of Physical Optics, School of Physics, A28
University of Sydney, NSW 2006, Australia.
email:lukins@physics.usyd.edu.au

Abstract. Scanning tunneling microscopy of biological systems is generally difficult because of the low conductance, high mobility and complex topography of the specimens together with complications in interpretation of the images obtained. A method is presented for improved STM imaging and interpretation for biological specimens, particularly those containing light-sensitive components. This approach is demonstrated by application to the light-harvesting complex LHCII in photosynthetic membranes. These LHC II particles are imaged and the formation of light-induced delocalised states is characterised.

1 Introduction

Scanning tunneling microscopy (STM) [1] has proved to be a powerful technique for atomic-resolution imaging of the surfaces of conductive materials and has lead to a variety of scanning probe microscopies. However, STM has only rarely [2] been applied successfully to biological specimens because of their low conductance, softness, mobility and complications associated with image interpretation and optimisation of substrate binding properties. For these reasons, the standard constant-current and constant-height mode STM methods have so far been unable to yield meaningful STM images of biological material. Here, we demonstrate a quasi-constant-height mode (QCHM) operation [3,4] and show that the resulting current-height images can be interpreted using scanning tunneling spectroscopy (STS). Light-sensitive biological systems should be able to be studied using STM/STS by directly exploiting the photoconductive or other light-activated processes in these systems to obtain both a contrast mechanism for imaging and a means of probing electronic effects in these systems as they undergo light-induced changes. As an example, we apply the QCHM method to imaging photosystem II (PS II) membrane fragment particles and an investigation of photoexcitation of LHC II which is the most important light-harvesting chorophyll protein associated with PS II. Photosystem II itself is one of the two plant proteins which catalyse the conversion of light energy from the sun into electrochemical energy on which plants, and hence life on earth, depend [5]. In the LHC II / PS II complex, LHC II is involved in light capture and exciton transfer while PS II is involved in exciton capture and energy conversion.

2 Methods

PS II particles were prepared by the method of Berthold et al. [6]. STM and STS were carried out on a modified Park Scientific SPM system, with QCHM operation achieved by hardware and software modifications to enable precise control of STM feedback conditions particularly gain and frequency response. QCHM scanning tends to lead to an enhanced signal on the leading edge at a topographic feature together with a shadow effect on the trailing edge. This effect has proved very useful for edge detection and contrast improvement. Lateral resolutions were 0.02 nm for graphite and 0.3 nm for photosynthetic material while the vertical resolution was 0.01 nm in each case. STS studies involved measurement of the tunneling current I as the tip voltage V was ramped from -1 V to +1 V.

3 Results and Conclusions

Figure 1, which shows an STM image of two protein fragments, demonstrates the ability of the method to image biological material and illustrates how an image can be calibrated using the atomic separation of atoms in the substrate which appears in the background of the image. Biological material exhibits lower resolution than the atomically-resolved substrate because of the lower conductance of the protein and its greater topographic irregularity.

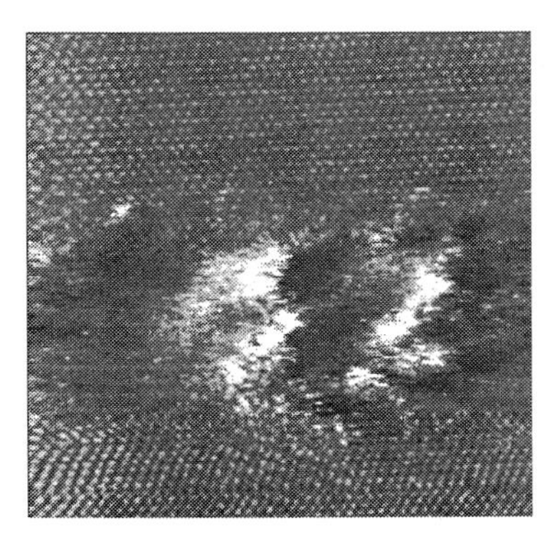

Fig. 1. An STM image of two small protein fragments together with background carbon atoms of the graphite substrate demonstrating the difference in resolution between flat conductive material and partially-conductive topographically-irregular biological material (13.8 x 13.8 nm scan size, 2 nA tunneling current, 0.2 V bias voltage).

Using a similar approach for PS II membrane fragments, we obtain the image in figure 2A which shows LHC II / PS II complexes with a dimeric two-fold symmetric structure and dimensions of 30.1 x 12.5 nm. By comparing these images with the images and subunit structures that we obtained in a recent STM/STS study of PS II core complexes [3], we propose the assignments shown in figure 2B for features seen in the STM images of the LHC II / PS II complex. Of particular importance here is the fact that the structure of the PS II core region in figure 2 closely resembles that for the isolated core complex [3] and that two LHC II units are located at the ends of the LHC II / PS II complex (region 3 in figure 2). Our observed shape and size for the LHC II (circular with diameter of 4.1 nm) [3] is in agreement with the electron crystallographic data of Kuhlbrandt et al. [7].

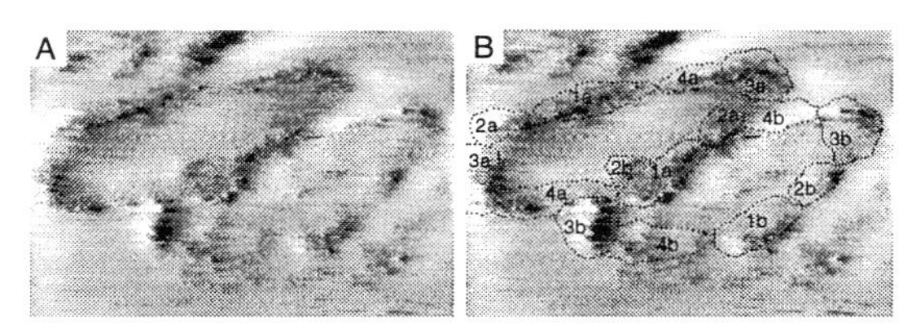

Fig. 2. Structure of PS II membrane fragments. (A) STM image of three fragments showing an anti-parallel two-fold symmetric structure with overall fragment dimensions of 33.5 x 15.9 nm (51 x 36 nm scan size, 1.1 nA tunneling current, 0.14 V bias voltage). (B) Outlines of the major topographic features with symmetrically corresponding regions similarly numbered. The raw dimensions of these features and the subunits to which they belong are: 1, 13 x 5 nm, D1/D2/cytb-559 reaction center complex; 2, 8 x 5.5 nm, CP43/CP47 proximal antennae; 3, circular diameter 4.1 nm, LHC II; 4, 10 x 5 nm, accessory linker proteins.

STS of the LHC II regions yield spectra (figure 3) which differ markedly for dark and white-light illuminated conditions. The spectrum for graphite under similar tunneling conditions is also shown for comparison. The LHC II spectra, which are symmetric about V=0, contain a small linear (ohmic) contribution in addition to the dominant second and third order contributions due to tunneling to delocalised LHC II states as well as hopping conduction and space charge limited current effects. The energy dependences of the density of states in the tunneling region can be calculated from these STS spectra and these are shown in figure 3B. We conclude that the main tunneling mechanism involves tunneling to the delocalised electronic states of the chlorophyll network comprising the LHC II. On photoexcitation, there is (a) a dramatic increase in the tunneling current due to increased conductance of LHC II surface structures in the excited state, (b) an increase in the density of states within the tunneling gap and (c) increased exciton and electron delocalisation in the bulk LHC II structure. We believe that the light-induced enhancement in the density of delocalised states is important for the efficient transfer of excitation energy to the reaction center.

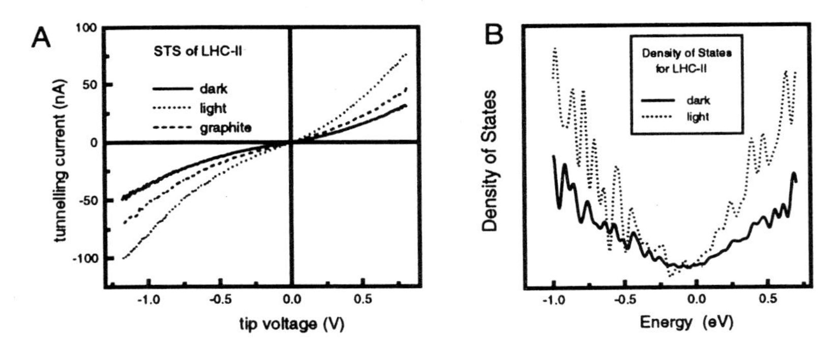

Fig. 3. Tunneling spectra (A) and the energy dependence of the density of states (B) for LHCII.

A quasi-constant height-mode scanning approach can, under a narrow range of conditions, allow STM imaging and STS electronic spectroscopy of certain biological systems including light-sensitive biomolecules. We have used these techniques to image LHC II from higher plants and describe the fundamental tunneling processes that occur during photoexcitation. We thank the Australian Research Council for finnancial support.

References

1. Binnig, G. and Rohrer, H. (1982) Scanning tunneling microscopy. Helv. Phys. Acta 55: 726-735
2. Engel, A. (1991) Biological applications of scanning probe microscopies. Ann. Rev. Biophys. Biophys. Chem. 20: 79-109
3. Lukins, P.B. and Oates, T. (1998) Single-molecule high-resolution structure and electron conduction in photosystem II by scanning tunneling microscopy and spectroscopy. Biochim. Biophys. Acta, in press
4. Lukins, P.B. and Oates, T. Biological STM using quasi-constant height-mode scanning and image interpretation by STS. To be published
5. Hansson, O. and Wydrzynski, T. (1990) Current perceptions of photosystem II. Photosynth. Res. 23: 131-162
6. Berthold, D.A., Babcock, G.T. and Yocum, C.A. (1981) A highly resolved oxygen-evolving photosystem II preparation from spinach thylakoid membranes. FEBS Lett 134: 231-234
7. Kuhlbrandt, W., Wang, D.N., Fujiyoshi, Y. (1994) Atomic model of plant light-harvesting complex by electron crystallography. Nature 367: 614-621

Optical Inspection of Surface Quality of Pharmaceutical Compacts

Kai-Erik Peiponen, Raimo Silvennoinen and Ville Hyvärinen
Department of Physics, University of Joensuu, P.O. Box 111, FIN-80101 Joensuu, Finland
Pasi Raatikainen and Petteri Paronen
Department of Pharmaceutics, University of Kuopio, P.O. Box 1627, FIN-70211 Kuopio, Finland

Abstract. Optical techniques for pharmaceutical compact inspection are presented. Surface statistics of porous compacts was investigated with aid of specular reflectance. Diffractive optical element was applied for surface quality inspection.

1 Introduction

When coherent laser radiation is incident on a porous material, and thereafter diffusely reflected, one usually observes generation of speckle pattern [1]. However, in some cases it is possible to distinguish the specularly reflected ligth component provided that the angle of incidence is high enough. Then it is possible to investigate the surface quality of porous materials by measuring the strength of the specularly reflected light as a function of the angle of incidence of laser beam. It is also possible to get rough estimate for optical surface roughness with the aid of classical theory for surfaces obeying gaussian statistics [2].

Another technique that can also be applied for inspection of porous surfaces[3-6] is based on the image information that can be recorded using a diffractive element based sensor.

One class of porous materials includes pharmaceutical compacts (compressed from powder materials). Often a medicine has a tablet form and its bulk and surface porosities can affect the rate of release and absorption of drug substance. Sometimes coating of a tablet is also needed for instance to control the release of the drug. For the purpose of correct dosing there is a desire to optimize the quality of the tablets.

In this paper we study the information of specularly reflected ligth from pharmaceutical compacts and also exploitation of diffractive element based sensor for surface quality assesment of pharmaceutical compacts.

2 Specular Reflection from Tablet' s Surface

When a plane wave front from a laser is incident on the the tablet' s surface part of the ligth can experience diffuse reflection due to surface roughness. In addition part of the ligth can penetrate the tablet and experience multiple back scattering. If the angle of incidence becomes high enough then specularly reflected ligth can usually be observed together with diffuse ligth. The specular component is expected to yield information about the surface roughness of the tablet. Provided that the surface statistics would obey normal distribution, we have a rough estimate for optical surface roughness that is based on the mathematical model of Beckmann and Spizzichino [2].

Of course there are many problems related to their model in context of porous pharmaceutical compacts. We mention that usually we have no information about the reflectance of perfect pharmaceutical compact, we face the problem of diffuse ligth that can be incident on the dectector and also that the classical model of Ref. [2] is best suited for materials with infinite conductivity. However, we may get information about the statistics of the surface and also estimate for the optical surface roughness using a modified version of the classical formula of Beckmann and Spizzichino. The modified version neglects the reflectance of perfect sample surface and replaces it with the output intensity of the laser. Then we can write

$$I = I_0 \exp\left(-16\pi^2 z^2 \frac{\cos^2\theta}{\lambda^2} \right) \tag{1.1}$$

where I is the intensity of the signal, I_0 is the output intensity of the laser, is the wavelength of the laser, θ is the angle of incidence and z is the optical surface roughness.

In Fig.1(a) is presented a schematic diagram of the measurement system. The laser source was a stable semiconductor laser $\lambda = 635\text{nm}$.

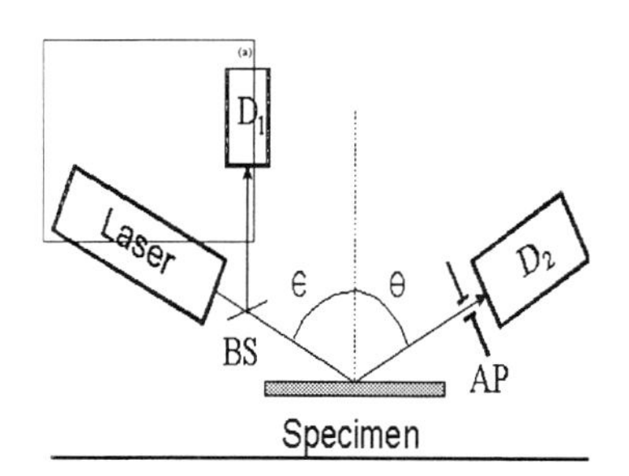

Fig. 1(a) Schematic diagram of measurement of specular reflectance

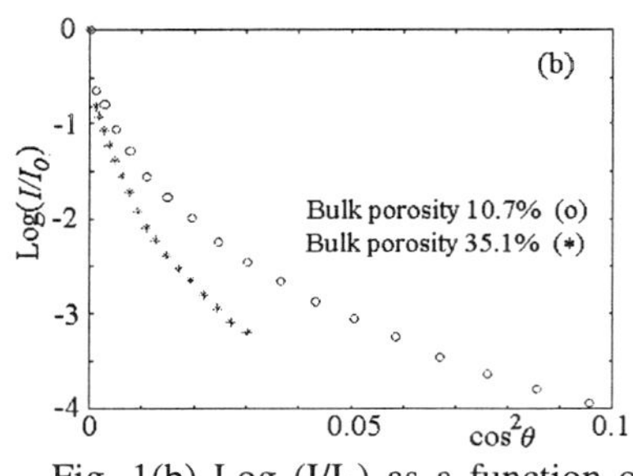

Fig. 1(b) Log (I/I_0) as a function of $\cos^2\theta$

The sample is in a goniometer. For this particular measurement we had to prepare slabs due to the fact that the angle of incidence, needed for detection of strong specular component, is rather high.The samples were compressed producing compacts with various bulk porosities. The pharmaceutical properties of starch asetate compacts were altered by changing the degree of acetate substitution of hydroxylic groups by asetate groups. When the substitution degree of asetate is increasing the surface roughness of the the compact is decreasing. In the Fig. 1(b) we show as an examples the plot of the logarithm of the intensity ratio, I/I_o, as a function of $\cos^2\theta$ for two tablets (bulk porosties 10.7% and 35.1%, and the same degree of asetate substitution).We can distinguish curves therefore we may state that gaussian statistics probably is not the correct choice. Note that in the case of smaller bulk porosity in Fig. 1 the signal could be detected for smaller angle of incidence. This is believed to result from lower surface roughness of the compact. The estimate of optical surface roughness, calculated from Eq. (1), was parctically speaking constant despite of the value of bulk porosity, which was ranging between 10-30% in our experiments. However, the degree of asetate substitution had a clear impact on the value of the optical surface roughness obtained from Eq. (1). As an average optical surface roughness of order of one micron could be calculated. Such a value is quite reasonable.

The surface quality of compacts and also of tablets can be based to a sligtly different scheme than that of Fig.1(a). If we replace the photodiode with a a CCD camera and use a fixed angle of incidence, we observe in far-field simultaneously a pattern that contains information about specular and diffuse reflection.

3 Diffractive Element Based Sensor

We have constructed a robust setup to analyse ligth reflected from the same compacts like above. The diffractive element was a binary amplitude-type element which was calculated using scalar diffraction theory.

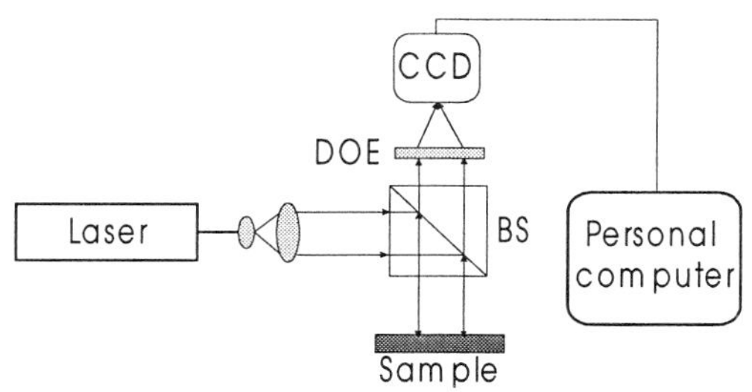

Fig. 2. Schematic diagram of diffractive element based sensor

The output of the element was chosen to be a regular array of a 4x4 light spot matrix. The idea of the 4x4 light spot matrix is to get a good spatial resolution when diffracted light front (the diffractive element is attached in the aperture of the CCD camera) is incident on the chip (there is no objective lens).The element was fabricated by e-beam lithography. The mesurement setup is presented in Fig. 2. Note that the present setup is well suited for the inspection of tablets, which can have also small size. The surface quality was estimated using processed image information. An example of image information obtained from a tablet is shown in Fig.3.

We have investigated also oval- shaped tablets that have coatings. These tablets contained new drug developed for Parkinson' s disease. The change of coating thickness was observed from the image data similar to that of Fig. 3.

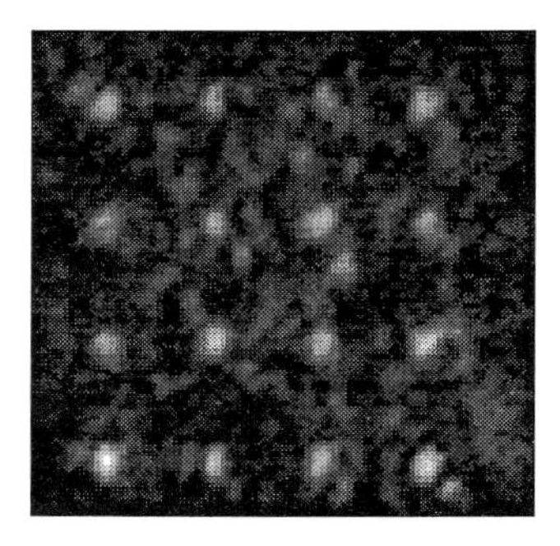

Fig. 3. Image data obtained from a flat surface starch tablet by the sensor of fig. 2.

4 References

[1] T. Asakura, Surface roughness measurement in Speckle Metrolgy ed. R. K. Erf (Academic, New York, 1978).

[2] P. Beckmann and A. Spizzichino, The Scattering of Waves from Rough Surfaces(Pergamon, Oxford, 1963).

[3] R. Silvennoinen, K.-E. Peiponen, P. Laakkonen, J. Ketolainen, E. Suihko, P. Paronen, J. Rasanen and K. Matsuda, Meas. Sci. Technol. 8, 550 (1997).

[4] K.-E. Peiponen, R. Silvennoinen, J. Rasanen, K. Matsuda and V. P. Tanninen, Meas. Sci. Technol.8, 815 (1997).

[5] R. Silvennoinen, K.-E. Peiponen, M. Sorjonen, J. Ketolainen, E. Suihko and P. Paronen, J. Mod.Opt.45, 1507(1998).

[6] R. Silvennoinen, K.-E. Peiponen, J. Rasanen, M. Sorjonen, E. J. Keranen, T. Eiju, K. Tenjimbayashi and K. Matsuda, Opt. Eng.37, 1482(1998).

Single Shot, Laser Plasma X-Ray Contact Microscopy of *Chlamydomonas*

A. C. Cefalas[1)], P. Argitis[2)], E. Sarantopoulou[1)], Z. Kollia[1)], T. W. Ford[3)], A. Marranca[3)] A. D. Stead[3)], C. N. Danson[4)], J. Knott[4)], D. Neely[4)]
National Hellenic Research Foundation, Theoretical and Physical Chemistry Institute, 48 Vas. Constantinou Av. Athens 11635 Hellas.
e - mail: ccefalas@eie.gr
2) Institute of Microelectronics, NCSR "Demokritos" 15310 Ag. Paraskevi, Hellas
3) School of Biological Sciences, Royal Holloway, University of London, Egham, Surrey TW20 0EX, UK.
4) Central Laser Facility, Rutherford Appleton Laboratory, Chilton, Didcot, Oxon OX11 0QX, UK

Abstract. Images of living green alga *chlamydomonas* were obtained using shoft X-rays in the water window (2.3–4.4 nm). The specimen was placed on a photoresist, which acts as the recording medium. Successful biological imaging is based on the different absorption caused by water and the carbon-rich proteins of the living biological specimens. Atomic force and scanning electron microscopy of the biological image clearly showed the flagella and flagella intersections of the motile green alga, *Chlamydomonas*, recorded in the photoresist suggesting a lateral resolution better than 150nm.

1 Introduction

The interest in using soft X-rays in imaging of living biological specimens [1-3],

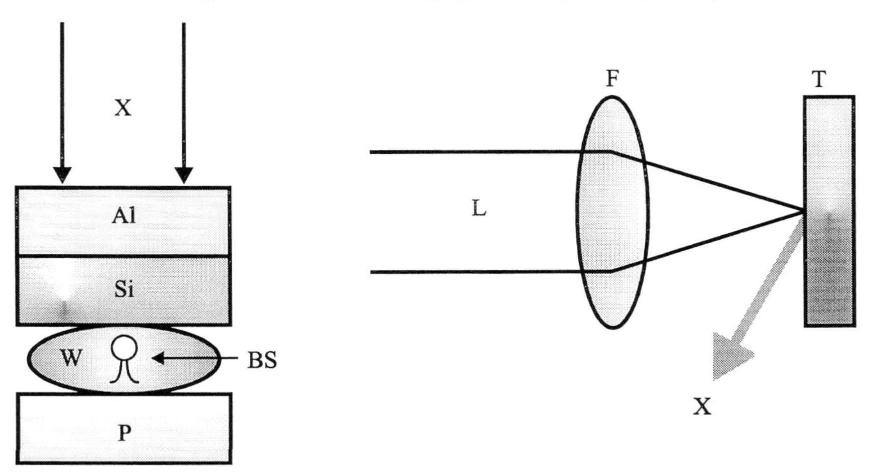

Fig.1 Experimental arrangement for obtaining the image of the living biological specimen motile green alga *chlamydomonas,* in the water window 2.3-4.4 nm. X: X-rays, Al: Aluminum filter 100nm thick, Si: Silica Nitride window 100 nm thick, BS: Biological Specimen, W: Water, P: Photoresist, L: Laser Beam, F: Lens T: Vanadium Target

is based on the different absorption of the carbon containing proteins and the water in the "water window", which is the region of the spectrum between the oxygen and the carbon K-absorption edges, i.e 2.3-4.4 nm. In this region the absorption of carbon is approximately 10 times higher than of oxygen and water. Therefore images of living biological specimens in their nutrient environment, can exhibit considerable high degree of contrast when they are exposed in contact with a photoresist. The image is then chemically developed, and is examined by a scanning electron microscope (SEM). Using X-rays from laser–produced plasmas, such images can be recorded over a very short period of time, of the order of 100 psec to 10 nsec. Therefore images of living and fast moving biological specimen can be taken. Despite the fact that the application of this method is limited only to large-scale national facilities, where synchrotron and high power laser sources are available, this method has certain advantages over the conventional methods of light and electron microscopy. Light microscopy has the advantage of simplicity and usually allowing living specimens to be viewed without the need of tissue preparation, but its resolution is limited by the wavelength of the light source to the best value of 250nm. On the other hand, living biological specimens cannot be examined with electron microscopy, and the need of preparation, dehydration and metal coating of the specimens, could destroy the tissue. Electron microscopy does not allow high contrast of the carbonated proteins as well, due to the fact that the scattering coefficient of the electrons for the carbon based proteins is very similar to that of water. In this work we report on the application of the soft X-ray contact microscopy (SXRCM) to record images of the living motile green alga *chlamydomonas*, using a negative chemically amplified epoxy novolac based photoresist [4]. This photoresist was two orders of magnitude faster than PMMA; the standard used photoresist in X-ray contact microscopy. The flagella of the motile green alga were clearly seen, suggesting a lateral resolution better than 150 nm. The photoresist was also capable to differentiate high features as small as 20 nm in atomic force microscope depth profiles, and to register the flagella intersection areas. These results suggest that the application of SXRCM to record images of living biological specimens could be used with less intense X-ray sources and could form the basis of developing a table-top X-ray contact microscope, using a small commercial laser.

2 Experimental

The experimental apparatus is indicated in Fig.1. It consists mainly of the laser source to generate the soft X-rays, and the vacuum chamber where the biological specimen was placed. The laser source was the VULCAN Nd: glass laser of the Rutherford Appleton Laboratory, UK. The rod chain output deliver 11 J at 1064 nm, in a 2 min shot cycle. The output from this laser was focused on an yttrium target with a lens of 40 cm focal length. With this arrangement and at 6 J of focused laser energy, 10^{15} photons per sterad in the water window were emitted and this value corresponds to the energy fluence of 4 mJ/cm^2.

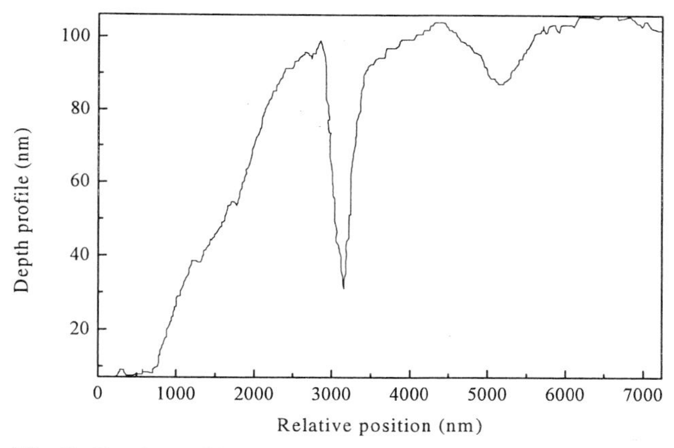

Fig.2. Depth profile of the image of *chlamydomonas* registered on a negative epoxy based novolac photoresist taken with an atomic force microscope. The overlapping of the two flagella is clearly seen with 80 nm depth difference.

The water solution containing the biological specimens was confined between the photoresist and a Si_3N_4 0.1 μm thick window. In order to screen out photons with energy lower than 400 eV, a thin alluminium foil of 100 nm thickness was used as a filter. Taking into consideration that the absorption coefficient of the Al foil and the Si_3N_4 window is ~ $2x10^4$ cm^{-1}, 2.5 mJ/cm^2 of X-rays in the water window are incident on the sample. Since the aim of soft X-ray contact microscopy is to capture the image of living specimens, the environment of the specimen during exposure must be designed to maintain its hydrated state. The imaging system is placed in position inside the vacuum chamber by a specially designed arm, which can be evacuated separately from the main chamber. When the pressure inside the arm and the main vacuum chamber are equal, the diaphragm that separates the vacuum chamber and the arm is removed. This step is critical because the silicon nitride window is thin and it has to remain intact surviving the pressure gradient, at least up to the moment of the image capture. For successful imaging the biological specimen has to maintain its hydrated state during exposure. Ideally, the specimen has to be wet even when the specimen holder is taking out of the chamber after exposure.

3 Results and Discussion

High contrast conditions such as the ones preferred in photolithography are not suitable for this kind of application, because they would not allow any depth differences coming from different degree of masking by the specimen on the relief image. Methyl isobutyl cetone was used as developer for giving lower contrast. The depth profile of an image as it was measured with an atomic force microscope (Burleigh ARIS 3300) is given in Fig. 2. It is possible to differentiate between a single flagellum (relative position at 5200 nm), with 20 nm depth difference, and those areas where two flagella are overlapping (relative position at 3000 nm), with 80 nm depth difference. Images of *chlamydomonas* cells were obtained successfully with electron microscopy as shown in Fig. 3. The diameter of the cell body was between 1 and 5 μm and the corresponding thickness of the flagella was between 150 and 300 nm.

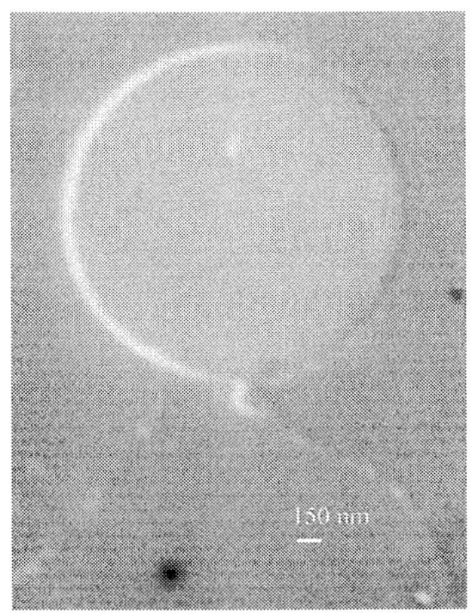

Fig.3. Scanning electron microscope image of *chlamydomonas*. The two flagella are clearly visible, suggesting a resolution better than 150 nm.

References

1. T. W. Ford, A. D. Stead and R. A. Cotton, Electr. Microsc. Rev., 4, 269, (1991)
2. S. Bollanti, P. Di Lazzaro, F. Flora, G. Giordano, T. Letardi, G. Schina, C. E. Zheng, L. Filippi, L. Palladino, A. Reale, G. Taglieri, D. Batani, A. Mauri, M. Belli, A. Scafati, L. Reale, P. Albertano, A. Grilli, A. Faenov, T. Pikuz and R. Cotton, Journ. X-ray Sch. Techn. 5, 261, (1995)
3. C. Cefalas, P. Argitis, Z. Kollia, E. Sarantopoulou, T. W. Ford, A. D. Stead, A. Maranca, C. N. Danson, J. Knott, and D. Neely, Appl. Phys. Letters 72, 3258, (1998)
4. P. Argitis, I. Raptis, C. J. Aidinis, N. Glezos, M. Baciocchi, J. Everett and M. Hatzakis J. Vac. Sci. Technol. B 13, 3030, (1995)

Image Formation in Optical Coherence Tomography and Microscopy

C.J.R. Sheppard and M. Roy

Department of Physical Optics, School of Physics, A28,
University of Sydney, NSW 2006, Australia.

Abstract. Three dimensional image formation in optical coherence tomography has been investigated theoretically. Imaging can be described by a three-dimensional (3-D) coherent transfer function (CTF), which has contributions from the different wavelength components. From this, two dimensional imaging can be described using the 2-D CTF obtained as a projection of the 3-D CTF. For very low numerical aperture axial imaging results from the limited coherence length of the light source. Interference microscopy results in an optical sectioning property similar to that in confocal microscopy. Thus for intermediate values of numerical aperture, axial imaging is a combination of coherence gating and confocal sectioning, for which a paraxial theory can be used. At very high numerical apertures it is necessary to use a full high aperture theory. These theoretical treatments can be used to model images of known structures, and to estimate expected imaging performance.

1 Introduction

Optical coherence tomography (OCT) is a form of broad-band interferometry. Early forms of this techniques were coherence-probe microscopy (CPM) [1], and coherence-domain reflectometry [2]. In CPM, an interferometer is constructed using a tungsten-halogen or arc lamp as source. Interference takes place only from the section of the sample located within the coherence length relative to the reference beam. This technique is usually used for profilometry of surfaces. A series of 2D images is recorded using a CCD camera as the sample is scanned axially through focus. For each pixel an interferogram is recorded, which can be processed to extract the location of the surface either from the peak of the fringe visibility, or from the phase of the fringes. Use of a broad-band source avoids phase-wrapping ambiguities present in laser interferometry with samples exhibiting height changes of greater than half a wavelength.

If a microscope objective of high numerical aperture is employed in CPM, interference occurs only for regions of the sample which gives rise to a wavefront whose curvature matches that of the reference beam. For this reason it has been termed phase correlation microscopy. The effect was first pointed out by Corcoran [3]. An optical sectioning effect results which is analogous to that in confocal microscopy. This phenomenon been used by Fujii [4] to construct a lensless microscope, and by Sawatari [5] to construct a laser interference microscope. Overall optical sectioning and signal collection performance is slightly superior to that of a confocal microscope employing a pinhole.

In order to extract the height information from the interferogram, a complete axial scan through the sample must be recorded. An alternative method is based on phase-shifting, in which the visibility at a particular axial position can be determined by three or more measurements for different values of reference beam phase. This avoids the necessity to scan through the complete depth of the image, and is particularly advantageous with thick structures. Phase shifting is usually performed with laser illumination, and with a broad-band source a problem is encountered in that the phase shift of the different spectral components, when reflected from a mirror, varies. This problem has been overcome by performing the phase shifting by use of geometric phase-shifting (GPS), in which polarisation components are used to effect an achromatic phase shift [6].

Although the principle of OCT is identical to that of CPM, the embodiment is usually quite different [7]. In OCT the signal, scattered from bulk tissue, is very weak . Hence the source is often replaced by a super-luminescent laser diode, which is very bright but exhibits a smaller spectral bandwidth than a white-light source. The source is used in a scanning geometry, in which the sample is illuminated point-by-point, so that considerable time is necessary to build up a full image. Instead of phase-shifting, the alternative techniques of heterodyning is used, which can result in shot-noise limited detection performance. In OCT optical sectioning results from the limited coherence length of the incident radiation, but if the numerical aperture of the objective lens is large, an additional optical sectioning effect identical to that in confocal imaging results from the correlation effect [8]. Both OCT and OCM are usually implemented using single mode optical fibres for illumination and collection, and the geometry of fibre spot influences the imaging performance of the system [9].

Either low-coherence light or ultra short pulses [10] can be used to measure internal structure in biological specimens. An optical signal that is transmitted through or reflected from a biological tissue contains time-of flight information, which in turns yields spatial information about tissue microstructure. The equivalence of these two approaches has been discussed [11]. Time-resolved transmission spectroscopy has been used to measure absorption and scattering properties in tissue, and has been demonstrated as a non-invasive diagnostic measure of haemoglobin oxygen in the brain. Optical ranging technique using femtosecond pulses have been performed of the eye and skin. Time gating by means of coherent, as well as non-coherent techniques, has been used to detect directly transmitted light and obtain transmission images in turbid tissue. In contrast to time domain techniques, OCT can be performed with continuous-wave light avoiding an expensive laser and a cumbersome system.

2 Significance of Transfer Functions

An optical system usually behaves as a low-pass filter, so that it only transmits the low spatial frequency information efficiently, but does not transmit high spatial frequencies representing very fine details. The efficiency with which the different spatial frequencies are transmitted is called the transfer function. In an optical system this usually goes to zero at a particular cut-off frequency. In general the bigger the range of spatial frequencies transmitted, the better the image. Also the transfer function should be a smoothly varying function, as otherwise ringing occurs, which is observed as fringing of the image. According to the optical system, transfer functions can be classified into two main groups: the optical transfer function (OTF) which is applicable for an incoherent system and the coherent transfer function (CTF) which is applicable for a coherent system. The difference between the OTF and CTF is that the OTF operates on the intensity variations in the object, whereas the CTF operates on the amplitude variations in the object.

3 3-D Imaging

For OCT, tranverse imaging results from the imaging properties of the focusing lens, while axial imaging is determined solely by the limited coherence length. Thus 3-D imaging can be desribed using a 3-D CTF which is a weighted itegration of 2-D CTFs over the spectral components. Three-dimensional imaging by a microscope objective in the brightfield reflection mode can be described and modeled using a 3-D CTF, which has been presented in the paraxial approximation for a cofocal system [12]. The CTF for monochromatic interference imaging is identical to this. If a broad-band source is used, the resultant CTF is given by a weighted integration over those for the individual spectral components. The result is the same as for time-resolved imaging with a short pulse of the same bandwidth [13]. Account must be taken of the change of spectral composition on propagation [14]. The image of a specimen which exhibits 3-D variations in refractive index can be modeled [15]. Image reconstruction can be based on these methods [16].

References

1. Davidson, M., Kaufman, K., Mazor, I. and Cohen, F. (1987) An application of interference microscopy to integrated circuit inspection and metrology. Proc. SPIE. **775**, 233-247
2. Danielson, B. L. and Whittenburg, C. D. (1987) Guided-wave reflectometry with micrometer resolution. Appl. Opt. **26**, 2842-2836
3. Corcoran, V. J. (1965) Directional characteristics in optical heterodyne detection processes. J. Appl. Phys.**36**, 1819-1825
4. Fujii, Y. and Takimoto, H. (1976) Imaging properties due to the optical heterodyne and its application to laser microscopy. Opt. Comm. **18**, 45-47

5. Sawatari, T. (1973) Optical heterodyne scanning microscope. Appl. Opt. **12**, 2768-2772
6. Roy, M. and Hariharan, H. (1995) White-Light geometric phase interferometer for surface profiling. Proc. SPIE **2554**, 64-72
7. Huang, D., Swanson, E. A., Lin, C. P., Schuman, J. S., Puliafito, C. A. and Fujimoto, J. G. (1991) Optical coherence tomography. Science **254**, 1178-118
8. Izatt,J. A., Hee, M. R., Owen,G. A., Swanson, E. A. and Fujimoto, J. G. (1994) Optical coherence microscopy in scattering media. Opt. Lett. **19**, 590-592
9. Gu, M., Gan, X. and Sheppard, C. J. R.(1991) Three-dimensional coherent transfer functions in fibre optical confocal scanning microscopes. J. Opt. Soc. Am. A **8**, 1019-1025
10. Fujimoto, F. G., De Silvestri, S., Ippen, E. P., Puliafito, C. A., Margolis,R. and Oseroff, A. (1986) Femtosecond optical ranging in biological systems. Opt. Lett. **11**, 150-152
11. Caulfied, H. J., Hirschfield, T. and Williams, A. D. White-light interferometry and phase-locked pulse analysis. Opt. Lett. **1**, 56-57
12. Sheppard, C. J. R., Gu, M., Mao, X. Q. (1977) Three-dimensional coherent transfer function in a reflection-mode confocal scanning microscope. Opt. Commun. **81**, 281-284
13. Gu, M. and Sheppard C. J. R. (1995) Three-dimensional image formation in confocal microscopy under ultra-short laser-pulse illumination. J. Mod. Opt. **42**, 747-762
14. Sheppard, C. J. R. and Gan, X. (1997) Free-space propagation of Femto-second light pulses. Opt. Commun. **133**, 1-6
15. Sheppard, C. J. R., Connolly, T. J. and Gu, M.(1995) The scattering potential for imaging in the reflection geometry. Opt. Commun. **117**, 16-19
16. Sheppard, C. J. R. (1998) Imaging of random surfaces and inverse scattering in the Kirchhoff approximation. Waves in Random Media **8**, 53-66

In vivo Measurement of the Optical Properties of Human Tissues in the Wavelength Range 610-1010 nm

R. Cubeddu, A. Pifferi, P. Taroni, A. Torricelli and G. Valentini
INFM-Dipartimento di Fisica and CEQSE-CNR, Politecnico di Milano, Italy

Abstract. *In vivo* absorption and scattering spectra of different human tissues were obtained using a system for time-resolved reflectance measurements in the wavelength range from 610 nm to 1010 nm, every 5 nm. The system is based on a dye laser and a Ti:Sapphire laser as light sources and an electronic chain for time-correlated single-photon counting for detection. Measurements were performed on the breast, the arm, and the abdomen of healthy volunteers. The scattering spectra are decreasing upon increasing the wavelength, while the absorption spectra show the spectral features of oxy- and deoxyhemoglobin, of water, and of lipids.

1 Introduction

The efficient use of optical radiation for both diagnostic and therapeutic techniques [1, 2] relies on the knowledge of tissue optical properties. Moreover, it is of basic importance to assess the optical properties of biological tissues *in vivo*. In fact, due for example to blood deoxygenation, significant changes may occur when measures are performed *post mortem*, and even more if tissues need to be manipulated to produce thin sections, as generally required for the application of conventional techniques [3]. However, up to now *in vivo* data have been reported only at single wavelengths or in limited spectral intervals [4-8].

In this paper we report on an *in vivo* study for the evaluation of absorption and reduced scattering spectra of biological tissues (breast, arm, and abdomen), by means of time-resolved reflectance measurements.

2 Materials and Methods

A synchronously pumped mode-locked dye laser (Mod. CR-599, Coherent, Ca) and an actively mode-locked Titanium:Sapphire laser (Mod. 3900, Spectra-Physics, Ca) were used as the illumination sources in the wavelength range 610-700 nm, and 700-1010 nm, respectively. Laser light was delivered to and collected from the sample by a couple of 1 mm diameter plastic-glass fibers (Mod. PCS1000W, Quartz et Silice, France), placed at a relative distance of 2 cm in a reflectance geometry. A double microchannel plate photomultiplier (Mod. R1564U, Hamamatsu, Japan) and an electronic chain for time-correlated single-photon counting were used for the detection of time-resolved reflectance curves. Overall, the instrument transfer function was <150 ps FWHM. A small fraction of the incident beam was coupled to a 200 μm fiber (Mod. PCS200W, Quartz et Silice, France) and fed directly to the photomultiplier to account for any eventual time drift of the instrumentation.

In order to allow measurements to be carried out *in vivo*, the system was fully automated and the analysis and display of the measured spectra were performed in real time. The overall measurement time for data acquisition and system adjustment was of 8-10 s/wavelength with an incident power <10 mW.

The experimental time-resolved reflectance curves, collected every 5 nm, were fitted with the solution of the transport equation in the diffusion approximation for a semi-infinite homogeneous medium with the extrapolated boundary condition [9].

A more detailed description of the experimental set-up and of the fitting procedures can be found in [10] and [11], respectively.

Data were acquired from breast, arm, and abdomen of 5 healthy volunteers. At each site, two measurements were repeated on close positions.

3 Results

Figure 1.a reports the absorption spectra of breast. In the red region of the visible spectrum, μ_a progressively decreases from 0.15 cm^{-1} to 0.05 cm^{-1} upon increasing the wavelength. In the near infrared, a small absorption peak is detected around 740-760 nm. For longer wavelengths, a weak peak is observed at 830-840 nm. However, the spectrum is dominated by a major absorption peak around 970 nm. Depending on the volunteer, a further spectral feature can be detected around 930 nm as a shoulder or as a peak comparable to the one at longer wavelength.

A similar spectral behavior is observed for the muscular tissue of the arm (Fig. 1.c), even though the absorption values are several times higher. In the red region of the visible spectrum μ_a progressively decreases from 0.41 cm^{-1} to 0.15 cm^{-1} upon increasing the wavelength. A sharper peak, centered around 760 nm, is present in middle wavelength range, while in the long wavelength range, like for the breast, the dominant absorption feature is the peak centered around 970 nm.

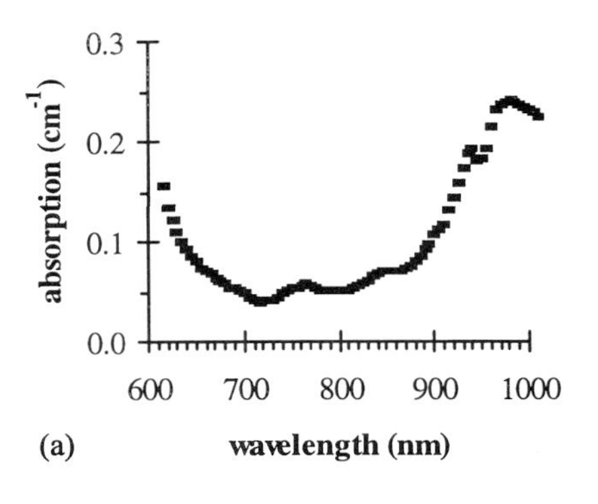

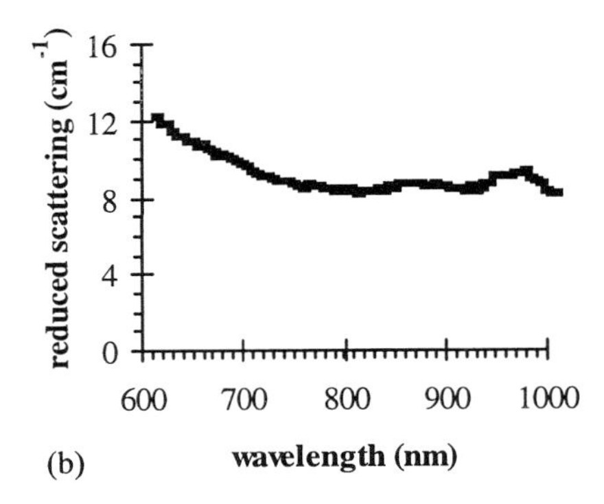

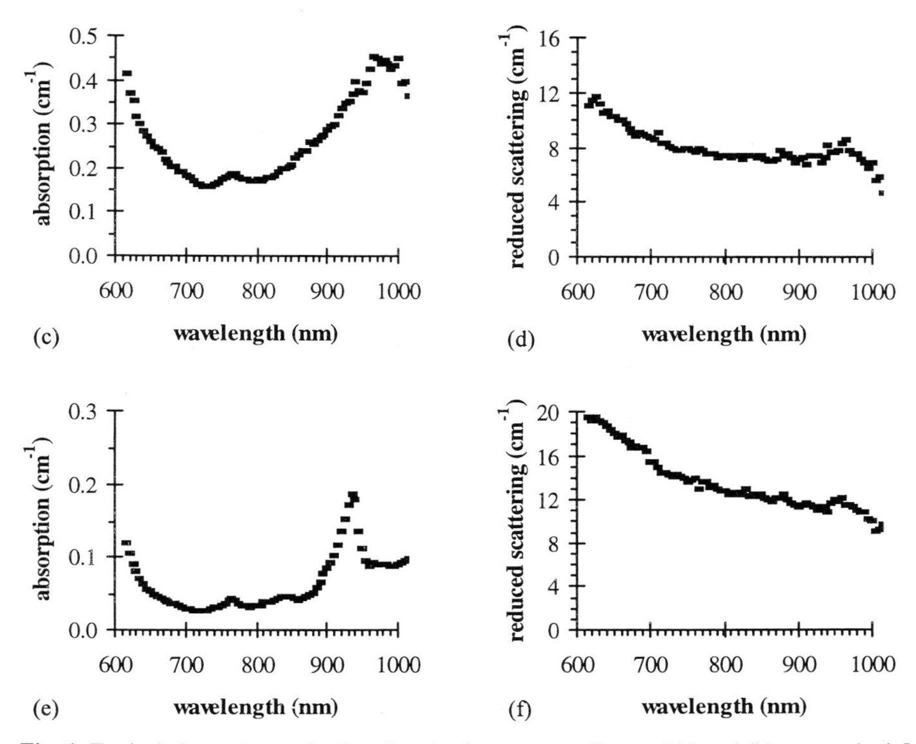

Fig. 1. Typical absorption and reduced scattering spectra of breast [(a) and (b), respectively], arm [(c) and (d)], and abdomen [(e) and (f)].

For the adipose tissue of the abdomen (Fig. 1.e), the decrease is steeper at short wavelengths, while the absolute values of the absorption coefficient are comparable to or slightly lower than the ones obtained for the breast. A peak, centered around 760 nm, is present even in the spectra of the abdomen. The infrared absorption is characterized by a sharp peak at 930 nm, similar to what observed in some measurements on the breast.

The reduced scattering as a function of wavelength (Fig.1.b, Fig.1.d Fig.1.f for breast, arm, and abdomen, respectively) presents a progressive decrease for all the tissue types considered, with no detectable spectral features either in the red or in the near infrared. Similar values of μ_s' are detected in breast and arm (8-12 cm^{-1}), while the abdomen seems to be more strongly diffusive (12-20 cm^{-1}).

4 Discussion

If we take into account the absorption spectra of the main tissues components and we compare the results obtained from different tissue types, we can derive some information on the origin of the measured line shapes.

The rapidly decreasing absorption in the red region is likely influenced by the tail of the visible absorption of oxy- and deoxyhemoglobin. The absorption spectrum of deoxyhemoglobin is characterized also by a peak around 760 nm, which seems to be detected in all tissue types.

A contribution at these wavelengths may also come from lipids, as suggested by the fact that significant absorption is detected in the case of lipid-rich abdomen. Lipids are clearly detected also around 930 nm, as a peak or shoulder in the case of abdomen and breast, respectively.

The presence of oxyhemoglobin is not easily identified in the near infrared absorption spectra, since it is characterized by a broad absorption band centered around 900 nm, but by no sharp spectral features. However, its contribution is detected, especially at long wavelengths, as a high background in the strongly vascularized muscular tissue of the arm.

The dominant peak around 970 nm can be attributed to water, which is likely responsible also for the small peak or shoulder around 835 nm. Where the water content is higher (i.e. breast), even its absorption peak around 740 nm can be detected, as it is slightly blue-shifted with respect to deoxyhemoglobin and lipids (at 760 nm) and broadens the absorption feature observed in the range 740-760 nm.

Work is presently in progress to confirm the results obtained so far with a more accurate evaluation of the inter-subject variability and of the effects of the measurement position, and subsequently to extend the study to the assessment of the optical differences between pathologic areas and surrounding healthy tissues.

References

1. *Photon propagation in tissue III*, D.A.Benaron, B.Chance, and M.Ferrari, Editorss, Proceedings of SPIE, vol.3194, (1988).
2. *Advances in Optical Imaging and Photon Migration*, Technical Digest, Optical Society of America, Washington DC, (1988).
3. B.C.Wilson, W.Patrick, and D. M.Lowe, Photochem. Photobiol. 42, 153 (1985).
4. K.A.Kang, B.Chance, S.Zhao, S.Srinivasan, E.Patterson, and R.Troupin, SPIE Proc. 1888, 487 (1993).
5. H.Heusmann, J.Kölzer, and G.Mitic, J.Biomed. Opt. 1, 425 (1996).
6. C. af Klinteberg, R.Berg, C.Lindquist, S.Andersson-Engels, and S.Svanberg, SPIE Proc. 2626, 149 (1995).
7. K.Suzuki, Y.Yamashita, K.Ohta, M.Kancko, M.Yoshida, and B.Chance, J.Biomed. Opt. 1, 330 (1996).
8. S.J.Matcher, M.Cope, and D.T.Delpy, Appl. Opt. 36, 386 (1997).
9. R.C.Haskell, L.O.Svaasand, T.-T.Tsay, T.-C.Feng, M.S.McAdams, and B.J.Tromberg, J.Opt.Soc.Am.A 11, 2727 (1994).
10. A.Pifferi, R.Cubeddu, P.Taroni, A.Torricelli, and G.Valentini, in *Photon Migration in Tissues*, K.G.Tranberg, L.O.Svaasand, J.M.Brunetaud and A.Katzir, Editorss, Proceedings of SPIE, vol. 3566, in press (1998).
11. R.Cubeddu, A.Pifferi, P.Taroni, A.Torricelli, and G.Valentini, Appl. Opt. 35, 4533 (1996).

Depth Estimation of an Absorbent Embedded in a Dense Medium Using Diffused Wave Reflectometry

T. Iwai[1], K. Tabata[1], G. Kimura[1], and T. Asakura[2]

[1] Research Institute for Electronic Science, Hokkaido University, N12 W6, Sapporo 060-0812, Japan

[2] Faculty of Engineering, Hokkai-Gakuen University, S26 W11, Sapporo 064-0926, Japan

Abstract. The purpose of this research is to propose a new and simple method in measuring the depth of an absorbent embedded in a dense scattering medium by using a diffused wave reflectometry. The estimation of the depth of the absorbent is based on the assumption that the total intensity over the backscattering plane equals to the probability that the intensity profile is formed by the contribution of the waves with the path-length shorter than a certain maximum path-length. To confirm the principle, the Monte Carlo simulations and experiments are repeated for the absorbents with various depths and shapes and the various positions of the beam incidence. Finally, we demonstrate the validity of the proposed principle and the possibility to apply the new method to the surface profiling of the absorbent, i.e. the diffused wave topography.

1 Introduction

The techniques based on the light scattering have been playing an important role in the optical diagnosis and therapy for human tissues. The analysis of the tissue structure with thickness over a few millimeters is still an open problem. The technique for the such the thickness of tissues will be powerful for diagnoses of a breast cancer and a tumor near the skin layer, the analysis of the blood flow dynamics near the surface layer of the brain and so on.

For such the thick tissue, the diffused wave is dominant in the light scattering. The purpose of the research is to propose the new and simple method based on the diffused wave reflectometry to estimate the depth of the absorbent embedded in the dense scattering medium. The proposed method is based on the relation between the probability density function of the optical path-length and the total intensity integrated spatially over the backscattering plane. The principle of the depth estimation is successfully confirmed by Monte Carlo simulations and experiments and applied to the surface-profiling method of the absorbent.

2 Principle of Depth Estimation

The absorbent restricts the contribution of scattered waves to the total backscattered intensity within those with the path-length shorter than the certain maximum path-length. Assuming that the total intensity over the backscattering plane equals to the probability that the waves with the path-length shorter than the maximum path-length L contribute to the formation of the intensity profile, the assumption is formularized as

$$P(L)=\int_0^L p(s)ds \Big/ \int_0^L p(s)ds \approx \int_\Sigma I_a\, dS \Big/ \int_\Sigma I_0\, dS \tag{1}$$

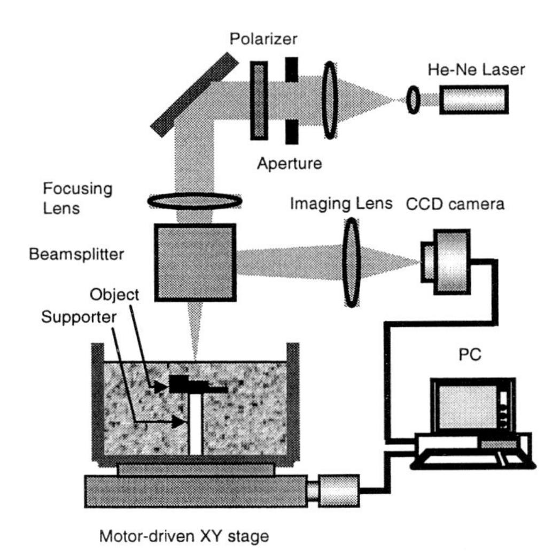

Fig.1. Experimental setup. The focused beam emerging from the He-Ne laser with the 633nm-wavelength is incident to the dense scattering medium which is composed from the 1%-suspension of the 460nm-polystrene particles. The black- and mat-painted absorbent is supported by a thin rod in the scattering medium.

where p(s) is a probability density function of the optical path-length when the absorbent does not exist, I_a and I_0 the intensity profiles produced from the media with and without the absorbent, respectively, and Σ the area over the backscattering plane. This principle corresponds to renormalizing the variation in p(s) due to the restriction of the scattered waves by the absorbent to the maximum path-length L. In Eq.(1), the probability density function p(s) can be obtained by the Monte Carlo simulation or as the analytical solution of a diffusion equation of photons in advance and Ia and I0 are observable values in the experiment.

To estimate the depth from the backscattering plane to the surface of the absorbent, it is assumed that the maximum path-length L is directly proportional to the depth d of the absorbent and defined by

$$L=\alpha d\ . \tag{2}$$

The constant α is optimized by Monte Carlo simulations in such the way that the squared error between the estimated and the give depth of absorbent is minimized for various depths of the absorbent. For 1%-suspension of 460nm-polystrene latex particles, $\alpha=3$ was obtained as the optimum value.

3 Experiment Setup

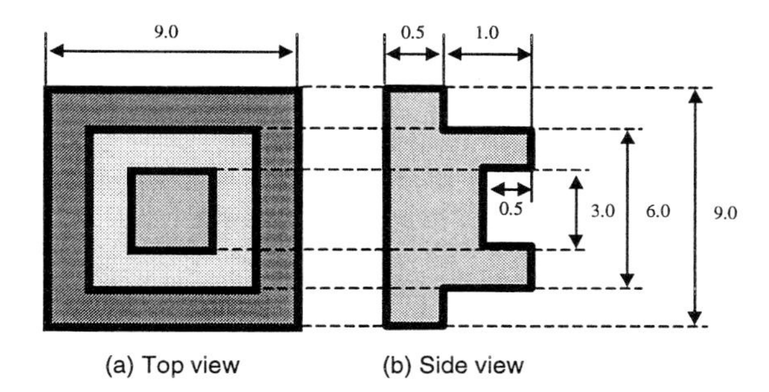

Fig.2. (a) Top view and (b) side view of the absorbent used in the experiment, which is painted black and mat.

The experimental setup is schematically shown in Fig.1. The laser light with the wavelength of 633nm is focused to the backscattering plane. The absorbent shown in Fig.2 is supported by the thin rod in the 1%-suspension of 460nm-polystrene latex particles that the 5cm×5cm×1cm-vessel is filled with. The backscattered intensity profile is formed by the light waves traveling along the long trajectories inside the medium, and imagined in a CCD plane of the digital camera with 1/5 magnitude. The digital image is fed into the microcomputer and integrated spatially over the backscattering plane. The incident position of the laser beam is two-dimensionally moved by a motor-driven XY stage.

4 Experimental Results

Figures 3 (a) and (b) show the intensity profiles of backscattered light by the latex particle suspensions without and with the absorbent, respectively. The comparison between the figures shows that both the peak and extension of the profile decrease due to the ray restriction by the absorbent. As is mentioned above, the variation of the probability density function of the path-length due to the absorbent is reflected to the intensity profile in the backscattered plane.

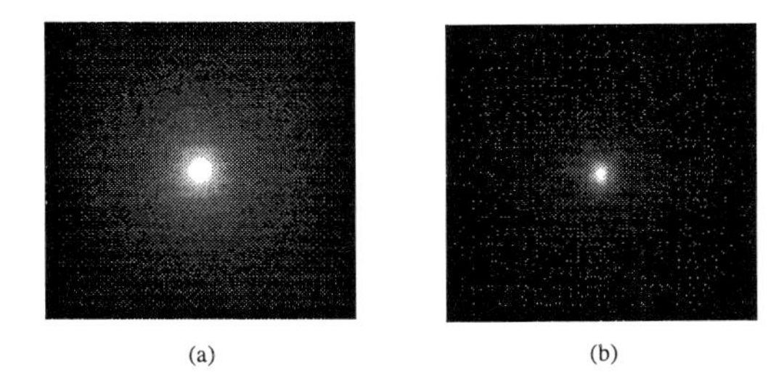

Fig.3. Intensity profiles of backscattered light by the dense scattering medium (a) without the absorbent and (b) with the absorbent. The intensity profile is quantized to 12 bit gray scales. To show the photographs, the intensity near the incident position is saturated.

Figure 4 shows the reconstruction result of the absorbent shown in Fig.2, in which the topographical image and the line along the horizontal axis at the center of

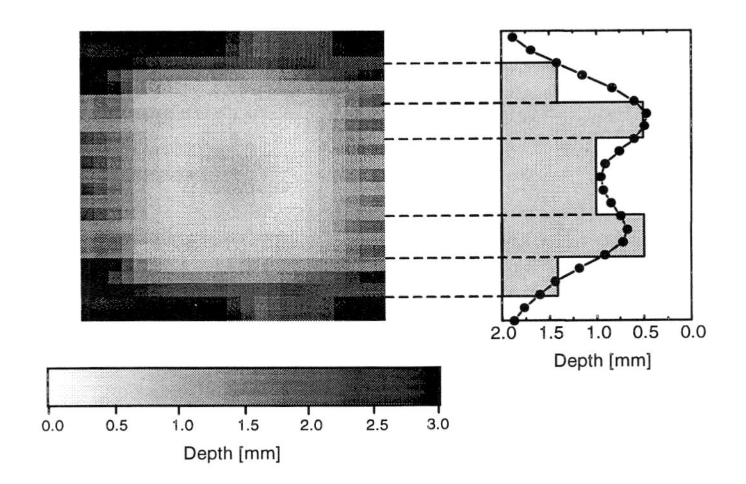

Fig.4. Reconstructed absorbent using the proposed method based on the diffused wave topography. The left and right figures show the two-dimensional surface profile of the absorbent and the line scanned along the vertical axis at the center of the horizontal axis, respectively.

the vertical axis. The gray scale indicates the distance from the backscattering plane to the surface of the absorbent, in which the depth increases from the white to the black. It is shown that the surface profile is successfully reconstructed by the proposed method though the edge is blurred.

5 Conclusion

In this study, the depth of the absorbent embedded in the dense medium was successfully estimated by the diffused wave reflectometry. Although the new principle of the measurement is quite simple, the reconstruction of the topographical image of the absorbent shape could be attained.

Acknowledgements

We thank R. Meike and K. Hasegawa of Division of Technical Saffs in Research Institute for Electronic Science for making the parts of the experimental setup and the absorbents. This research is financially supported by a Grant-in-Aid for Scientific Research from the Ministry of Education, Science, Sports and Culture of Japan (10650396).

In vitro Optical Characterization of Female Breast Tissue with Near Infrared fsec Laser Pulses

G. Zacharakis[a], V. Sakkalis[a], G. Filippidis[a], A. Zolindaki[b], E. Koumantakis[b], T. G. Papazoglou[a]

[a] Foundation for Research and Technology-Hellas, Institute of Electronic Structure and Laser, Laser and Applications Division, P.O. Box 1527, 71110 Heraklion, Crete, Greece
Fax# ++ 30 81 391318, e-mail: zahari@iesl.forth.gr

[b] Gynecology Department, University Hospital of Heraklion, Crete, Greece

Transmission measurements of ultra-short laser pulses through tissue obtained from female breast during tumor extraction operation or biopsy were performed. The study concerned tissue from twenty cases in total. Twelve of them were macroscopically characterized as fat deposited and eight of them as fibrous. All tissues were obtained half an hour after excision and were returned after the experiment to the Department of Gynecology of the University Hospital for histologic analysis. The thickness of the samples varied from 4 to 30 mm and they were placed on a X-Y translation stage in order to take measurement from more than one point of each tissue. The distance of the back side of the tissue to the front face of the detector was kept constant at 1 cm for all the experiments. Thus, the observation angle was kept constant during all measurements.

The source of the fsec pulses is a mode-locked Ti:Sapphire laser emitting at 800 nm, pumped by a Spectra Physics Ar^+ laser operating with a power of 9W. The repetition rate of the Ti:Sapphire was 82 MHz with an average power of ~1W. A Hamamatsu C 5680 Streak Camera with a temporal resolution of 2 psec was used as the detector. A CCD camera operating at -50°C for reduced dark noise was used for the recording of the signals. The resulting signal was fed to a PC for further processing. Two different signals were simultaneously recorded. The diffused pulse emerging from the sample and a small portion of the original pulse which provided the time reference. This part of the pulse was coupled into an optical fiber and directed to the entrance slit of the camera. The position of the coupler was adjusted so that the two signals arrive at the same time without the sample. The width of the entrance slit of the streak camera was 28 mm and during the experiments it was opened 5-10 μm in order to have the maximum temporal resolution. A Hamamatsu PIN photodiode provided the optical trigger to the streak camera using a very small part (few mW) of the original pulse. The diameter of the beam was about 3 mm while the samples were irradiated with a power of 100 mW.

The signal was recorded by the streak camera in the form of three dimensional images. Time, space and intensity in the form of pseudocolors, were recorded. Time profiles are obtained in order to extract the time information. Intensity versus time graphs can then be constructed (figure 1).

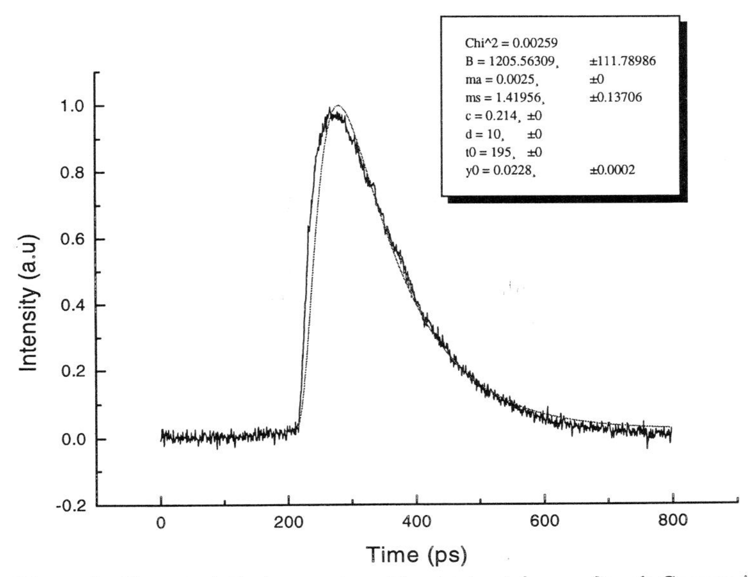

Figure 1: Characteristic temporal profile obtained from a Streak Camera image. The solid curve represents the theoretical fitting.

The optical parameters (absorption and scattering coefficient) of the samples were calculated by fitting the diffusion curves with the Patterson analytical expression[1] (derived from the diffusion theory), modified to fit the data from the streak camera:

$$T(d,t) = y_0 + B(4\pi Dc)^{-1/2} t^{-3/2} \exp(-\mu_a ct)$$
$$\left[(d-z)\exp\left\{\frac{-(d-z)^2}{4Dct}\right\} - (d+z)\exp\left\{\frac{-(d+z)}{4Dct}\right\} + (3d-z)\exp\left\{\frac{-(3d-z)^2}{4Dct}\right\} - (3d+z)\exp\left\{\frac{-(3d+z)}{4Dct}\right\} \right]$$

where,

$t = t - t_0$, $z = \ell_t = ((1-g)\mu_s)^{-1} = 1/\mu_s'$ and $D = \left(3\left(\mu_a + \left(1/z\right)\right)\right)^{-1}$

The fixed parameters of the equation were the thickness of the sample (d), the time corresponding to the position of the maximum of the reference pulse (t_o) and the speed of light (c). The value of the latter was calculated for water (0.214×10^9 m/sec, with $n_{water} = 1.4$)[2]. B and y_o, were parameters that had to do with specific characteristics of each curve and are important for the adjusting of the theoretical curve to the data. The output of the calculation was the absorption and the inverse reduced scattering coefficient (μ_a and $1/\mu_s'$ respectively). Since the scattering coefficient can not be calculated independently of the anisotropy factor g, $1/\mu_s'$ was used for the characterization of each type of tissue.

The absorption coefficient was not taken into account since it did not affect significantly the fitting of the theoretical curve and furthermore, its values were at least two orders of magnitude smaller and had a large deviation for the same tissue. Thus it could not be connected to the characterization of the tissue.

The difference of the mean values of these coefficients were used in order to discriminate the different types of tissue. This was possible since a different range of values corresponded to the two types of tissue studied here. For the lipid tissue the values lid in the interval [0.43 - 0.64] mm^{-1} and for the fibrous tissue in the interval [0.71 - 1.1] mm^{-1}. Values which were between the two distributions corresponded to tissue with different percentage of fat and fibrosis. These results were then compared with the corresponding from the histologic analysis. The same method could also be used for the discrimination of the other ingredients of the breast. The desired result is the discrimination between malignant and benign tumor with non-invasive techniques, that may eventually replace biopsy, which requires the extraction of tissue sample from the patient.

The "holy grail" of this study is the clinical application of optical tomography with in - situ characterization of tumors without biopsy. The construction of compact laser systems, such as femtosecond diode lasers which are under development, will assist towards the realization of this goal. In conclusion, the use of ultra - short laser pulses is very promising for non - invasive imaging inside the human body.

References

1. M. S. Patterson, B. Chance and B. C. Wilson : "Time resolved reflectance and transmittance for the noninvasive measurement of tissue optical properties", Applied Optics Vol.28, No. 12, 2331,1989
2. D.T Delpy, M. Cope, P. van der Zee, S. Arridge, S. Wray, J. Wyatt : "Estimation of optical pathlength through tissue from direct time of flight measurement", Selected papers on tissue optics. Applications in medical diagnostics and therapy, Valery V. Tuchin Editor, SPIE Milestone Series, Volume MS 102

Bio-Speckle Phenomena for Blood Flow Measurements: Speckle Fluctuations and Doppler Effects

Yoshihisa Aizu [1] and Toshimitsu Asakura [2]
[1]Muroran Institute of Technology, Muroran 050-8585, Japan, e-mail:aizu@muroran-it.ac.jp
[2]Hokkai-Gakuen University, Sapporo 064-0926, Japan, e-mail:asakura@eli.hokkai-s-u.ac.jp

Abstract. Both laser speckle and laser Doppler techniques can be used to measure human blood flows, but their interrelation is still unclear. This paper studies their differences, similarities, and relations in the moving rough surface and particle flow models, which are summarized in the table. The dominance of either speckle fluctuations or Doppler effects depends mainly on the coherence condition, particle concentration, and velocity distribution.

1 Introduction

Laser speckle and laser Doppler techniques are both based on dynamic light scattering phenomena. However, what is detected is originally different due to their different principles: the former is random intensity fluctuations and the latter is periodic beat signals. In spite of this, the terms "speckle fluctuations" and "Doppler effects" are carelessly mingled when they are applied to blood flow measurements. In fact, we have already reported bio-speckle flowmetry [1,2] for blood flow measurements, which is based on the speckle technique but very often it is considered to be the Doppler technique. The two techniques have been well established, but their interrelation is still unclear [3,4]. We discussed their differences, similarities, and mutual relations in the moving rough surface and particle flow models, and summarized results in the table.

2 Single Particle and Rough Surface Models

2.1 Single Particle

For the case of a single particle or very dilute particle suspension with a constant moving velocity, the laser Doppler technique works quite well. When the coherence condition is satisfied, the heterodyne mixing produces truly periodic Doppler beat signals with a Gaussian envelope due to the transit time τ_T. There is no speckle produced in this case because of a single or very few scattering centers.

2.2 Rough Surface

Single Beam Illumination. A moving rough surface illuminated by a single laser beam produces time-varying speckle fluctuations in the both diffraction and imaging fields. But, all the scattered waves from the surface are Doppler-shifted by different frequencies each other. No use of a reference beam means conventionally "*homodyne* mixing" which shows a monotonically decreasing spectrum. Thus, detected signals are truly speckle fluctuations and can be possibly considered homodyne components.

Dual Beam Illumination. When the differential-type Doppler arrangement is applied to a moving rough surface [5], detected components are Doppler beat signals with an envelope of speckle fluctuations. Then, the Doppler effects and speckle fluctuations coexist, but are not equivalent because of different frequency components. When the intersection angle θ of the two beams is reduced to zero degree, the Doppler heterodyne beat component disappears but the speckle fluctuation still remains.

Michelson Interferometer. Briers' model [4] using the Michelson interferometer in which one of reflecting mirrors is replaced by a rough surface moving along the optical axis corresponds to exactly the case of "Dual Beam Illumination" with θ = 90 (deg). Thus, detected signals are again the Doppler heterodyne beat with the speckle fluctuation envelope. The in-plane movement in this model yields no Doppler effect and no heterodyne beat even with the reference beam. Detected signals are truly speckle fluctuations which are very similar to homodyne signals.

3 Particle Flow Models

3.1 Single-Beam Doppler Velocimetry (LDV)

Moderate Concentration. In the single-beam Doppler technique studied by Riva *et al*[6], the Doppler-shifted wave from a moving red blood cell is heterodyned with the non-shifted wave reflected back from the vessel wall. Moderate concentration of particles may produce time-varying speckle fluctuations while it maintains dominant Doppler components. Thus, detected signals are possibly Doppler beat with the speckle fluctuation envelope. Figure 1(a) shows a typical recorded signal and the corresponding power spectrum measured for 1-μm diam acrylic particles flowing with water (1% concentration) in a cylindrical glass tube under He-Ne laser beam illumination. The signal shows periodic beat components with the envelope of speckle-like random fluctuations, and the power spectrum demonstrates the heterodyne cut-off characteristics with some broadening as we expected.

Receiving Cone Angle Enlarged & Flow in Orthogonal Direction. When the receiving cone angle is enlarged, the coherence condition tends to be corrupted due to range of scattering angles. Thus, the heterodyne components are degraded and speckle fluctuations become dominant. When the illumination and detection are both made in the direction orthogonal to the flow, no Doppler effect is produced. Then, speckle fluctuation components only remain and make the broadened spectrum centered at the zero frequency, which may be referred to as the homodyne spectrum.

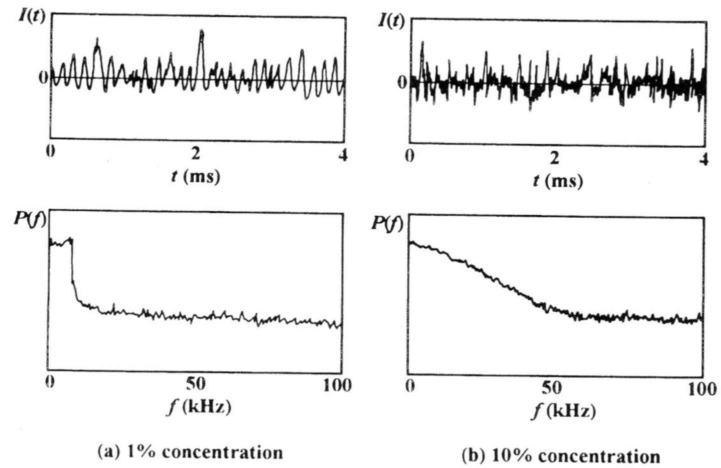

Fig.1. Typical signals $I(t)$ and power spectra $P(f)$ obtained from particles within water flowing in a glass tube for the particle concentration of (a) 1 % and (b) 10 %.

3.2 High Concentration and Bio-Speckle Flowmetry

Increased concentration of particles causes multiple scattering, corresponding to the case of blood flow measurements. In fact, bio-speckle flowmetry has been developed for this case. Numerous random directions of illuminating and scattered waves and random velocity distribution cause randomly distributed Doppler-shifted frequencies. Random positions of particles result in the randomness of optical path length for scattering waves. All of these effects randomize the phases and, therefore, the phase-consistent time or the speckle correlation time is significantly shortened. Bio-speckle flowmetry employs both the extended illuminating spot and the enlarged receiving cone angle which enhance the speckle fluctuations. The Doppler heterodyne components are degraded significantly and finally wiped out by speckle fluctuations.

Figure 1(b) shows a typical signal and its power spectrum measured for 10% concentration of particles in water. The signal shows random speckle fluctuations, and the power spectrum demonstrates monotonically decreasing components, which wiped out the heterodyne cut-off characteristics. The mixing of various Doppler-shifted waves with themselves may also be interpreted as the homodyne, and possibly be referred to as the intensity correlation (or fluctuation) spectroscopy (ICS or IFS), light beating spectroscopy (LBS), and photon correlation spectroscopy (PCS).

4 Conclusion

The relations of speckle fluctuations and Doppler effects were discussed and results are briefly summarized in Table 1. Surely the two phenomena are not equivalent, but in some situations they coexist or are mixed. The homodyne components are quite similar to the speckle fluctuations. Whether the laser Doppler velocimeter covers only the heterodyne technique or also the homodyne technique is a problem of terminology and probably due to a difference of interpretation.

Table 1. Relations of speckle fluctuations and Doppler effects in various situations.

		Speckle technique	Doppler technique	Relations & remarks
Single particle (very dilute suspension)		--- (no speckle)	O	Doppler beat only
Rough Surface	Single beam	O	--- (no reference)	speckle only [possibly homodyne]
Rough Surface	Dual beam	O	O	envelope + beat signal [coexisting, not equivalent]
Rough Surface	Michelson interferometer [movement along axis]	O	O	envelope + beat signal [coexisting, not equivalent]
Rough Surface	Michelson interferometer [in-plane movement]	O	--- (no shift)	speckle only [similar to homodyne]
Flowing particles	Moderate concentration [single or dual beam]	O	O (broadened)	envelope + beat signals (various frequencies) [coexisting, not equivalent]
Flowing particles	Large receiving cone angle	O (enhanced)	Δ (degraded)	speckle dominant [possibly homodyne]
Flowing particles	Flow in orthogonal direction	O	--- (no shift)	speckle or homodyne
Flowing particles	High concentration [single or dual beam]	O (higher freq. fluctuations)	--- (wiped out)	speckle or homodyne, [also ICS, IFS, LBS, and PCS]

Acknowledgement

Part of this work was supported by the research grant of Nakatani Electronic Measuring Technology Association of Japan.

References

1 Y. Aizu, K. Ogino, T. Sugita, T. Yamamoto, N. Takai, and T. Asakura, Appl. Opt. **31**, 3020 (1992).
2 Y. Aizu, T. Asakura, K. Ogino, T. Sugita, Y. Suzuki and K. Masuda, Bioimaging **4**, 254 (1996).
3 H.M. Pedersen, Opt. Acta **29**, 105 (1982).
4 J.D. Briers, J. Opt. Soc. Am. A **13**, 345 (1996).
5 J. Ohtsubo and T. Asakura, Optik **52**, 413 (1978).
6 C.E. Riva, B.L. Petrig, R.D. Shonat, and C.J. Pournaras, Appl. Opt. **28**, 1078 (1989).

Layered Gel-Based Phantoms Mimicking Fluorescence of Cervical Tissue

Svetlana Chernova*, Alexander Pravdin*, Yury Sinichkin*, Valery Tuchin*, Sandor Vari**
*Saratov State University, Department of Optics, Astrakhanskaya 83, 410026 Saratov, Russia
** Laser- und Medizin-Technigie GmbH, Kramerstrasse 6-10, D-12207 Berlin, Germany

1 Introduction

In recent decades autofluorescence spectroscopy has proved its potential to distinguish between neoplasia and tumors and normal human tissue. When combined with endoscopic techniques and instrumentation the method has become an effective tool for noninvasive diagnosis of neoplasia and early tumors in human epithelium and subepithelium which are the tissues easily accessible to endoscopic optical probing [1 - 7].

World Health Organisation reported about 500,000 cases of cervical cancer diagnosed every year [8], and early detection of cervical intraepithelial neoplasia (CIN) could be capable to reduce the mortality associated with this type of cancer. Colposcopy - the endoscopic method of visual examination of cervix surface, which is widely used in the routine of CIN diagnosis, could be improved by implementation of autofluorescence spectroscopy technique. But the quantification of spectral differences between normal and pathological tissues because of wide intrapatient and interpatient variability, overlapping of emission bands of endogenous fluorophores, effect of inner filter, and unreproducible alterations of the conditions of spectral data collection requires an employment of correlation technique. Nevertheless, in deciding on the form of correlation algorithm relating spectral features to tissue status it would be well to lean upon the relationship of spectral manifestations with definite biochemical and morphological changes in tissue.

To avoid the difficulties of direct search for such dependencies in experiments in actual biotissues we propose to employ tissue-like phantoms – structures mimicking optical and spectral characteristics of the natural object and reproducing morphological structure and fluorophore/chromophore composition of it.

Unlike actual biotissue, in phantom we are free to vary the structural parameters and chemical composition in an arbitrary way and follow the linked changes in fluorescence spectrum.

2 Materials and Methods

Following the stratified texture of actual cervical tissue we have proposed a three-layer planar architecture of phantom. Outer layer representing epithelium contains scatterers, in order to reproduce lightscattering in actual tissue, and, in series of samples, endogenous fluorophore - nicotinamide adenine dinucleotide reduced form (NADH). Gelatin gel that serves as a mechanical base of the layer moderately fluoresce under UV irradiation; this imitates a weak emission from keratinized epithelium. Such emission has been observed *ex vivo* in epidermis at UV excitation [9].

In cervical tissue, epithelium and stroma are separated by basal membrane, the main constituent of it being collagen IV. In the phantom the basal membrane is modeled by a thin film of dried gelatin. It counts in favor of this substitution that fluorescence spectra of gelatin used and collagen IV obtained with 365± 20 nm excitation reveals rather close similarity.

Cervical stroma, which incorporates capillary network, was represented in the phantom by inner layer. Besides gelatin gel and scatterers, this third layer contained human hemoglobin solution

Gelatin gel, which was used as a base of phantom layers, was prepared from gelatin (Sigma) and scatterer suspension in water. The proportion was: one gram of gelatin to nine milliliters of suspension, what gave the volume content of gelatin equal to one in 10% per weight water solution. The problem was to find out neutral scatterers that had no or low fluorescence under UV irradiaton. After some preliminary experiments we had chosen two: irregularly shaped quarts particles of micrometer and submicrometer dimensions and freshly precipitated $CaCO_3$.

Gel layers were prepared by charging melted gelatin gel with proper additives between nonadhesive plates separated by 300, 500 and 1100 mcm spacers. Then samples were allowed to harden in refrigerator for 10 – 15 min. Upon hardening, layers were cut into around 5 x 10 mm plates which were stuck together to form 3-layer phantom. To perform spectrophotometric measurements the layered structures were placed on quartz plates.

The additives used were: 4mM solution of NADH in borate buffer (pH 8.3) and 24.4 mg/ml water solution of hemoglobin; they were added to gel base just before charging in order to prevent thermal damage during melting. Dried gelatin film mimicking basal membrane was prepared using 500 mcm 10% gelatin gel layer which after 24 hours drying underwent nearly exact tenfold shrinkage, giving 50 mcm film. Radiation of either N_2 laser (337 nm) or Xe arc lamp (filtered by 360 ± 20 nm band pass filter) was used for excitation of phantom fluorescence. Spectral data was collected using grating monochromator with PMT and was stored on PC.

3 Results and Concluding Remarks

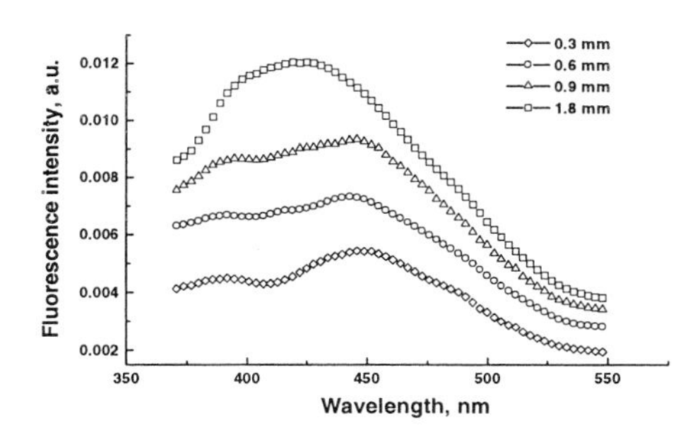

Fig. 1. Fluorescence of the phantom with different thickness of the "epithelial" layer (excitation 337 nm)

In phantom development we tried to reproduce actual dimensions of the tissue. Regarding to squamous epithelium we based on the assumption that its thickness is about 300mcm. Hystological data [10] did not significantly differ from this estimation.

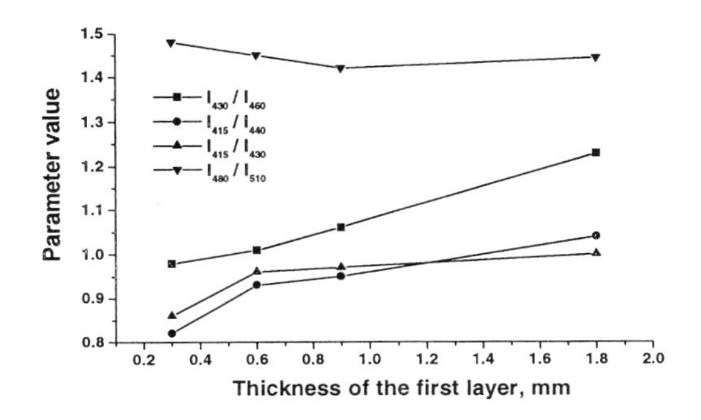

Fig. 2. Algorithms for spectral dependency on "epithelium" thickness

As there is some evidence that neoplasia is associated with epithelium thickening, we prepared a series of phantom samples with different "epithelium" thickness starting from "normal" value of 300 mcm. Concentration of hemoglobin in the "stromal" layer was the same in all samples and coresponded to 4.5 % p.v. of tissue blood content. Fluorescence spectra of these phantoms at 337 nm excitation are shown in Fig. 1.

Emission from phantom results from the fluorescence of gelatin of all three layers, but emission from the "stromal" layer is modulated by hemoglobin absorption. This is also true for the portion of emission of two upper layers that is remitted by the scattering within the hemoglobin containing third layer. This reabsorption manifests itself in fluorescence spectra as an through at 405 – 420 nm (Soret band of hemoglobin). The increase of "epithelium" thickness leads to attenuation of excitation radiation within "stromal" layer and thus to the decrease of the fraction of modulated emission from the third layer in total fluorescence escape. As a consequence, the thickening of epithelium manifests in smoothing of the spectrum, and in the case of 1.8 mm outer layer the through practically is not observed. To quantify this spectral dependency we tried four algorithms – all being the ratio of fluorescence intensity at two chosen wavelengths (see Fig. 2).

Analyzing these results we are inclined to give the preference to I_{430} / I_{460} algorithm for its linearity in "epithelium" thickness and reasonable slope.

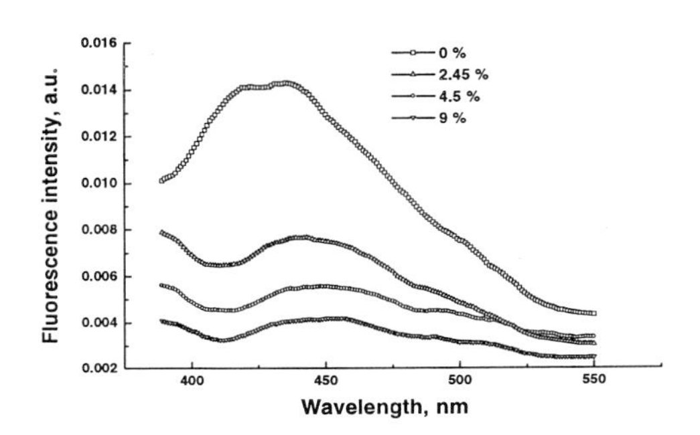

Fig. 3. Fluorescence of the phantoms with different blood content in "stromal" layer (excitation 337 nm, per volume percentage of blood is shown)

The development of cervical neoplasia may be accompanied by vascularisation of the tissue, that leads to increase of tissue blood content. Also, the blood content may rise in inflammatory processes. We have made an attempt to follow the

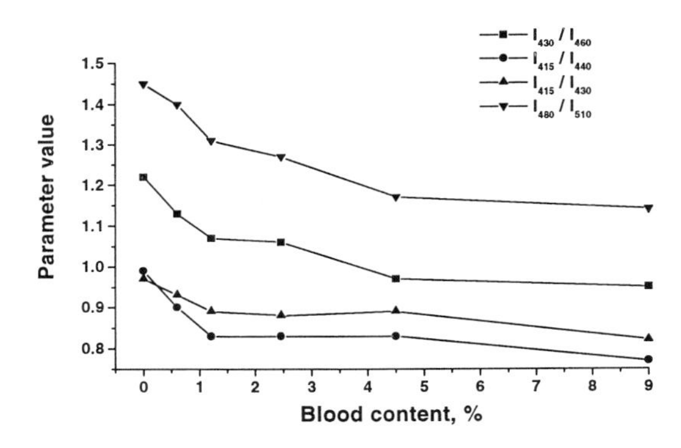

Fig. 4. Algorithms for spectral dependency on blood content in "stroma"

manifestation, in fluorescence spectra, of the variation of hemoglobin content in "stromal" layer. Fluorescence spectra obtained are shown in Fig. 3.

As in the previous case, we tried for algorithms four spectral differences quantification (see Fig. 4.).

Here, two algorithms, I_{430} / I_{460} and I_{480} / I_{510}, may be suppoused for the estimation of tissue blood content from fluorescence spectra date. But the preference to one of them may be given only after the study of more complicated model accounting for more exogenous fluorophores.

One of the main spectral features of the spectra in Fig. 1 and Fig. 2 is practically the same wavelength of the peaks, but the fluorescence spectra of actual tissue in pathology [11] show the shift of peak position to higher wavelength. This behavior can not be modeled on the base of fluorescence reabsorption only. Keeping in mind that increased NADH fluorescence may be considered as a maker of neoplasia, we prepared the series of phantoms with different amount of NADH solution added to "epithelial" layer. An evident longwavelength shift of the fluorescence peak if the samples with NADH added can be easily seen from Fig. 5. The very shape of spectral curve of NADH containing samples more closely reproduses spectra of actual cervical tissue with pathology.

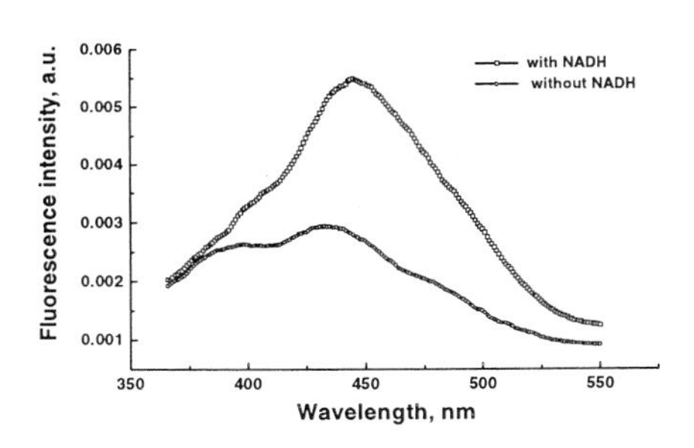

Fig. 5. The influence of NADH on the shape of phantom fluorescence spectra (excitation 337 nm)

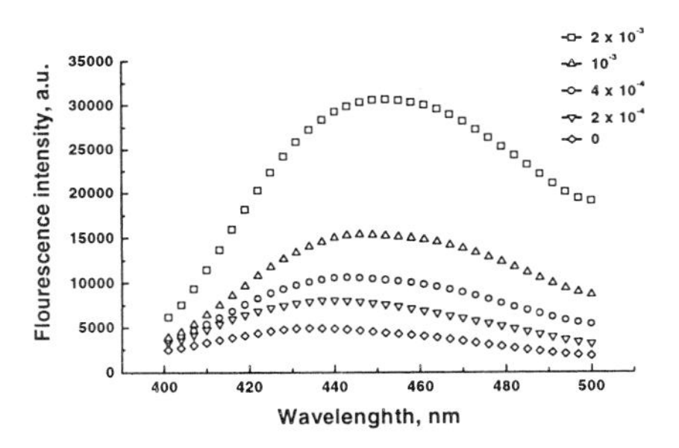

Fig. 6 illustrates the shift of phantom fluorescence peak wavelength (365 nm excitation) linked with the increase of NADH content in "epithelium".

Acknowledgements

This work was supported by EC in the frame of INCO-COPERNICUS Program: Cooperation in Science and Technology with Central and Eastern European Countries, grant PL 965117.

References

1. G.Zonios, R.Cothren, J.Arendt, J.Wu, Fluorescence spectroscopy for colon cancer diagnosis, Proc. SPIE **2324**, 9 – 13 (1994).

2. M.L.Harries, S.Lam, C.MacAulay, J.Qu, B.Palcic, Diagnostic imaging of the larynx: autofluorescence of laryngeal tumors using the helium-cadmium laser, J.Laryngol. Otol. **109** (2), 108 – 110 (1995).

3. F.Koenig, F.G.McGovern, A.F.Althausen, T.F.Deutsh, K.T.Schomacker, Laser induced autofluorescence diagnosis in bladder cancer, J.Urol. **156** (5), 1597 – 1601 (1996).

4. M.Panjehpour, B.F.Overholt, T.Vo Dinh, R.C.Haggitt, D.H.Edwards, F.P.Buckley, Endoscopic fluorescence detection of high grade dysplasia in Barrett's esophagus, Gastroenterology **111** (1), 93 – 101 (1996).

5. J.K.Dhingra, D.F.Perrault, K.McMillan, E.E.Rebeiz, S.Kabani, R.Manoharam, I.Itzkan, M.S.Feld, S.M.Shapshay, Early diagnosis of upper aerodigestive tract cancer by autofluorescence, Arch. Otolaryngol. Head Neck Surg. **122** (11), 1181 – 1186 (1996).

B.W.Chwirot, S.Chwirot, W.Jedrzejczyk, M.Jackowski, A.M.Raczynska, J.Winczakiewicz, J.Dobber, Ultraviolet laser – induced fluorescence of human stomach tissues: detection of cancer tissues by imaging techniques, Lasers Surg. **21**, 149 – 158 (97).

7. N.Ramanujam, M.F.Mitchell, A.Manadevan, S.Thomsen, E.Silva, R.Richards-Kortum, Fluorescence Spectroscopy: A Diagnostic Tool fro Cervical Intraepithelial Neoplasia (CIN), Gynecologic Oncology **32**, 31 – 38 (1994).

8. Ya.V.Bokhman. Manual on oncogenicology. – Leningrad: Medicine, 1989 (in Russian).

9. S.Chernova, A.Pravdin, E.Bukatova, Autofluorescence of human epidermis under UV-irradiation in vitro, Proc. SPIE **3053**, 152 – 159 (1996).

10. Dr. B.Sipos. Private communication.

11. N.Ramanujam, M.F.Mitchell, A.Manadevan, S.Warren, S.Thomsen, E.Silva, R.Richards-Kortum, In vivo diagnosis of cervical intraepithelial neoplasia using 337 nm – excited laser-induced fluorescence, Proc. Natl. Acad. Sci, USA, Medical Sciences, **91**, 10193 – 10197 (1994).

Interferometrical Microscope for Investigations of Biological Objects

D. Tontcheva,I. Sainova, N. Metchkarov*, V. Sainov*

*Central Laboratory of Optical Storage and Processing of Information to the Bulgarian Academy of Sciences, Sofia 1113, P.O.Box 95, e-mail: vsainov@optics.bas.bg; nikola@optics.bas.bg

Abstract. In the present work the results for multidimensional visualisation of cells and chromosomes by computerised phase-stepping laser interferometry are presented. The set-up used is based on a Mach-Zender interferometer with projection microscope (magnification up to 600×) in one of the arms and a reference five-steps shifted beam in the other. High resolution CCD camera in measurement mode with 256 grey levels is used. The recordings are performed with different wavelengths in the red (He-Ne laser), green and blue (Ar^+ laser) spectral regions. The distortions of the optical elements are omitted by subtracting images of the objects and/or reference plane. Essential features of the system are the optico-electronical feed-back and the specialised software for recording, digital processing and visualisation of the images.

1 Introduction

A pressing task for the modern biomedical investigations down to cellular and chromosome's level is the multidimensional (3D or even 4D) analyses of the objects. Different approaches such as confocal microscopy, scanning laser microscopy, computerised thomography, digital image processing and others have been successfully applied [1]. The confocal microscopy offers several advantages over the conventional one. The blurring of the image from light scattering is reduced, the signal to noise ratio is improved, the effective resolution is higher and x-y scan can be made over wide areas for measurements. The new developments based on lasers scanning beam and digital processing of the images give new understanding to cellular and chromosomes structures and functions. The observation of structural components in three dimensional space provides better information about the objects and finds large practical applications. The progress is astonished but not enough to meet update requirement of biomedical investigations for in-live and fast measurement of microobjects with uncertainty close to the uncertainty of the electronic and atomic force microscopes. The new scientific tasks for measurement in the visual spectral range are connected with improvements of sensitivity, accuracy and lateral resolution. Due to the technological advancements the Rayleigh resolution limit may be overcome [2-4]. The supper resolution and some otherwise unresolved details of the objects could be obtained using a given apriori information about the input signal such as shape, phase distribution above the investigated object and others. As a first step of solving this problem the aim of the present work is to apply phase-stepping laser interferometry for visualization and microscopic investigation of cells and chromosomes. The main goal is to obtain information about the object's shape and phase distribution at different wavelengths.

2 Experimental setup and results

The experimental setup for microscopic cells and chromosomes investigation by phase-stepping laser interferometry is shown in Fig.1. It is based on a Mach-Zender interferometer with projection microscope (magnification up to 600×) in one of the arms and a reference five-steps shifted beam in the other.

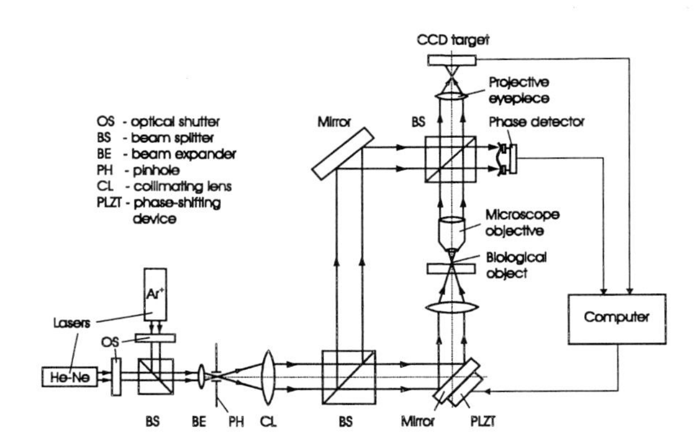

Fig.1. Experimental setup for microscopic cells and chromosomes investigation by phase-stepping laser interferometry.

High resolution CCD camera in measurement mode with 256 grey levels is used. The recordings are performed with different wavelengths λ in the red (He-Ne laser), green and blue (Ar^+ laser) spectral regions. The distortions of the optical elements are omitted by subtracting images of the objects and/or reference plane. Essential features of the system are the optico-electronical feed-back and the specialised software for recording, digital processing and visualisation of the images. The objects are phased - leucocytes and human chromosomes, dispersed on a usual for microscopic investigations glass substrates. Omitting refraction of the light the optical path changes $\Delta\Phi(x,y)$ over the objects in normal (z) direction could be presented as:

$$\Delta\Phi(x,y) = \int [n(x,y,z) - n_0] dz = N\lambda \qquad (1)$$

where n_0 is the refractive index of the medium surrounded the object ($n_0 = 1$ for air, without immersion liquid), $n(x,y,z)$ is the refractive index of the objects and N is a number.

In the case of illuminations with different wavelengths λ_1 and λ_2 the phase distribution $\varphi_{1,2}(x,y)$ of the objects light can be express as:

$$\varphi_{1,2}(x,y) = \frac{2\pi}{\lambda_{1,2}} \int_0^L n(x,y,z)dz\,, \qquad (2)$$

where L is the size of the object in z direction.

In the phase stepping laser interferometry the phase differences could be measured with uncertainty up to $\lambda/100$. In our experiments the wavelengths used are λ_1 = 632,8 nm (He-Ne laser), λ_2 =488 nm and λ_3 = 514,5 nm (Ar^+ laser). From the measured optical path changes in known dimensions of the object, the distribution of the refractive index over the objects, and vice versa in known refractive index - the dimensions of the object could be estimated.

The experimentally obtained results are presented in Fig.2 to Fig.4.

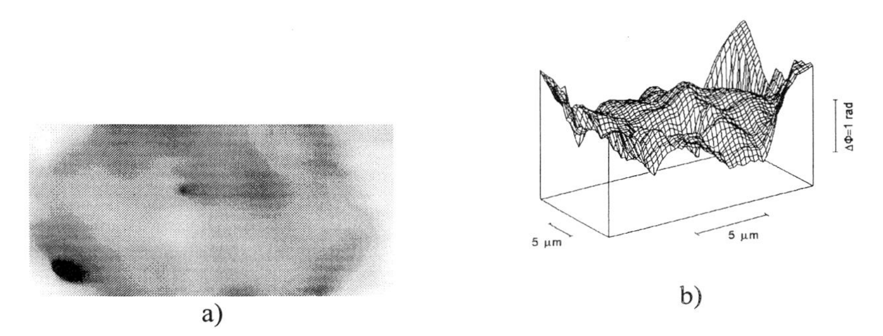

Fig.2. Phase map modulus 2π - a) and 3D-visualization - b) of a leucocyte at illumination by He-Ne laser

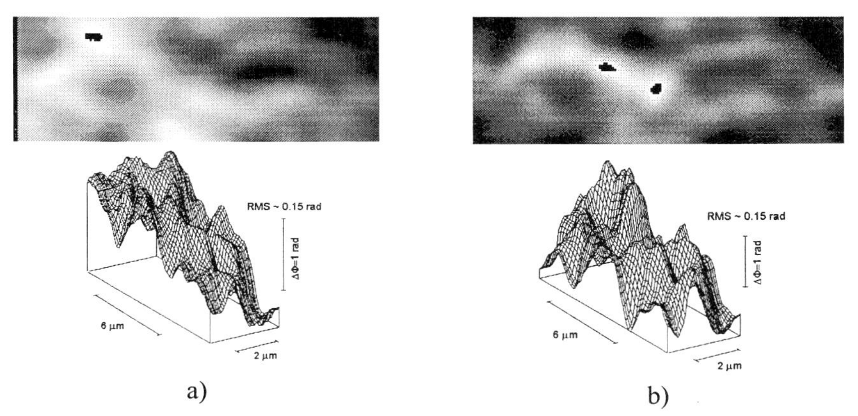

Fig.3. Phase map modulus 2π and 3D-visualization of a chromosome at illumination by Ar^+ laser (λ=488 nm) a) and a difference phase map modulus 2π at consecutive illumination of the same object by He-Ne and Ar^+ laser (λ=488 nm) b).

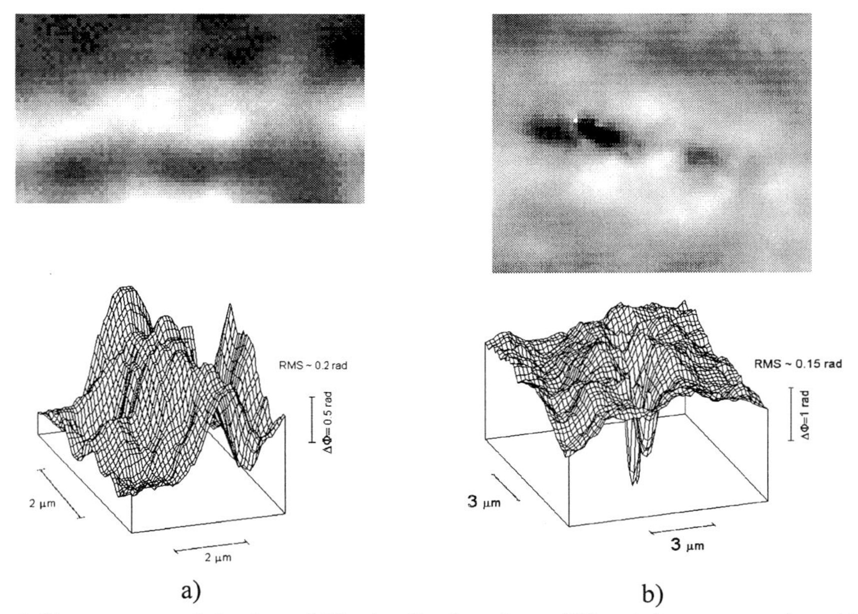

a) b)

Fig.4. Phase maps modulus 2π and 3D-visualization of two different chromosomes (a and b) at illumination by He-Ne laser.

3 Conclusions

The promising results for 3D visualization and measurement of optical path changes in illumination of cells and chromosomes with coherent (laser) light have be obtained. The use of phase-stepping laser interferometry and digital processing of interference patterns could be applied for investigation of biological micro objects. The new additional information about the structures offer new and better possibilities for the biomedical investigations.

4 Acknowledgements

This work is funded through contracts F-613 and F-638 with National Scientific Foundation in Bulgaria and international scientific collaboration.

References

1. H. Sickinger, J. Schwider; G. Häusler at al., Optik Ann. Rep. (Univ. Erlangen-Nürnberg, 1997) 10, 64
2. A.I. Kartashev, *Optical systems with enhanced resolving power*, Opt. Spectr. **9**, 204-206 (1960)

3. V. Tychinski, *Wavefront dislocations and registering images inside the Airy disk*, Opt. Comm. 81, 131 (1990)
4. O.J.F. Martin, A. Dereux, C.Girard, *Iterative scheme for computing the total field propagating in dielectric structures of arbitrary shape*, J.Opt. Soc. Am. 11, 1073 (1994)

Realization of a Double-Exposure Interferometer with Photorefractive Crystals for Biomedical Applications

M. Weber, F. Rickermann, and G. von Bally

Laboratory of Biophysics,
Institute of Experimental Audiology, University of Münster,
Robert-Koch-Str. 45, D-48129 Münster, Germany
Tel.: (+49) 251-83-56888, Fax: (+49) 251-83-58536,
e-mail: maweber@uni-muenster.de

Abstract. A holographic double-exposure interferometer for deformation and vibration analysis of biomedical objects is presented. Utilizing two ns pulse-lasers and a photorefractive sillenite-type crystal to store the holograms, double-exposure interferograms can be recorded and visualized with high repetition rates. The limitations of the energy density for irradiation of human tissue defined by the German national laser safety regulations (VBG93) are taken into account. To achieve a quasi real-time inspection of dynamic processes (e.g. in medical diagnostics), a progressive scan camera and a fast frame grabber system are used to digitize the read out interferograms. A visualization of the object movement in video frame rate (25 Hz) is obtained. Using a spatial-heterodyne technique, object deformations can be calculated quantitatively from each interferogram.

1 Introduction

Holographic interferometry is a well established, highly sensitive, contactless, and non-destructive metrology in the medical and technical domain [1]. Here, photorefractive crystals have certain advantages compared to conventional holographic recording materials, e.g. they do not need any development or sensitizing processes and the holograms are erasable by homogeneous illumination. Several promising applications of photorefractive crystals in holographic interferometry, especially in real-time or quasi (video) real-time interferometry, have been demonstrated [2,3]. The latter uses fast sequences of double-exposure interferograms to investigate object movements between the two exposures. Thus, interferometric stability is required only between the two corresponding exposures, and the measurement becomes insensitive against environmental disturbances which are slow with respect to the time delay between the two light pulses. Nanosecond pulses from Q-switched lasers are classical means to avoid phase variations during in-vivo recording. For an automated digital evaluation spatial-heterodyne techniques are preferable, because from each recorded holographic interferogram optical phase calculation is possible without time consuming additional readouts of phase shifted interferograms [4].

2 Experimental Methods

A cadmium-doped (1.4 wt% Cd) bismuth titanate ($Bi_{12}TiO_{20}$) single crystal is used (BTO:Cd) as rewritable holographic recording material. The crystal was grown at the Kurnakov Institute of the Russian Academy of Sciences in Moscow [5]. The sample is cut along the crystallographic planes [110], [001], and [$1\bar{1}0$]. This orientation yields the highest diffraction efficiency in holographic experiments [3]. The size of the polished surface is 8.2 mm × 8.8 mm and the sample is 1.9 mm thick. The experimental setup for a spatial-heterodyne double-exposure interferometer is shown in Fig. 1.

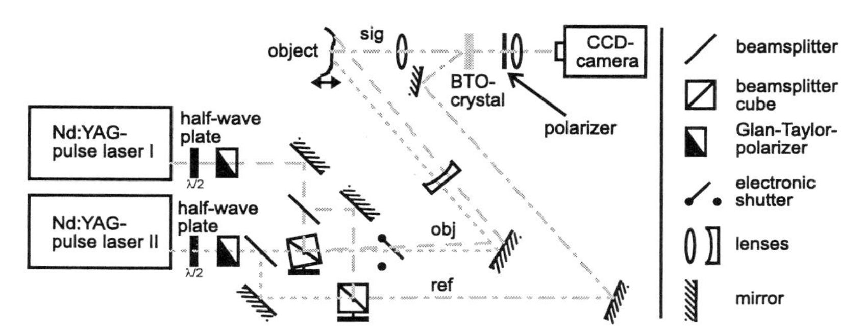

Fig. 1. Schematic diagram of the double-exposure spatial-heterodyne interferometer. Two holograms of an object are recorded with two frequency-doubled Q-switched Nd:YAG lasers in a BTO:Cd crystal. Each hologram is recorded by a signal beam (sig) reflected from the object and a reference beam (ref) from one laser. After blocking the object illumination (obj) with an electro-mechanical shutter, both holograms are reconstructed simultaneously with one reference beam. The resulting interferogram is digitized by a CCD-camera (for details see text).

To achieve the required short writing times as well as a short time delay between the double-pulses, two frequency-doubled Nd:YAG lasers (Coherent Infinity 40-100: wavelength 532 nm, pulse width 3 ns, maximum energy 250 mJ/pulse, repetition rate 0.1–100 Hz, beam diameter 10 mm (flat-top)) are used to record and read out the holograms. Two identical Q-switched Nd:YAG-lasers are necessary to ensure the recording with the reference wave at the same spatial position, and the pulse delay between the investigating double pulse can be adjusted down to 10 ns.

In the first step, two image-plane holograms of the object are stored with a short time delay in a BTO:Cd crystal. Here, the first hologram is recorded with a pulse from laser I, and after a time delay, the second hologram is written with a pulse from laser II. Each pulse is divided into a signal (sig) and a reference (ref) beam to record one hologram of the respective object state. Afterwards, the object beam (obj) is blocked by an electro-mechanical shutter and both holograms are read out simultaneously by one pulse of the reference beam. The resulting interferogram is digitized by a CCD-camera and a frame grabber for further phase evaluation. To digitize a full video-frame with one nanosecond-pulse, a progressive scan camera is necessary. The video blank signal is used to synchronize the electro-mechanical shutter and the laser pulses. The interferograms are visualized and digitized in high resolution with a video frequency of 25 Hz. Another pulse of the reference beam erases the holograms in the crystal [3]. A special spatial-heterodyne technique is used for quantitative evaluation of the object deformation [4]. By means of a permanent tilt between the object beam of the first and the second laser, a spatial carrier frequency is generated. The advantage of this approach is that no moving mechanical element, which could cause phase errors due to inaccurate repositioning, is needed. Thus, only one digitized interferogram is necessary and high repetition rates can be achieved.

3 Experimental Results

The spatial-heterodyne technique allows an automatic phase evaluation, and the exact determination of the movement direction is possible [4]. Fig. 2 presents an example sequence recorded with this system. The upper row shows digitized interferograms of a loudspeaker membrane (24x24 mm^2) vibrating with 3 kHz. The lower row displays the corresponding object movement obtained by using a Fourier-transform evaluation.

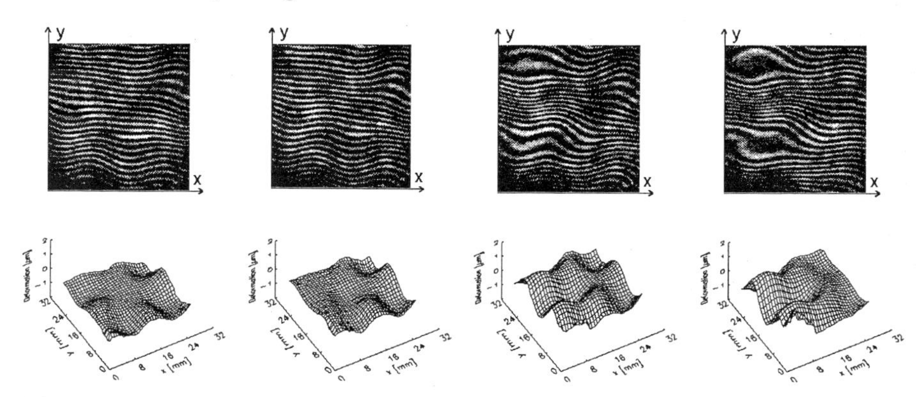

Fig. 2. Upper row: double-exposure interferograms taken from a loudspeaker membrane (24x24 mm^2) vibrating with 3 kHz. The interferograms are recorded at a frequency of 12.5 Hz (delay between recording pulses: 50 ms, spatial carrier frequency: 39 lines/cm). Lower row: object movement calculated from these interferograms.

When investigating human tissue, the permitted irradiation with laser light is limited in energy density as well as exposure time and rate by national laser safety regulations (e.g. in Germany by VBG93). To obtain fringe contrast in the interferograms suitable for digital evaluation, it is sufficient to have a signal beam energy density of 0.02 mJ(cm)$^{-2}$ at the crystal surface. Taking into account the losses within the setup and the reflectivity of the

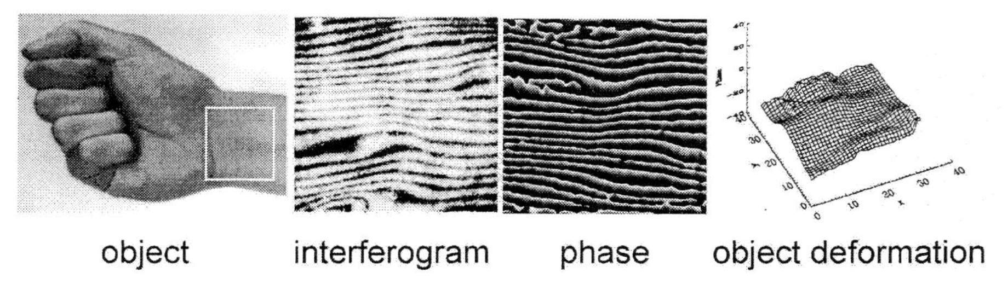

Fig. 3. Example for in-vivo recording of human tissue movements due to heartbeat: spatial carrier frequency: 18 lines$(\text{cm})^{-1}$; delay between recording pulses: 1 ms; repetition rate: 8 Hz; energy at the object: 5 mJ$(\text{cm})^{-2}$. The modulation frequency is that of the heartbeat and the deformation is caused by the blood pressure variation.

object (e.g. tissue), the energy density of the object beam has to be larger than 1 mJ$(\text{cm})^{-2}$.

Within the limits of the legal requirements of German national laser safety regulations a repetition rate of 8 Hz has been achieved in-vivo . An example is shown in Fig. 3. To obtain higher repetition rates, e.g. for visualizing movements in video frame rate (25 Hz), the sensitivity of the recording material has to be further increased.

References

1. G. von Bally, H.I. Bjelkhagen, eds., Optics for Protection of Man and Environments Against Natural and Technological Disasters, Vol.2 of Series in Optics Within Life Sciences (Elsevier Science, Amsterdam, 1993)
2. J.P. Huignard and J.P. Herriau, Real-time double-exposure interferometry with $Bi_{12}SiO_{20}$ crystals in transverse electrooptic configuration, App. Opt. **16** No. 7, (1977), 1807–1809.
3. F. Rickermann and S. Riehemann and G. von Bally, Utilization of Photorefractive Crystals for Holographic Double-Exposure Interferometry with Nanosecond Laser Pulses, Opt. Commun., **155**, (1998), 91–98.
4. M. Takeda, Spatial-carrier fringe-pattern analysis and its applications to precision interferometry and profilometry: An overview, Ind. Metrology, **1**, (1990), 79–99
5. S. Riehemann, F. Rickermann, V.V. Volkov, A.V. Egorysheva und G. von Bally, Optical and photorefractive properties of BTO crystals doped with Cd, Ca, Ga, and V, J. Nonlinear Opt. Phys. Mater. **6** (1997) 235–249.

Financial support of the Deutsche Forschungsgemeinschaft (SFB 225, project D7) is gratefully acknowledged.

Comparative Studies of Laser Induced Fluorescence and Intravascular Ultrasound for the Human Coronary Artery Diagnosis of Atherosclerosis

A. Manolopoulos[a1], A. Vasileiou[a], V. Kokkinos[b], G. Athanasopoulos[b], E. Agapitos[c], N. Kavantzas[c] and D.Yova[a]

[a] Biomedical Optics & Applied Biophysics Lab, Department of Electrical and Computer Engineering, National Technical University of Athens, Greece.

[b] Onassis Cardiac-Surgery Center, Athens , Greece.

[c] Department of Pathology, Medical School, University of Athens, Greece.

Abstract. We have acquired Intravascular Ultrasound (IVUS) images and recorded Laser Induced Fluorescence (LIF) spectra from postmortem human coronary artery samples (58 specimens, 18 healthy, 32 fibrous and 8 atheromatic, according to histology) in order to improve the efficacy of diagnosis of atherosclerosis by combining the information derived from each method. The studies were performed by a dual catheter, which was developed for simultaneous LIF and IVUS signal recording from the same cross section of the artery lumen. The dual catheter was consisted of a flexible 3.5 Fr phased array, intravascular catheter and a 6.5 Fr lateral firing quartz fiber, joined together through a biocompatible thermoplastic sheath.Fluorescence spectra were collected with the use of a LIF device, with a Nitrogen Laser (337nm), as the excirtaion source, coupled to the lateral firing quartz fiber. The arterial wall spectra were collected by the same fiber; displayed and analyzed by an optical multichannel analyzer. LIF characterisation was made by selection and implementation of empirical dimensionless ratio algorithms A7=I(380)/I(540) and A9=I(450)/I(420) combined in a binary diagnostic scheme, introducing a subsequent index of discrimination among healthy, fibrotic and atheromatic coronary arterial specimens. IVUS characterisation was made by an experienced echo-cardiologist according to image vessel wall thickness, echo-density and smooth or irregular distribution of echoes. Thickness was assessed as normal or increased; density as normal, increased or decreased; and configuration as smooth or irregular. The implemented LIF diagnostic scheme managed to provide adequate discrimination between normal and fibrotic specimens with sensitivity reaching 0.94 and 0.85 respectively, whereas specificity, accuracy and predictive value were over 0.88. Interpretation of the statistical results shows that LIF & IVUS are complementary & co-operative intravascular diagnostic methods, that can be combined towards improved diagnosis.

1 Introduction

Atherosclerosis produces a complex histological pattern, composed of fibrotic thickening, lipids, calcification and/or necrosis. No presently available diagnostic method allows this process to be completely characterized in human coronary arteries. Moreover early diagnosis is very important, since the implementation of therapeutic procedures is then easier and more effective. The most common diagnostic technique is contrast coronary angiography, which provides an image of the longitudinal lumen cross section but it gives no information about the volume and the composition of the atheroma present [1]. On angiography, a mild degree of lumen narrowing may actually represent a large atheroma volume [2].

Intravascular Ultrasound (IVUS) images coronary artery stenosis with greater detail than contrast angiography [3]. IVUS imaging can also identify dense calcification [4] and to a limited degree, lipid filled zones and regions of fibrous composition [5]. Accordingly, this modality is being utilized more frequently in clinical practice, as a means of assessing vessel wall architecture.

Laser Induced Fluorescence Spectroscopy (LIFS) is a technique under investigation, for tissue characterization. LIF ability to discriminate diseased from normal artery wall has been verified by several researchers [6, 7, 8]. Furthermore, classification algorithms based upon spectral features have been implemented on peripheral, aortic or coronary arterial spectra in order to discriminate among lesion types [9, 10, 11, 12].

Since different techniques base their diagnosis on different information revealing different lesion characteristics, the attempt to combine them in order to improve the efficacy of diagnosis, is considered important. The combination of LIFS and IVUS has the potential of providing information for both the biochemical and structural information of tissue.

This work describes a combined diagnostic system which utilizes ultrasound to provide a cross-sectional image of the artery while at the same time autofluorescence spectra are recorded through a suitable side-firing optical fiber adjacent to the ultrasound transducer. Using this dual catheter data from 58 coronary artery specimens were wecorded and diagnosis from both techniques is presented and confirmed by histopatology. The statistics of the combined diagnosis is then presented revealing a significant increase in all the statistical parameters.

2 Materials and methods

2.1 Tissue samples

Fifty-eight (58) post-mortem human coronary artery samples were obtained from several autopsies and after excision, immediately stored at -70°C (in liquid N2). During the experiments, samples were left to thaw at room temperature normally, rinsed and immersed in saline. The dual LIF & IVUS catheter was inserted by the echocardiologist into the lumen, imitating the catheterization procedure. The specimen lesions of interest were marked externally and sent for histopathologic examination, which has shown that, 18 of them were healthy, 32 fibrous and 8 atheromatic.

2.2 Experimental apparatus

The experimental set up system is shown in Figure 1, and specifically it is consisted from the following subsystems.

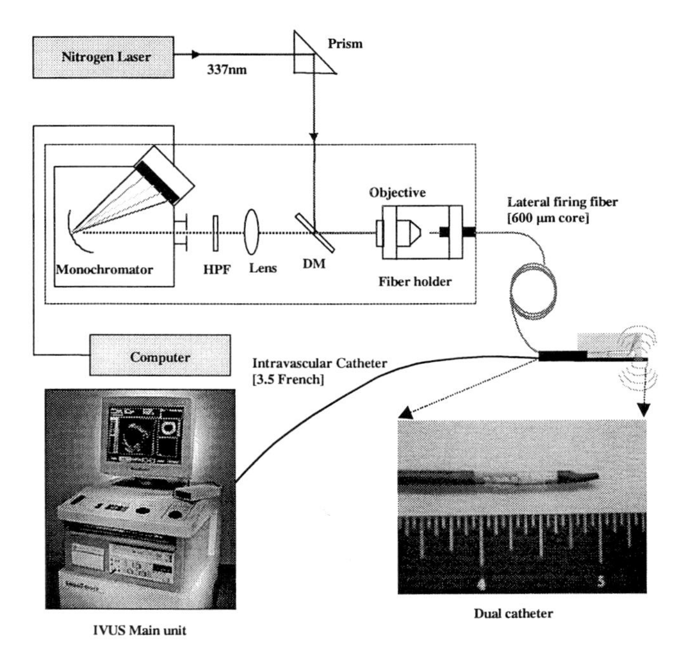

Figure 1. Experimental apparatus, including the LIF system, IVUS main unit and the dual catheter.

Laser Induced Fluorescence System. A pulsed nitrogen laser (Laser Science VSL337NDS, $\lambda = 337$ nm, $E_{pulse} = 240$ μJoule, t_{Pulse}=7 ns) was used for excitation. The output of the laser was coupled to the LaserSonics UltraLine Catheter (6.5 fr, 600 nm core, lateral firing quartz fiber) of the dual LIF & IVUS catheter. Fluorescence emission was collected by the same fiber and focused at the entrance slit (100 μm) of a 0.25m spectrograph. Data acquisition and analysis were performed via an Optical Multichannel Analyzer (OMA) employing a diode array detector. The signal from the diode array detector was fed to a PC for A/D conversion and further processing. A high pass filter was placed in front of the spectrograph entrance to eliminate the reflected laser light. The spectrum obtained during the experiment represented the time-integrated fluorescence of the underlying tissue for 75 laser pulses.
Intravascular Ultrasound System. IVUS images were obtained by a flexible 3.5 French intravacular catheter Endosonics-Visions® Five 64, (phased array catheter, f =20 MHz). The image display system (Endosonics S7700470 - INV) provided a

two dimensional 360° cross sectional 64 grayscale images of the vessel with 68 dB dynamic range.

Dual LIF & IVUS catheter. A dual catheter was developed for simultaneous LIF and IVUS signal recording from the same cross section of the artery lumen. The dual catheter was consisted by the Endosonics-Visions® Five 64 intravascular catheter and the LaserSonics UltraLine lateral firing quartz fiber joined together through a biocompatible thermoplastic sheath. The dual catheter operation was verified for both diagnostic methods. The overlapping regions of simultaneous LIF and IVUS signal recording was between "hours" eleven and one of the IVUS image periphery, whereas the body of the LIF catheter was creating a "shadow" around "hours" eight and ten. This defect was overwhelmed by manual rotation of the dual catheter from the echocardiologist (a common act for everyday intravascular catheterization).

2.3 Data analysis

IVUS characterisation was made by an experienced echo-cardiologist according to image vessel wall thickness, echo-density and smooth or irregular distribution of echoes. Thickness was assessed as normal or increased; density as normal, increased or decreased; and configuration as smooth or irregular.

LIF characterisation was made by selection and implementation of empirical dimensionless ratio algorithms, based on spectral intensity features at 390, 420, 450 and 540 nm. These spectral regions are proposed as reflecting the relative tissue composition of fluorophores [11,12,13]. Analytically, the selected algorithms were A7=I(380)/I(540) and A9=I(450)/I(420) combined in a binary diagnostic scheme, introducing a subsequent index of discrimination among healthy, fibrotic and atheromatous coronary artery specimens.

3 Results and Discussion

IVUS images and LIF spectra were used to independently classify the specimens as normal, fibrous and atheromatic. Typical examples of IVUS image and corresponding LIF spectra are shown in Fig.5 which demonstrates the ability of the dual catheter to provide information for differences in arterial wall fluorescence even in the same section. This is considered important since atherosclerotic lesions are mostly located eccentrically to the lumen axis without covering the whole lumen circumferentially.

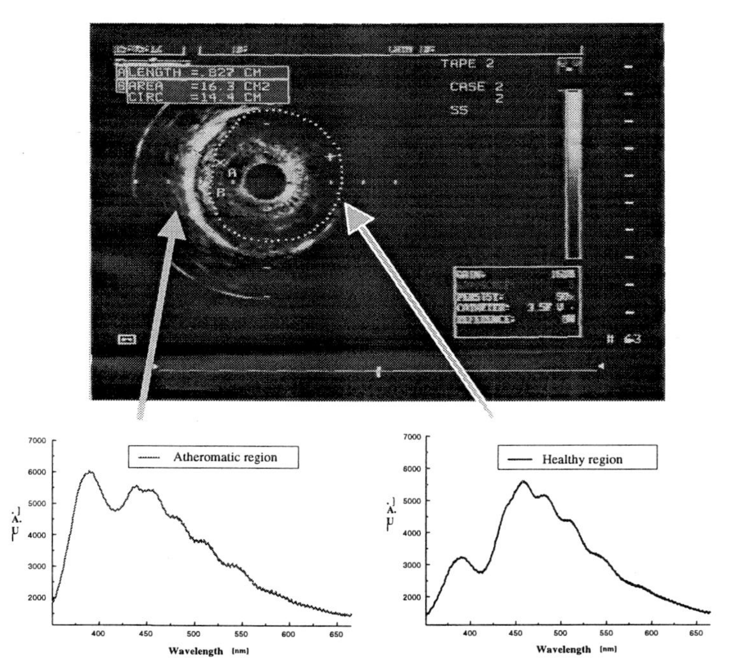

Figure 2. Characteristic IVUS image of coronary artery section comprising both healthy and atheromatic regions with corresponding LIF spectra.

IVUS results: The statistical interpretation of the ultrasound results is shown in Table 1. IVUS appears to be a highly sensitive and specific technique when assessing atheromatic specimens (with values over 0.9 for all statistical parameters). On the other hand for normal specimens although it provides a moderate 0.85 for specificity, it appears to be no sensitive (0.5, only 9/18 true positives). This fact has to be evaluated taking into concideration that there were not many false positives (6). This reveals a preliminary observation, that there is a tendency for classification of normal samples as fibrotic, explained by the fact that atherosclerosis is a gradually evolving disease and early fibrotic lesions do not differentiate enough from normal artery wall.

LIF results: Algorithm A7 focuses on the region of maximum collagen fluorescence (390 nm), and the region where the elastin fluorescence tail prevails (540 nm) compared to that of collagen. The second empirical Algorithm A9 uses the ratio of the maximum observed spectrum intensity (450 nm), with the intensity at 420 nm. The two algorithms were combined in a binary diagnostic scheme towards improved diagnosis (Table 1). The decision line A7 = 0.47 separates the majority of the fibrous specimens, while the A9=2 line separates the atheromatic from the normal ones. These lines were selected in order to maximize classification accuracy. Statistical results are shown in Table 1. Classification of the 18 normal specimens appears especially effective since values for all statistical parameters are over 0.89. Worth mentioning is the fact that 17 out of 18 specimens were classified correctly, while only 2 non-normal were included. The

32 fibrotic specimens were classified with specificity 0.89 and above 0.85 for all the other parameters. While the specificity and accuracy of the method is up to 0.9 for atheromatic specimens, the sensitivity drops to 0.63 dragging down the predictive value at 0.63, as well.

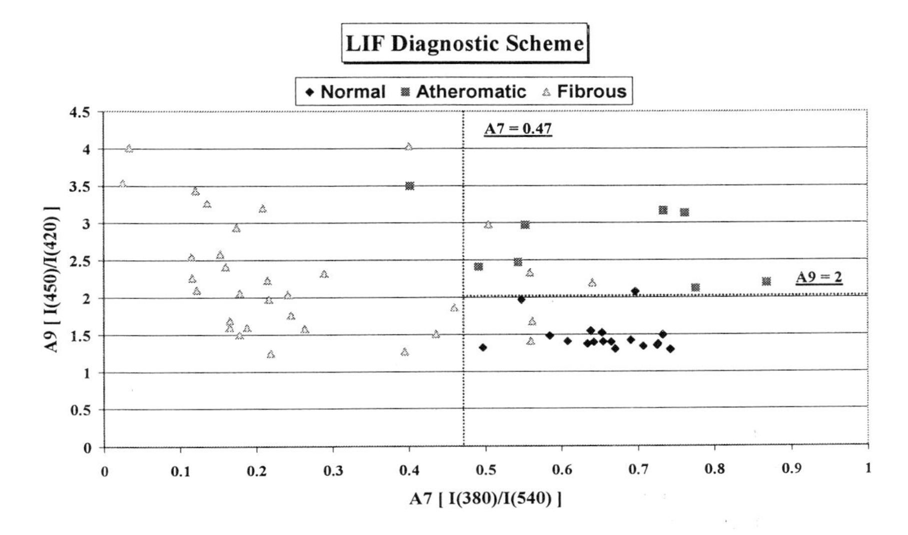

Figure 6. Classification of LIF spectra from normal, fibrous and atheromatic coronary artery specimens, according to a binary diagnostic scheme.

The combination of LIF & IVUS results are shown in Table 1. All specimens were included as correctly classified, if there was agreement between the technique and histology. For atheromatic samples, all statistical parameters are above 0.92, which is mainly due to the ultrasound classification of these samples. Successful results are also shown in the case of fibrous samples, that posses a maximum predictive value as well. Results for healthy specimens are slightly better than LIFS, comprising more than 0.9 values for all the statistical parameters.

Table 1. LIF, IVUS and LIF&IVUS characterization statistical results (sensitivity, specificity, accuracy, predictive value) for normal (N), fibrous (F) and atheromatic (A) coronary artery speciments.

	LIF			IVUS			LIF & IVUS		
	N	*F*	*A*	*N*	*F*	*A*	*N*	*F*	*A*
Sensitivity	0.94	0.86	0.64	0.50	0.61	0.91	0.94	0.96	1.00
Specificity	0.95	0.90	0.91	0.85	0.66	0.98	0.97	1.00	1.00
Accuracy	0.95	0.88	0.86	0.74	0.63	0.96	0.96	0.98	0.93
Predictive Value	0.89	0.89	0.64	0.60	0.63	0.91	0.94	1.00	1.00

4 Conclusions

- The implemented LIF diagnostic scheme managed to provide adequate discrimination between normal and fibrotic specimens with sensitivity reaching 0.94 and 0.85 respectively, whereas specificity, accuracy and predictive value were over 0.88.
- Interpretation of the statistical results shows that LIF & IVUS are complementary & co-operative intravascular diagnostic methods that can be combined towards improved diagnosis.
- The diagnostic methods have showed especially high sensitivity, specificity, accuracy and predictive value.
- First experiments on the combination of the two methods through a new dual LIF&IVUS catheter for simultaneous diagnostic implementation have shown very promising preliminary results towards improvement of the efficacy of non-invasive intravascular characterization of the arterial wall. This implementation has also revealed that such an approach is technologically applicable and effective in spite of its embryonic development stage and more efforts on this direction are considered well justified.
- Evaluation through preclinical studies is the next significant "step" to be made aiming to simultaneous implementation of both methods in vivo.

References

1 J.H.Eusterman, R.W.P. Achor, O.W. Kincaid, A.L.Jr.Brown, "Atherosclerotic disease of the coronary arteries: A pathologic-radiologic study", Circulation 26, 1288-1295 (1962).

2 Z. Vlodaver, R. Frech, R.A. Van Tassel, J.E. Edwards, "Correlation of the antemortem coronary arteriogram and the postmortem specimen", Circulation 47, 162-169 (1973).

3 P.G.Yock, P.J.Fitzerald, D.T.Linker, and B.A.J.Angelsen "Intravascular ultrasound guidance for catheter based coronary interventions" JACC, 17, 6, 39B-45B, (1991).

4 B.N.Potkin, A.L.Bartorelli, J.M.Gessert, R.F.Neville, Y.Almagor, W.C.Roberts and M.B.Leon " Coronary artery imaging with intravascular high frequency ultrasound" Circulation, 81, 1575-1585 (1990).

5 R.J.Siegel, M.Ariani, M.C. Fishbein, J.S.Chae, J.C. Park, G. Mourer and J.S. Forrester "Histopathologic Validation of Angioscopy and Intravascular Ultrasound" Circulation, Vol.84, No1 (1991).

6 M.Sartori, R. Sauerbrey, S. Kubodera, F.K. Tittel, R.Roberts, and P.D. Henry, "Autofluorescence Maps of Atherosclerotic Human Arteries - A New Technique in Medical Imaging", IEEE J. Quant Electr., QE-23(10), 1794-1797 (1987).

7 D. Yova, H. Gonis, C. Politopoulos, E. Agapitos, N. Kavantzas, S. Loukas, "Interpretation of Diagnostic Implications of Fluorescence Parameters for Atherosclerosis in Fibrous, Calcified and Normal Arteries" Tech. Health Care, 3:101-109 (1995)

8 R. Richards-Kortum, R.P. Rava, M. Fitzmaurice, L. Tong, N.B. Ratliff, J.R. Kramer, and M.S. Feld "A One Layer Model of Laser Induced Fluorescence for Diagnosis of Disease in Human Tissue: Applications to Atherosclerosis", IEEE Tran. Biomed. Eng. (1989).

9 S. Andersson-Engels,J. Johansson, U.Sternam, K.Svanberg and S. Svanberg, “Malignant tumor and atherosclerotic plaque diagnosis using laser induced fluorescence”, IEEE J. Quant. Electr. Vol.: 26, 2207-2217 (1990).

10 L.I. Deckelbaum, J.K. Lam, H.S. Cabin, K.S. Clubb, and M.B. Long, "Discrimination of Normal and Atherosclerotic Aorta by Laser Induced Fluorescence", Lasers in Surgery and Medicine, 7, 330-335 (1987).

11 S. Andersson-Engels, A. Gustafson, J. Johansson, U. Sternam, K.Svanberg and S. Svanberg, "Laser-indused Fluorescence used in localizing atherosclerotic lesions" Lasers Med.Sci., 4, 171-189 (1989).

12 D. Yova, H.Gonis, S. Loukas, K. Kassis, E. Koukoutsis, C. Papaodysseus, "Comparison of Classification Algorithms Based on Fluorescence Data for the Diagnosis of Atherosclerosis", Progress in Biomedical Optics, 2623, 436-447 (1995).

13 D. Yova, C.Politopoulos, H.Gonis, C.Papaodysseus "A Generalized Diagnostic Algorithm Based on Fluorescence Data for the Atherosclerosis", Med. Biol. Engin. Computing, 34, 297-29845 (1996).

Temporal and Spectral Narrowing of Sub-picosecond Laser-Induced Fluorescence of Polymeric Gain Media

G. Zacharakis, G. Heliotis, G. Filippidis, T. G. Papazoglou
Foundation for Research and Technology-Hellas, Institute of Electronic Structure and Laser, Laser and Applications Division, P.O. Box 1527, 71110 Heraklion, Crete, Greece, Fax# ++ 30 81 391318, e-mail: zahari@iesl.forth.gr

The narrowing effects of scatterers on the lifetime and spectral width of laser-induced fluorescence of dye imbedded PMMA polymer sheets were studied using a 500 femtosecond pulsed dye laser emitting at 496 nm. The samples were prepared by mixing PMMA and dye solution, in a 1cm x 1cm x 2cm Teflon cuvette, among with the scattering particles. The scatterers were TiO_2 nanoparticles (~400 nm diameter, R-900 Ti-Pure from E.I. du-Pont de Nemours & Co., Inc.). The evaporation of the solvent (dichloromethane - CH_2Cl_2) resulted to the creation of the 1 mm thick polymer sheet. The dyes used were R101, R6G, DCM and Py2 (Lambda Chrome part numbers LC 6400, LC 5900, LC 6500 and LC 7300 respectively). Samples of different dye ($3x10^{-3}$ M to 10^{-2} M) and scatterer (10^{12}-10^{14} cm^{-3}) concentrations were prepared and during the experiments they were irradiated with different excitation energies varying from 0.5 - 30 μJ/pulse[1-3].

The produced fluorescence signal was collected by the appropriate optics and focused at the entrance slit (100μm) of a 0.10 m spectrograph employing a 450 grooves mm^{-1} holographic grating. Wavelength calibration was performed with a mercury lamp. After spectral analysis the signal was focused at the entrance slit of a Hamamatsu C 5680 Streak Camera. A CCD camera, operating at -50°C for reduced dark noise, was used for the recording of the signals. The resulting signal was fed to a PC for further processing.

The goal of the experiments was the extensive study of the dependence of the spectral and temporal characteristics of the recorded fluorescence signal on the excitation energy and on the concentration of the scatterers and the dye. Furthermore, the different behavior of the liquid solution compared with the solid polymer can also be studied. Thus, the effects of the different embedding environment of the dye on the fluorescence characteristics can be investigated as well as the different characteristics of different dyes in the same environment.

Time and spectral profiles of the recorded 3D images were taken in order to extract the corresponding information (figure 1). The temporal evolution of the fluorescence can be extracted after fitting the time profiles with the appropriate function. Single exponential was used for the solutions and double for the solid polymer sheets, suggesting the different decay mechanism caused by the different properties of the liquid or solid medium. As far as the spectral features are concerned the width of the spectrum is compared with the spectral FWHM of the laser pulse.

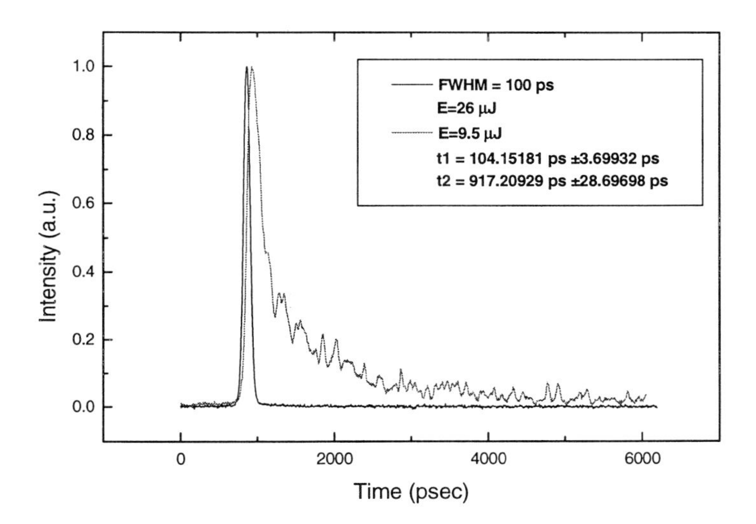

Figure 1 : Characteristic time profiles obtained from the images recorded with the Streak Camera on polymer sheet.

In both domains a sigmoidal curve is constructed depicting the life-time or the spectral FWHM in dependence on the excitation energy. From this curve the threshold point where the values are dramatically reduced can be seen (figure 2).

In both time and wavelength domains the width of the profile is dramatically reduced when the excitation energy is higher than a threshold value resulting to stimulated emission. This value is around 18 μJ. This leads to fluorescence lifetime of the same order as the measured width of the laser pulse (~ 20 psec) and much narrower wavelength curve - the FWHM is reduced to about one fourth of the standard value.

The time evolution of the decay of the fluorescence is different in the two types of samples. The solutions present a simple while the polymer sheets present a more complex decay mechanism, suggested by the double exponential fitting. The

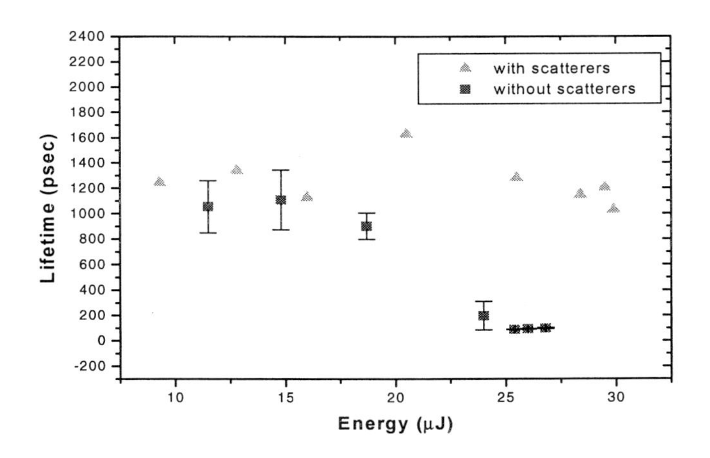

Figure 2 : Dependence of fluorescence lifetime on excitation energy, recorded on R6G polymer sheet with and without scatterers.

different behaviour is also shown by the longer fluorescence lifetime of the solutions in comparison with the polymers. A typical value for the R6G solution is 3 nsec while for the polymer sheet is 1 nsec.

The spectral and temporal narrowing can occur in wavelengths other than the maximum of the fluorescence curve. This can be caused by the different gain curve of the cavity that is formed by the scatterers inside the polymer sheet, due to the different concentration and distribution of the scatterers inside the sample.

The observed narrowing of the emission spectra of the sample with high dye concentration owes to the fact that the increased number of the dye molecules prevents the excitation photons to insert deeper and thus reacting with a great number of molecules. Since the excitation area in the sample is minimised and well-defined the possibility of self-absorption is reduced, resulting to narrow fluorescence spectra. The stimulated emission observed when the excitation energy is above the necessary threshold value, leads also in this direction.

In conclusion this is a study of the use of the spectral and temporal narrowing of fluorophore emission in a highly scattering environment. Similar studies are underway when various fluorophores are embedded in tissue[4]. The final goal is to take advantage of this effect towards a more spatially and spectrally confined agent in Photodynamic Therapy of target tissue lesions on skin or other types of superficial lesions.

References

1. V. S. Letokhov : "Generation of light by a scattering medium with negative resonance absorption", Sov. Phys. JETP **26**, pp. 835, 1968
2. R. M. Balachandran, D. P. Pacheco and N.M. Lawandy : "Laser action in polymeric gain media containing scattering particles", Appl. Opt. **V. 35**, No. 4, pp. 640, 1996
3. R. M. Balachandran and N.M. Lawandy : "Theory of laser in scattering gain media", Opt. Letters **V. 22**, No. 5, pp. 319, 1997
4. M. Siddique, Li Yang, Q. Z. Wang, R. R. Alfano : "Mirrorless laser action from optically pumped dye-treated animal tissues" Opt. Comm. **No. 117**, pp.475, 1995

Laser Induced Fluorescence in Atherosclerotic Plaque with Different Excitation Wavelengths

M. Makropoulou, H. Drakaki, N. Anastassopoulou, Y.S. Raptis and A.A. Serafetinides
National Technical University of Athens, Physics Department, Zografou Campus, 15 780, Greece, E-mail: mmakro@central.ntua.gr
A. Paphti, B. Arapoglou and P. Demakakos
University of Athens, Medical School, Department of Surgery, Areteion Hospital, Greece

Abstract. Laser induced fluorescence spectra were obtained from atherosclerotic plaque samples, in order to discriminate areas in the plaque of different composition. Different excitation wavelengths were used: 488nm and 476nm, provided by a c.w. Ar^+ laser, 457.9nm, provided by a c.w. Kr^+ laser, and 337.1nm, provided by a pulsed N_2 laser. All samples were histopathologically examined and showed three areas of different composition: fibrous tissue, lipid constituents and calcified plaque. Spectra corresponding to these areas were analyzed and mathematical fittings were performed, in order to reproduce each spectrum as a correlation of multiple Gaussian peaks. An effort was made to discriminate the composition of the plaque from the information obtained by the corresponding spectra.

1 Introduction

In the last few years there has been an increasing interest in laser induced fluorescence (LIF) as a diagnostic tool in laser angioplasty [1]. Various laser excitation wavelengths have been used in order to obtain the arterial fluorescence [2], during steady state [3] and time-resolved [4] experiments. Although discrimination between normal arterial tissue and atherosclerotic plaque has been repeatedly reported [5], there has not been found an accurate classification for all subtypes of plaque.

In this work, a discrimination of the composition of atherosclerotic carotid plaques is attempted, by activating tissue chromophores with visible c.w. and pulsed ultraviolet laser light. The histologic evaluation of the irradiated sites was used to determine the plaque type and any relevance of the recorded spectra with the histologic diagnosis of the plaque type is discussed.

2 Materials and Methods

Different lasers were used as an excitation source: an Ar^+ laser at 488nm and at 476nm as well as a Kr^+ laser at 457.9nm. The power output of both continuous wave, visible lasers was ranging from 5 to 15mW. The ultraviolet pulsed excitation laser was a nitrogen laser, emitting at a wavelength of 337.1nm, with 7ns pulse width. The energy output of the laser was <0.5mJ per pulse, while the pulse to pulse laser energy fluctuations were less than 5%, inducing relatively small fluctuations of the fluorescence signal intensity. The beam was focused onto the sample, which was located on a micro-adjustable mount. Before every measurement the sample was moistered with water. Atherosclerotic plaque samples from occluded carotid tissues were obtained from patients undergoing angiosurgery and kept refrigerated in aqueous solution. The samples were cut transversely and could be moved by steps of approximately 1mm vertically and horizontally. The emitted fluorescence was collected and analyzed either by a double monochromator with a photomultiplier tube, or by a single spectrograph (SPEX 1681 C), equipped with a holographic grating of 300 l/mm and a 1064-element photodiode array with a resolution of ~0.25nm/diode. The signal was converted from analog to digital and it was stored and processed by a personal computer. The experimental set-up is shown in previous work [6]. The samples were fixed in buffered formalin, after the spectra acquisition, until their histopathology examination in the hospital. The histopathology examination procedure was the conventional one: paraffin sections were stained with Hematoxylin and Eosin and all sections were evaluated by light microscopy.

3 Results

Many spectra were obtained from each sample in order to characterize different components of the plaque. In figure 1 are demonstrated spectra obtained from four different plaque samples, each one of them excited at a different wavelength. The histopathological examination showed that all spectra correspond to calcified plaque.

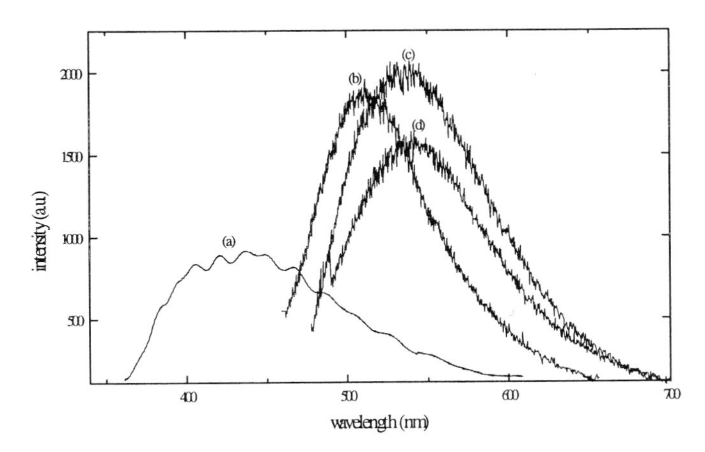

Fig. 1. Spectra obtained from calcified plaque samples excited at (a) 337.1nm, (b) 457.9nm, (c) 476nm and (d) 488nm

In figure 2 are presented fluorescence spectra obtained from two different areas of the same carotid plaque sample, which was excited at 337.1 nm. The macroscopic examination of the sample revealed a plaque composed mainly by lipid constituents. This area corresponds to the spectrum indicated with (a). An area with few calcium deposits was also observed on this sample, corresponding to the spectrum indicated with (b). In figure 3 is shown the respective photograph.

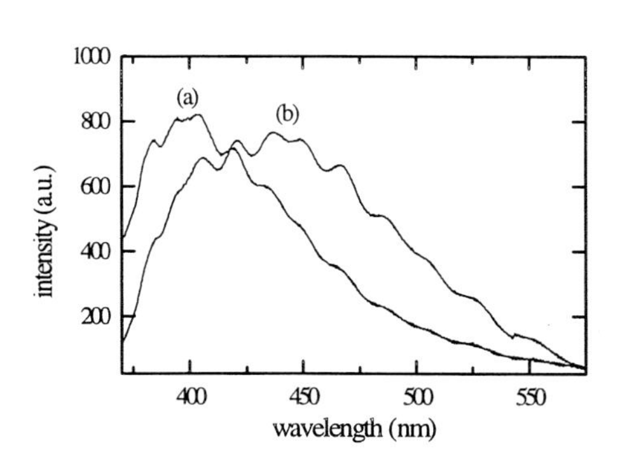

Fig.2. Fluorescence spectra obtained from a plaque sample, excited at 337.1nm, corresponding to (a) lipid constituents and (b) calcium

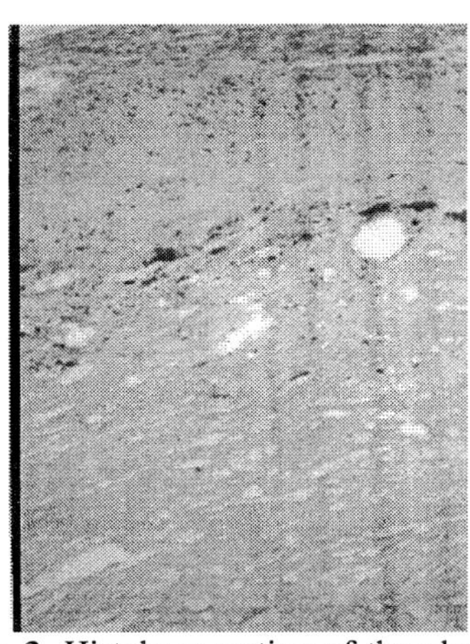

Fig. 3. Histology section of the plaque, corresponding to the spectrum shown in figure 2. Lipid constituents and few calcium deposits, on the upper side of the photograph, are observed

It is obvious that the overall lineshape of each spectrum represents the superposition of individual lineshapes from the laser induced fluorescence signal of tissue chromophores. By applying multiple gaussian fitting algorithms, we tried to analyze the overall spectra in one-peak curves. Figure 4 illustrates an example of the gaussian fitting analysis from a spectrum corresponding to a calcified plaque specimen.

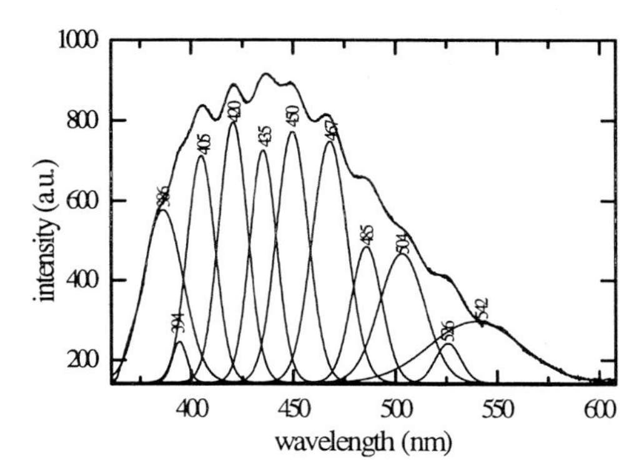

Fig. 4. Multiple gaussian fitting in an LIF spectrum of a calcified plaque, excited at 337.1nm. The wavelength of the peak of each gaussian is shown on the graph

4 Discussion

The histopathological examination of the samples revealed three areas of different composition: fibrous tissue, lipid constituents and calcified plaque. An effort was made to distinguish the composition of the atherosclerotic plaques from its spectral characteristics. Spectra obtained from calcified areas presented maximum at the same wavelengths (±3nm), depending on the excitation wavelength. At 457.9nm excitation wavelength, the LIF spectra presented maximum around 520nm. At 476 and 488nm excitation wavelengths, the spectral peaks were observed within the range from 540 to 550nm. Calcified plaques excited at 337.1nm presented maxima in the area of 520-550nm.

LIF spectra corresponding to lipid plaque did not present maximum in a specific wavelength range. This is probably due to the different components, such as cholesterol, composing a lipid plaque. The same behavior was also observed for areas composed of fibrous tissue. In tha case of samples excited at 337.1nm, peaks appearing at <440nm and >575 nm were considered as emitted mainly by cholesterol type plaque.

Mathematical fittings were performed in order to reproduce each spectrum as a correlation of multiple Gaussian peaks. Spectra obtained from samples excited at 337.1nm, demonstrated peaks mainly at 417±1nm, 432±4nm, 450±3nm, 464±3nm, 506±3nm, 521±1nm, 525±2nm, 550±3nm and 575±3nm. No specific information was obtained from fittings performed to spectra corresponding to samples excited at the visible wavelengths.

To summarize, the autofluorescence of the atherosclerotic carotid tissue depends on the laser excitation parameters, such as wavelength, pulse width and pulse energy. The discrimination of the different components of the plaque is limited by the overlapping spectra of the tissue chromophores.

References

1. Deckelbaum, M.L.Stetz, K.M. O'Brien, F.W. Cutruzzola, A.F.Gmitro, L.I.Laifer, and G.R.Gindi, "Fluorescence Spectroscopy Guidance of Laser Ablation of Atherosclerotic Plaque", *Lasers Surg. Med.*, Vol.9, pp.205-214, 1989.
2. Baraga, R.P. Rava, P. Taroni, C. Kittrell, M. Fitzmaurice, M.S. Feld, "Laser induced fluorescence spectroscopy of normal and atherosclerotic human aorta", *Lasers Surg. Med.*, Vol.10, pp.245-261, 1990.
3. Papazoglou, T. Papaioannou, K. Arakawa, M. Fishbein, V.Z. Marmarelis and W.S. Grundfest, "Control of excimer laser aided tissue ablation via laser induced fluorescence monitoring", *Appl. Opt.*, Vol.29, pp.4950-4955, 1990.
4. Stavridi, V.Z. Marmarelis and W.S. Grudfest, "Spectro-temporal studies of Xe-Cl excimer laser-induced arterial wall fluorescence", *Med. Eng. Phys.*, Vol.17, 595-601, 1995.
5. Kittrell, R.L. Willett, C. de los Santos Pacheo, N.B. Ratliff, J.R. Kramer, E.G. Malk and M.S. Feld, "Diagnosis of fibrous arterial atherosclerosis using fluorescence", *Appl. Opt.*, Vol.24, pp.2280-2281, 1985.
6. M. Makropoulou, H. Drakaki, N. Anastassopoulou, Y.S. Raptis, A.A. Serafetinides, A. Pafiti, B. Tsiligiris, B. Arapoglou and P. Demakakos, "Laser induced autofluorescence for discrimination of atherosclerosis", to be published

In vitro Laser-induced Fluorescence Measurements of Human and Lamb Heart Tissue

G. Filippidis[a], G. Zacharakis[a], G.E. Kochiadakis[b], S.I. Chrysostomakis[b], P.E. Vardas[b], T.G. Papazoglou[a]

[a] Foundation for Research and Technology-Hellas, Institute of Electronic Structure and Laser, Laser and Applications Division, P.O. Box 1527, 71110 Heraklion, Crete, Greece Fax# ++ 30 81 391318, e-mail: ted@iesl.forth.gr

[b] Cardiology Department, University Hospital of Crete, Greece

Laser-induced fluorescence spectra were recorded during the exposure of human and lamb heart tissue to Argon-ion radiation (457.9nm). The samples were irradiated an hour after the excision. Tissue from twenty (20) lamb and two (2) human hearts was investigated. Spectra from the left and right atria and ventricles, the myocardium, the epicardium, and the aorta, were recorded. The experimental apparatus is shown in Fig. 1. The average power irradiating the samples during the experiments was of the order of 10 mW. The spectra obtained during the experiments represented the time-integrated fluorescence of the underlying tissue. After the end of each experiment the hearts were stored in formalin (10%). The samples were irradiated again after forty eight (48) hours, in order to investigate the spectral difference that appear due to formalin conservation.

Simple algebraic algorithms based on the spectral intensity variation were developed in order to discriminate the different chambers of the heart. Each spectrum was processed in the following manner: the average intensities of the spectral regions 547-552 nm (PE 1), 556-561 nm (VA 1), 566-571 nm (PE2), 580-585 nm (VA 2), 595-600 nm (PE 3), 616-621 nm (VA3), and 666-671 nm (PE4) were calculated. Using these parameters twelve different simple algebraic algorithms were constructed. These were implemented on the acquired spectra. The twelve algorithms are shown in the table below:

A1= PE1-PE2 / VA1	**A7= PE3 / PE4**
A2= PE1-PE2 / VA2	**A8= PE2 / PE4**
A3= PE3-PE4 / VA3	**A9= PE1-PE3 / VA2**
A4= PE1 / PE2	**A10= PE2-PE4 / VA3**
A5= PE1 / PE3	**A11= PE1-VA1 / PE2-VA2**
A6= PE2 / PE3	**A12= PE2-VA2 / PE3-VA3**

It was verified that the intensity of the fluorescence spectra from aorta and left atrium of human heart was by far greater than those of the other parts of the heart. In addition the fluorescence spectra which were recorded from the myocardium of human heart exhibited significant changes in their spectral shape in comparison with spectra recorded from other cardiac compartments (Fig 2).

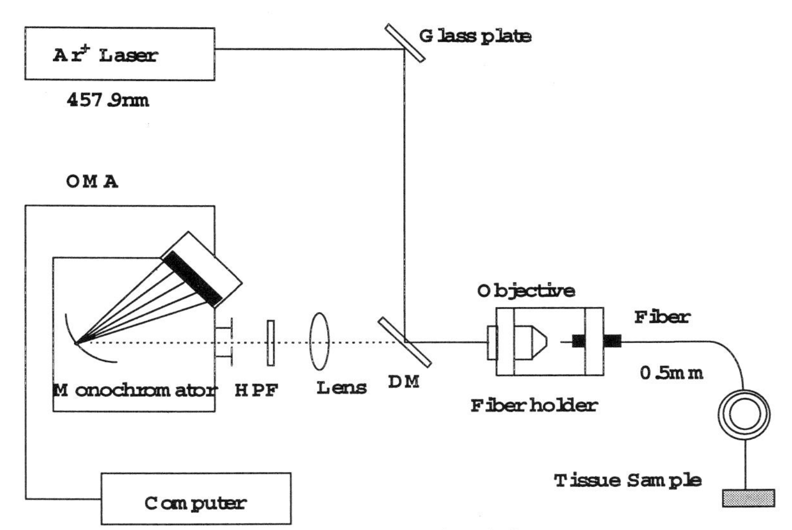

Fig 1: Experimental apparatus for laser induced fluorescence measurements of human and lamb hearts tissues. All optical components, except the pumping laser (Ar^+), are located in a diagnostic module. HPF: high pass filter, DM: dichroic mirror, OMA: optical multi-channel analyzer.

All the measurements taken from the same chamber of healthy lamb hearts were similar (Fig 3). Thus, an alteration in fluorescence spectra could then be linked to a lesion. The fluorescence spectra from the aorta and atrium (especially the left one) in lamb heart tissue usually gave higher fluorescence signal in relation with the other parts. This was in agreement with what was observed with the human heart tissue. In lamb heart tissue the algorithms seemed to have a good rate of success in discriminating between atrium and ventricle and between aorta and ventricle but on the other hand they seemed to have a very low rate of success in discrimination between aorta and atrium.

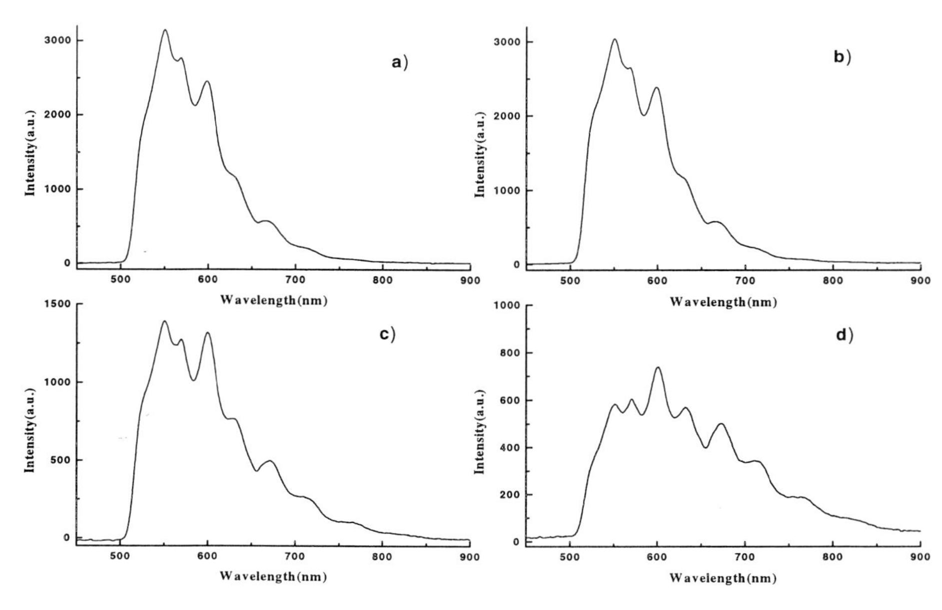

Fig. 2: Fluorescence spectra of human heart tissue, a) from the aorta b) from the left atrium c) from the left ventricle d) from the myocardium.

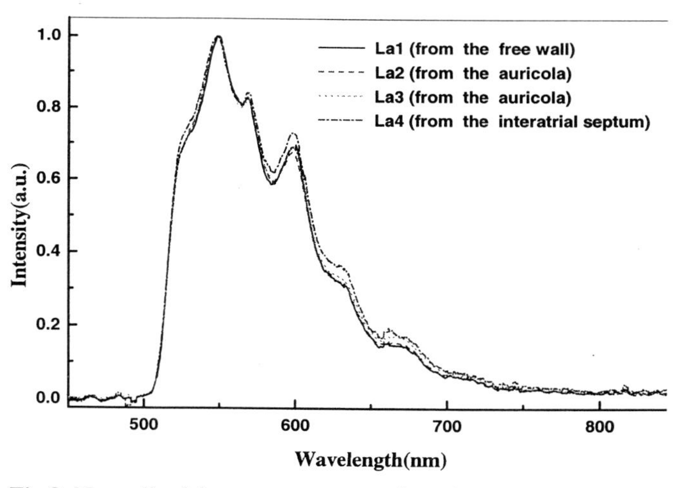

Fig 3: Normalized fluorescence spectra from four spots of left atrium of lamb heart tissue.

Moreover, in both human and lamb heart after the conservation in formalin for forty eight (48) hours, some differences in spectral distribution appeared especially in the region between 550-600nm (Fig 4).

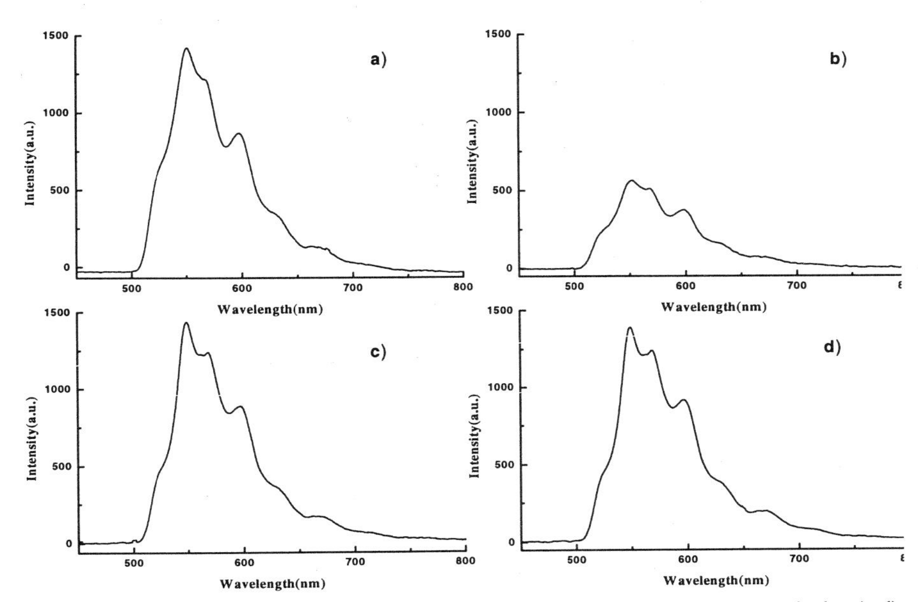

Fig 4: a), b), fluorescence spectra from left atrium and ventricle of lamb heart respectively. c), d), spectra from left atrium and ventricle after the conservation in formalin for 48 hours. The Ar^+ (457.9nm) laser was used as the excitation source.

References

1. T.G. Papazoglou, W.Q. Liu, A. Katsamouris and C, Fotakis, "Laser induced fluorescence detection of atherosclrotic deposits via their natural emission and hypocrellin (HA) probing", *J. Photochem. Photobiol. B: Biol.*, **22** 139-144 (1994)
2. A.M.K. Nilsson, D. Heinrich, J. Olajos, S. Andersson-Engels, "Near infrared diffuse reflection and laser-induced fluorescence spectroscopy for myocardial tissue characterisation", *Spectrochimica Acta Part A*, **53** 1901-1912 (1997)

Eye Model Using a CCD Camera for Observing the Images Constructed by IOLs

Kazuhiko Ohnuma[1], Yasuhiko Shiokawa, Norio Hirayama[3] and Qi Hua[4]
[1]Graduate School of Science and Technology, Chiba University 1-33 Yayoi-cho,Inage-ku,Chiba 263-8522 Japan
[2]Faculty of Engineering,Chiba University
[3]HOYA Healthcare Corporation
[4]HOYA Corporation

Abstract: The eye model that consists of a very high resolution CCD camera, a photographic lens, and an IOL put in water cell is constructed. The images are observed directly on a monitor and they can be compared and analyzed by a personal computer. The MTF of the eye model including CCD response are analyzed and it is found that the MTF resembles the total MTF of human eyes including the response of retina to brain. Here, we present the images constructed by two kinds of IOLs and the analyzed results using a resolution chart and a contrast chart.

1 Introduction

Eye models which can show us images of object around us directly on time, have been constructed[1,2]. But the images have low contrast in the high spatial frequency, because the images are observed byusing an objective lens with low MTF. So, we have developed a new eyemodel with a high-resolution CCD camera instead of an objective lens. The MTF of this eye model is calculated and is expected to be nearly same with the MTF of human eye including the MTF of retina to brain. Here we show the developed eye model and the images obtained by IOLs.

2 Eye Model

The eye model developed here is shown in Fig.1 . This eye model consists of photographic lens with 35mm of focusing length, an IOL in a water cell, aperture stop, and a very high resolution CCD camera. The 35mm photographic lens is selected because the incident angle to an IOL from this lens is nearly equal to the incident angle to crystalline lens from a cornea and this lens is used in MTF measurement of IOLs of Valdemar Portney[3]. The CCD camera is MegaPlus of Kodak with one pixel size of 6.8 μm which corresponds to 150 lines/mm in spatial frequency. The image is displayed on a monitor of a personal computer.

Similarity of MTF between Eye model and Human eye

Two MTF for 587.56nm (d line) are calculated under the same optical construction of a biconvex 20D IOL, aspheric cornea and 5mm aperture stop diameter. The total MTF of human eye is obtained by multiplying MTF of human eye and the MTF of retina to brain reported by [4]. And also, the total MTF of the eye model is obtained by multiplying the MTF of CCD array. The similarity of these two MTFs is shown in Fig.2.

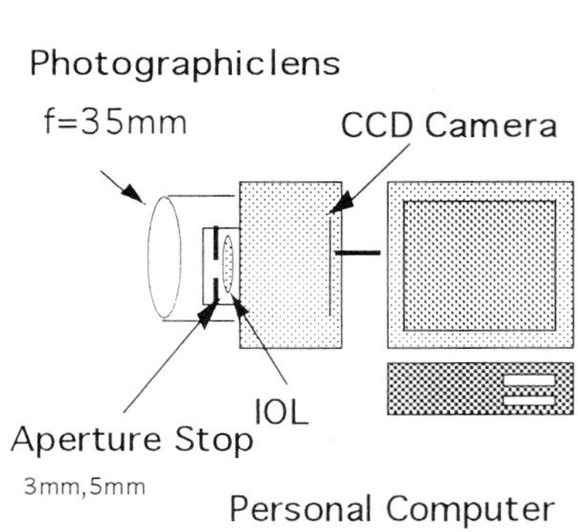

Fig.1 Eye model system

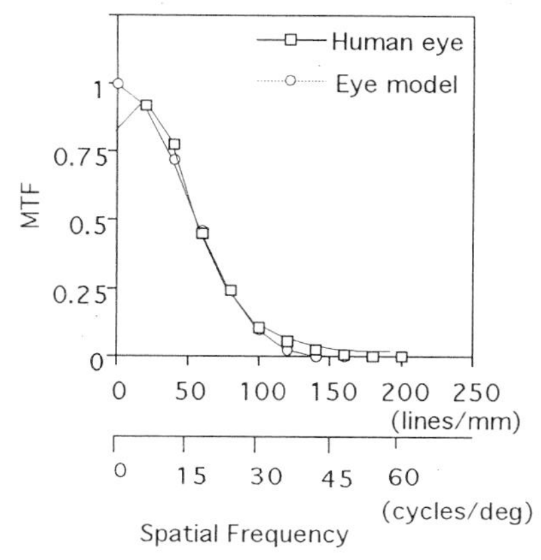

Fig.2 MTF of the Eye model and Human eye

3 Observation of images and the estimation

Contrast chart of VCTS6500 are put at 3m from the eye model and five resolution charts are set at 0.5m to 5m. The focus of eye model with a mono-focal IOL and /or multi-focal IOL is set at 5m and their images through eye model are displayed on a monitor. A man decides the contrast sensitivity and the resolution limitation from the images. These depend on the sensitivity of CCD camera that is able to detect 0.8\% difference of radiance. This sensitivity is higher than the sensitivity of human retina with 1% a little bit. So, the contrast sensitivity of human eye is predicted from the value obtained here. In the experiment, infrared cut filter and green filter are used to change the wavelength sensitivity of CCD to the wavelength sensitivity of the human eye.

4 Results of observation

The image of the resolution charts taken by using the mono-focal plano-convex lens with the power of 20D and focusing at 5m is shown in Fig.3. The resolution limit at each distance is shown in Fig.4. These two figures indicate that resolution decreases rapidly as the distance becomes near. The image taken by using the multi-focal diffraction lens with the power of 20D(4D added) and the resolution limit are shown in Fig.5 and Fig.6. These indicate that the resolution becomes lower at the distance of 2m. This is because two burred images are overlapped each other.

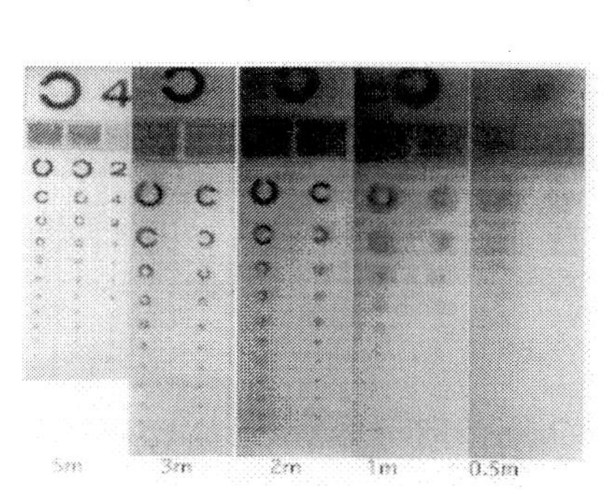

Fig.3 Image of mono-focal lens.

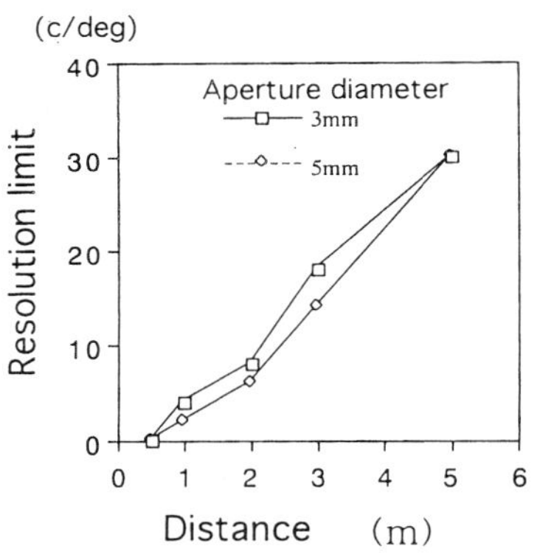

Fig.4 Resolution limit of mono-focal lens

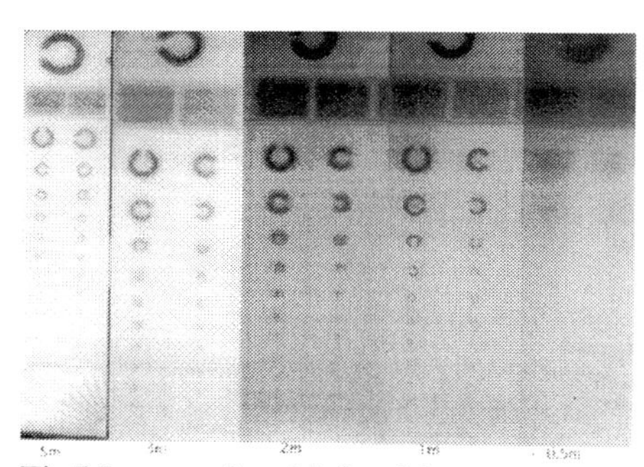

Fig5 Image of multi-focal lens.

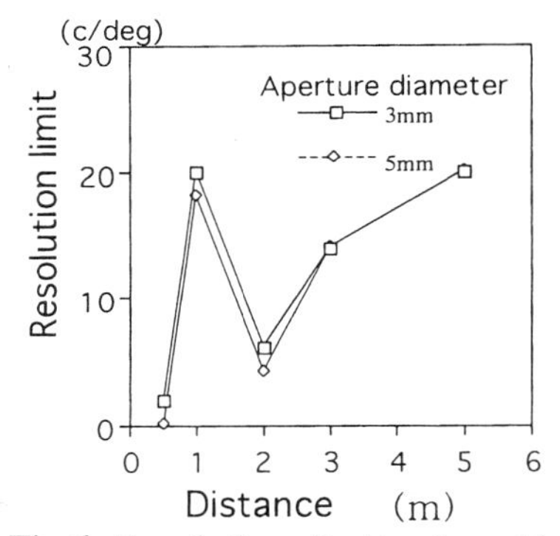

Fig.6 Resolution limit of multi-focal lens. (Diffraction lens)

The images of the contrast chart put at 3m taken by using the multi-focal diffraction lens are shown in Fig.7. And also, the contrast sensitivities derived from the images of mono-focal lens and the multi-focal are indicated in Fig.8. The contrast of the mono-focal lens is better especially in the high frequency.

5 Summary

The eye model with high resolution CCD camera has been developed for observing the images of IOLs. Here, the MTF similarity between the eye model and a human eye are shown. And also, the images of two different types of IOLs are shown. The merit of this eye model is that it presents us the image on time. So, we hope that it will be used for informed consent.

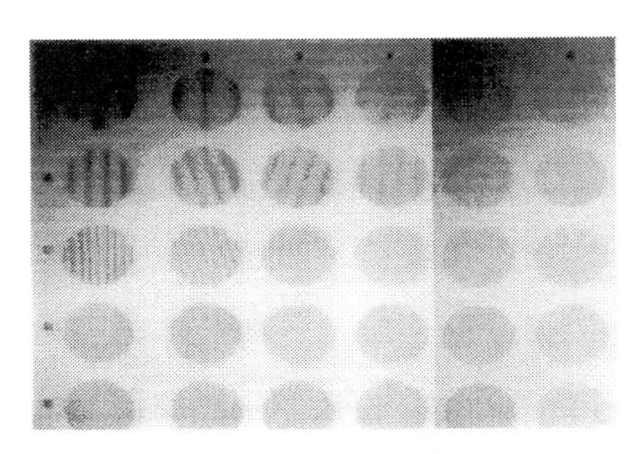

Fig.7 Image of multi-focal lens. (Diffraction)

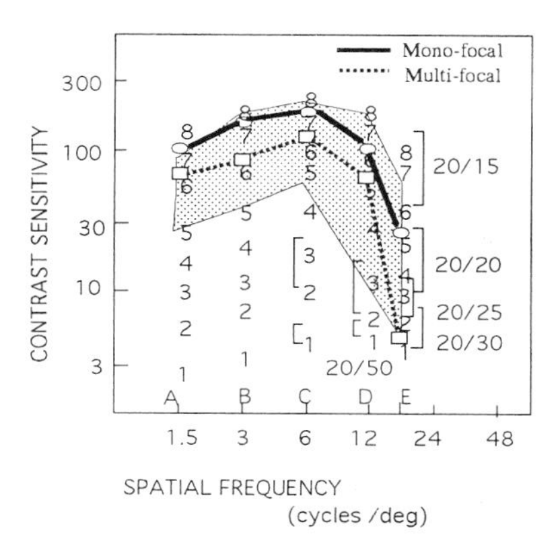

Fig.8 Contrast sensitivity of mono-focal ens and multi-focal lens.

References

1. Y. kora, T. Nagano, S. Yaguchi, T. Kosawa, and M. Kaneko: The retinal Image of three mulitfocal lenses through an eye model. J. of Japanese Opthalmological Society pp1091-1096 (1994)
2. S. Narai, R. Okura, Y. kora, S. Yaguchi, T. Kozawa, S. Okamoto, and H. Murakami: A new Developed eye model for evaluating the pseudophakic eye. J. of the eye (Japanese) 10 pp657-661(1993).
3. Alan Lang, Valdemar Portney: Interpreting multifocal intraocular lens modulation transfer functions J. Cataract Refract SURG 19 pp505-512(1993)
4. H. Ohzu et al.,: Optical Modulation by the Isolated Human Fovia", Vision Res., 12, 231-251 (1972)

Optical Wavefront Sensing Using a 2-Dimensional Diffraction Grating: Application to Biological Microscopy

Hiroshi Ohba[1] and Shinichi Komatsu[1,2]

[1] Department of Applied Physics, Waseda University, 3-4-1 Okubo, Shinjuku, Tokyo 169-8555, Japan
[2] Materials Research Laboratory for Bioscience and Photonics, 3-4-1 Okubo, Shinjuku, Tokyo 169-8555, Japan

Abstract. We developed a new wavefront sensor using a 2-dimensional diffraction grating with which we successfully measured a light wavefront distorted by thermal turbulence. In this paper, we improve the lateral resolution of this wavefront sensor and discuss the application of the sensor to biological microscopy.

1 Introduction

Adaptive optics is a technique to reduce image degradation produced by atmospheric turbulence. An adaptive optical system is divided into two sections; a wavefront sensor that measures the wavefront on which the phase $W(x, y, z)$ is constant, and a spatial light modulator (SLM) that manipulates and corrects the wavefront.

Many types of these wavefront sensors have been proposed. For example, the Shack-Hartmann wavefront sensor,[1)] which measures the tilt of the wavefront (equivalently ∇W) at only small apertures, and the Roddier wavefront sensor,[2,3)] which measures only the curvature of the wavefront ($\nabla^2 W$).

We propose a new wavefront sensor using a 2-dimensional diffraction grating, and in this paper we improve the lateral resolution of the wavefront sensor. With this improvement we can measure ∇W at arbitrary points in the pupil, and obtain the smoother wavefront.

2 Principle

This technique is based on the fact that the shadow of a diffraction grating pattern is spatially shifted when the incident wavefront of the light passing through the diffraction grating is tilted.[1)]

For simplicity, let us consider a 1-dimensional case where the period of the diffraction grating is d. If we consider a small segmented part whose length is $D(= n_0 d(n_0 \in \mathbf{N}))$, the wavefront over the segment can be approximated with a plane wave tilted by θ

$$\theta = \tan^{-1}\left[\frac{\tan^{-1}\{\Im\{I(mn_0u_0)\}/\Re\{I(mn_0u_0)\}\}}{2m\pi\Delta z/d}\right], \tag{1}$$

where $I(u)$ is the Fourier transform of the intensity distribution on the detector which is placed at the distance Δz behind the grating, and u_0 equals to $2\pi/D$.

We extend eq.(1) to a 2-dimensional case,

$$\begin{cases} \theta_x = \tan^{-1}\left[\frac{\tan^{-1}\{\Im\{I(mn_0u_0,0)\}/\Re\{I(mn_0u_0,0)\}\}}{2m\pi\Delta z/d}\right] \\ \theta_y = \tan^{-1}\left[\frac{\tan^{-1}\{\Im\{I(0,mn_0u_0)\}/\Re\{I(0,mn_0u_0)\}\}}{2m\pi\Delta z/d}\right], \end{cases} \tag{2}$$

where θ_x and θ_y denote the tilts of the wavefront over the segment along the x axis and y axis respectively.[4)]

A piston configuration of all the segmented parts come to minimize the sum of the squares of the differences in height of adjacent segment-edge midpoints, the wavefront is reconstructed.[5)]
In contrast with the Shack-Hartmann wavefront sensor, the present sensor has the advantage that both the size and position of the each segment can be varied according to shape of the wavefront. The lateral resolution is improved by increasing the number of segmented area, while the accuracy of the tilt measurement is improved by increasing the segment size.

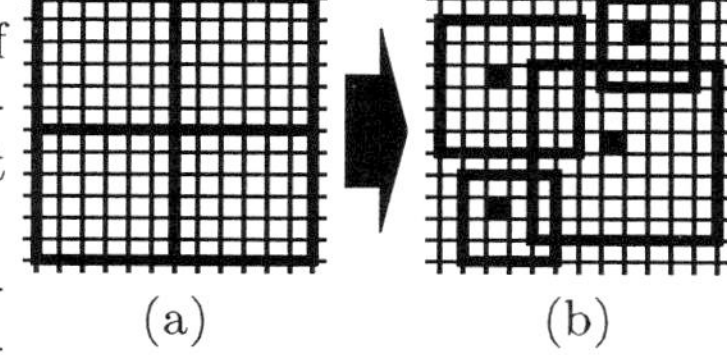

Fig. 1. The size and position of the grating segments. (a) Fixed segmentation. (b) Adaptive segmentation.

3 Experiments and Results

We measure the wavefront of a light wave distorted by thermal turbulence. As shown in Fig. 2, an expanded plane wave is distorted by a thermal turbulence caused by a soldering iron whose tip width is about 3.0 mm. A rectangular area (about 50 mm × 50 mm) 100 mm above the soldering iron is observed, whose reduced image is projected onto the grating plane with a lateral magnification of 0.20. A He-Ne laser, whose wavelength λ is 632.8 nm, is used as a light source. A diffraction grating (G: d=0.10 mm) is placed in front of a CCD detector (CCD: pixel size 25 μm × 25 μm) whose effective array size is 256 × 256 pixels. The distance Δz between G and CCD is 2.0 mm. We experiment on two cases. In the first case, *case A*, there is one soldering iron. In the second case, *case B*, there are two closely placed soldering irons, in which the heated range is twice as large as in *case A*.

Figures 3 and 4 show the experimentally obtained wavefronts for *case A* and *case B*, respectively. Figures 3(a) and 4(a) are the measured wavefronts obtained by using the fixed segmentation method, while Figs. 3(b) and 4(b) are those obtained by using the adaptive segmentation method.

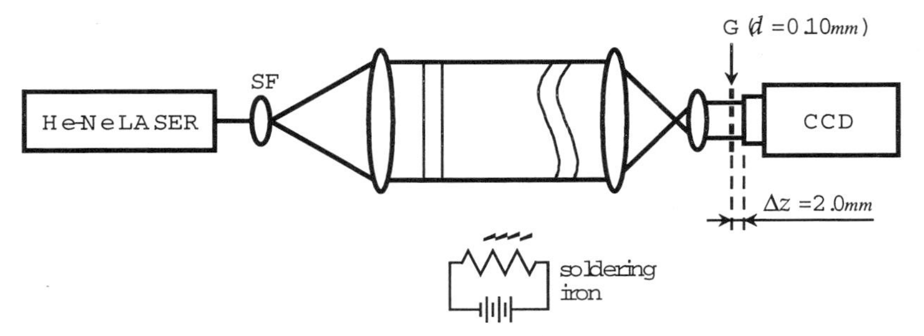

Fig. 2. Experimental set up. SF: spatial filter, G: 2-dimensional diffraction grating, CCD: CCD image detector.

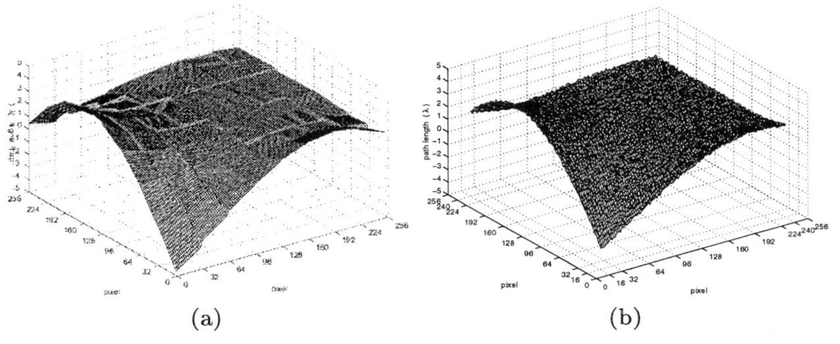

(a) (b)

Fig. 3. Measured wavefront of the light wave distorted by thermal turbulence (*caseA*) using (a) the fixed segmentation method and (b) the adaptive segmentation method.

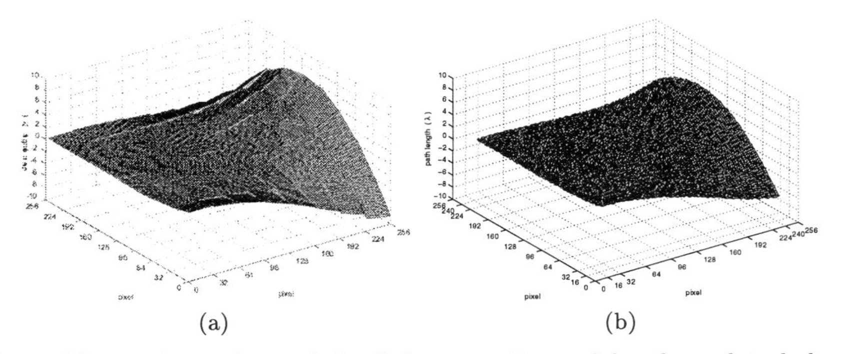

(a) (b)

Fig. 4. Measured wavefront of the light wave distorted by thermal turbulence (*caseB*) using (a) the fixed segmentation method and (b) the adaptive segmentation method.

4 Application to Biological Microscopy

The proposed wavefront sensor is not only simple and inexpensive, but also able to reconstruct the wavefront enough to be quickly applied to biological microscopy. Figure 5(a) shows the schematic diagram. By adopting, the magnified wavefront measurement shown in Fig. 5(b), higher resolution would be expected.

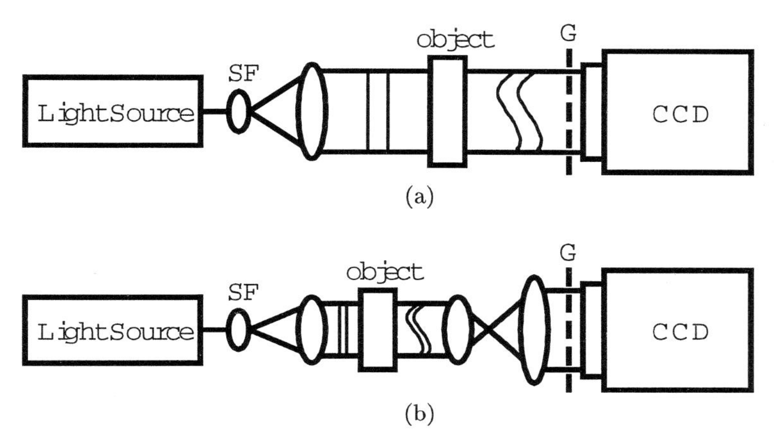

Fig. 5. Schematic diagram of biological microscopes using the wavefront sensor. (a) Direct wavefront measurement. (b) Magnified wavefront measurement.

5 Conclusion

We proposed a new technique that improve the lateral resolution of the wavefront sensor using a 2-dimensional diffraction grating. We successfully applied this wavefront sensor to the measurements of thermal turbulence. We also proposed the application of the sensor to biological microscopy.

This study was partly supported by a Grant-In-Aid (98A-586) from Waseda University as well as by the High-Tech Research Center Project by the Ministry of Education, Science, Sports and Culture.

References

1. F. Roddier, "Variation on a Hartmann theme," *Opt. Eng.*, **29**, 1239-1242 (1990).
2. F. Roddier, "Curvature sensing and compensation: a new concept in adaptive optics," *Appl. Opt.*, **27**, 1223-1225 (1988).
3. F. Roddier, "Wavefront sensing and the irradiance transport equation," *Appl. Opt.*, **29**, 1402-1403 (1990).
4. H. Ohba and S. Komatsu, "Wavefront sensor using a 2-Dimensional Diffraction Grating," *Jpn. J. Appl. Phys.*, **37**, 3749-3753 (1998).
5. S. Enguehard and B. Hatfield, "An exact analytic solution to segmented-mirror adaptive-optics control," *J. Opt. Soc. Am. A*, **11**, 874-879 (1994).

Measurement of Chlorophyll Distribution in a Leaf by Laser Induced Fluorescence

Kunio Takahash[1], Yasufumi Emori[2]
[1]Kisarazu National College of Technology, 2-11-1 Kiyomidai-higasi,Kisarazu, Chiba E-mail : ntakaha@minato, Kisarazu. ac.jp
[2]3-41-7 Kanamachi, Katsushika, Tokyo ,125 Japan.

Abstract: Information about the fluorescence light distribution in a leaf induced by ultra violet radiation is important for us to consider the photosynthesis of the vegetables . We have developed Micro-Fluorescence Imaging (MFI) apparatus that consists of a microscope system with a I:I installed CCD camera and the laser light. By utilizing this apparatus, we can obtain images of fluorescence distribution of a various intact leaves which have different cell structures and examine the relation between the fluorescence related closely with chlorophyll and chemically synthesized substance accumulated in leaves. We report that measurement of chlorophyll distribution in the leaf by laser induced fluorescence.

1 Micro-fluorescence Imaging System

This system is composed of three components; light source unit, image detector unit including a special sample holder and data processing unit as shown in Fig. 1-A. The light source unit has three types of laser; a cw blue color Ar^+ laser (488 nm), a cw green color Ar^+(514.8 nm) and a cw red color He-Ne laser (632.8 nm), which illuminate the sampled leaves by switching the reflectance mirror electronically. The detector unit has a metallurgical microscope ,CCD camera with a image intensifier (I.I.), a sliding filter holder and a specially designed sample holder. The filter holder is set between the microscope objective and the CCD camera , and is utilized for selecting out fluorescence of chlorophyll radiated from sampled plants.

A sample holder (Fig. 1-B) has a water pool for supplying water to the sampled leaf to prevent from the dehydration of the plant and keeping its water condition as normal. The sample leaf is vertically fixed on the holder and is cut it by a dual-edged razor so that its cross-section is observed 50 μm above the surface of the sample holder as shown in Fig.2. The stimulating lasers are incident perpendicularly on the upper surface of the sampled leaf and the cross-section of the leaf is illuminated by the transmitted lasers, and fluorescent lights are induced inside of the leaf. Fluorescence are consisted of scattered light of illuminating incident laser light ,chlorophyll fluorescence induced from the sampled leaf and another bleu-green color fluorescence which is induced from unknown pigment of the sample leaf.

The data processing unit consists of the image processor, computer, printer, plotter and video printer. The image captured by the CCD camera with the I.I. is converted to a 8-bit digital image of 512x512 pixels through an image processor.

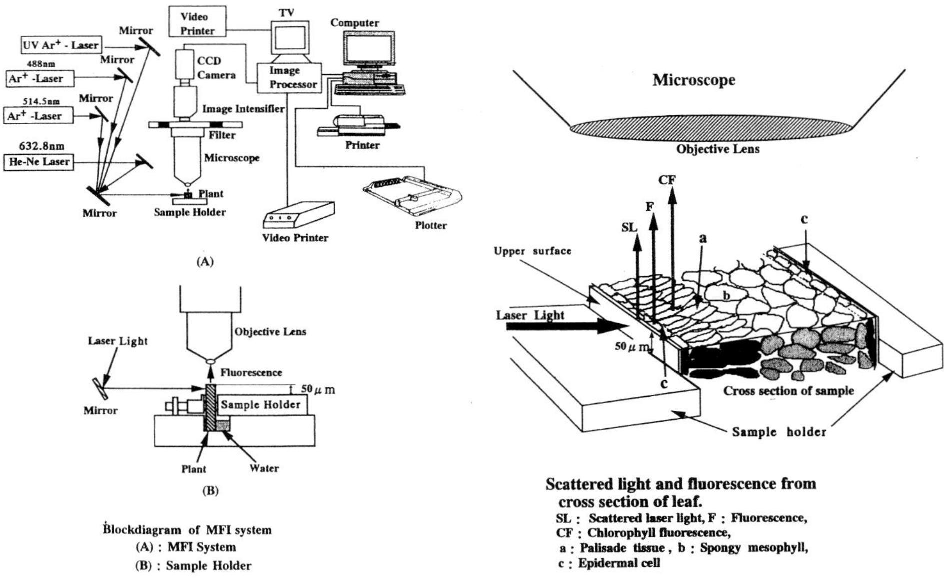

Fig .1 Diagram of MFI system

Fig. 2 Enlarged figure of sample holder

2 Test Sampled Leaves and Experiments

Two kinds of leaves; that is, Camellia Japonica that is seen in every temperate regions and Mangrove founded at the sea shore in the tropical arer, are selected. Distributions of induced fluorescence inside of the sample leaves are represented for the wavelength of 685 nm and 740 nm at which show maximum peak fluorescence intensities induced by the phto-system II and phto-system I respectively.

2.1 Camellia Leaf

Phot.1 and Fig.3 show a processed images for Camellia where visible lasers of 488 nm, 514.5 nm, and 632.8 nm stimulate on the front surface of Camellia ,and induced fluorescence patterns for three stimulating wavelength and fluorescence of two wavelengths of 687 nm and 741 nm induced by each sitimulated wavelengths are presented.

It is known from this figure that the more wavelength of stimulating radiation is longer ,the incident stimulating light scatters and expands deeper into the leaf, and maximum scattering is occurred at distance of 50 μm under the front surface of the leaf. The induced fluorescence distributions for chlorophyll and scattered light for incident light on the line of the thickness of the leaf are presented as like as Fig. 4. From these figures, we can know quantitatively the penetration differences and distribution of chlorophyll . Intensity ratio of induced fluorescence intensity at the wavelength of 741 nm (F 741) to that of 687 nm : F(741)/ F(687) is considered as a active factor which indicates re-absorption of fluorescence light by chlorophyll and utilization for photosynthesis in the leaf[1]. So F(741)/ F(687) as the function of the depth of the leaf is presented with a parameter of stimulating wavelength as shown in Fig. 5.

2.2 Mangrove Leaf

Mangrove leaf is analyzed by same method as that of Camellia. In Photo2, the up left image shows the cross-section of Mangrove leaf, up-right one is a scattered light in the leaf ,low-left and low-right ones indicate induced fluorescence for wavelengths of 687 nm and 741 nm respectively. and Fig, 6 shows intensity distrbution curves of each fluorescence light. From these curves, it is known that illumination light of 488 nm scatters only near the front surface of the leaf, and induced fluorescence lights are distributed at the center part of the leaf. This means that chlorophyll are distributed in the center part of the leaf so different from that of Camellia.

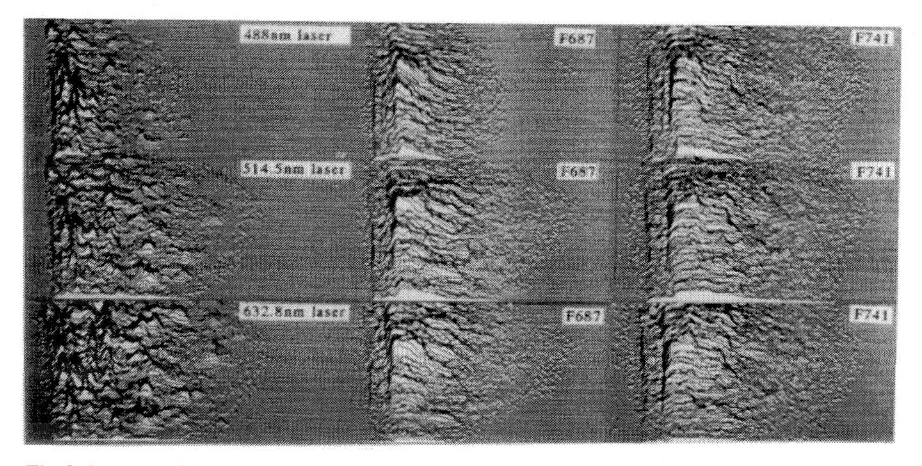

Fig.3. Processed images for fluorescence for Camellia

3 Irradiation of UV Radiation

As increasing the ozone hole, UV radiation may affect not only human life, but also plant. Here we have exposed UV radiation (354-361 nm) on Camellia and Mangrove and examined the relation between blue-green fluorescence induced by UV radiation and red fluorescence light with visible radiation.(phot. 3)

UV radiation is cut off at the upper layer of leaf of Camellia and does not penetrate into the inner layer, but in Mangrove, extended fluorescence area by included chlorophyll is found at the center part of the leaf. The active factor F (741) / F (687) in Mangrove becomes larger that of Camellia .

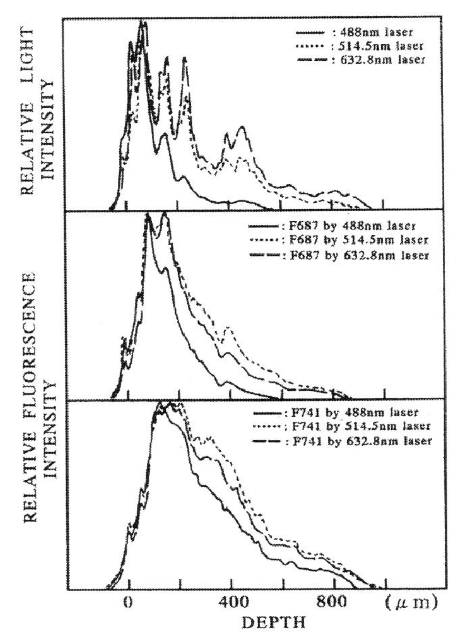

Fig. 4 Intensity distribution of Fluorescence and laser lights through the leaf thickness of Camellia

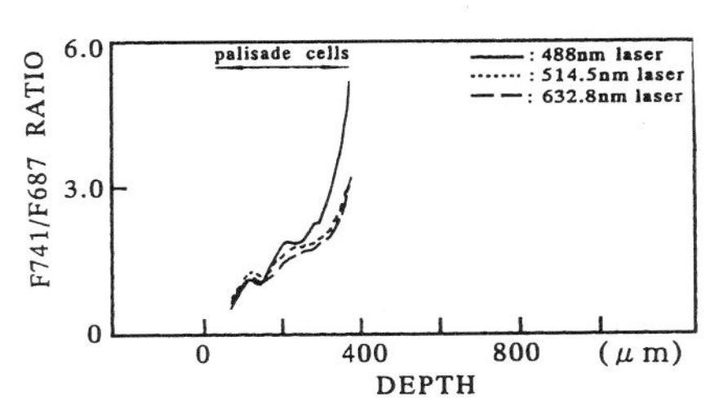

Fig.5 Active factor for Camellia

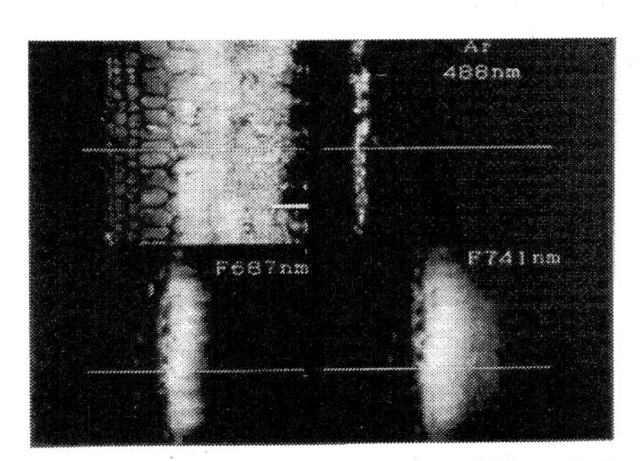

Phot.2 Fluorescence images and laser light for the leaf of Mangrove

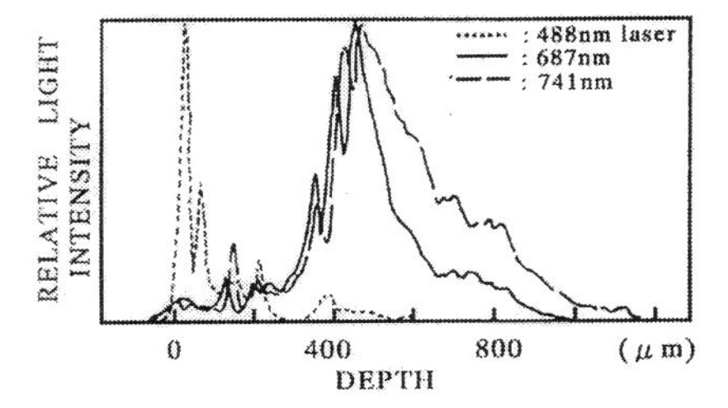

Fig. 6 Intensity distribution of fluorescence light through the leaf thickness for Mangrove

Phot. 3 Fluorescence and laser light images induced by UV irradiation for leaves of Camellia and of Mangrove.

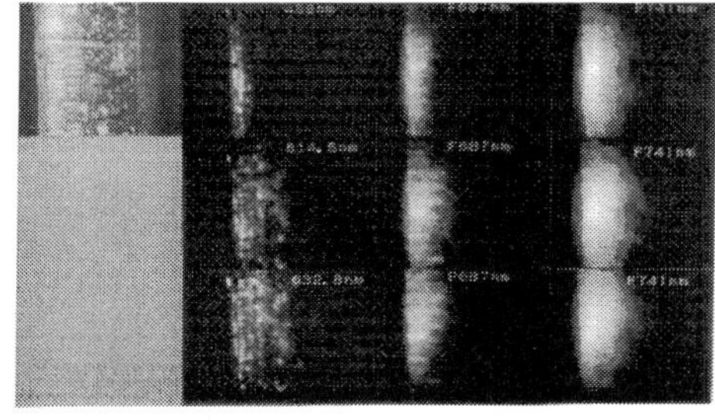

Phot. 1 Images of chlorophyll fluorescence and stimulating lights in the leaf of camellia

References

1. K Takahashi,K.Mineuchi,T.Nakamura,M.Koizumi:A system for imaging transverse distribution of scattered light and chlorophyll fluorescence in intact rice leaves, Plant ,Cell and Environment (1994) 17,105-110

High-Resolution Color Holography for Archaeological and Medical Applications

F. Dreesen, H. Delere, and G. von Bally

Laboratory of Biophysics, University of Münster,
Robert–Koch–Straße 45, D–48129 Münster, Germany

Abstract. New materials, and in particular the new laser recording technique developed at the Laboratory of Biophysics of the University of Muenster, enable the recording of high-resolution color holograms. The obtained (lateral) resolution for fine structures of scattering objects is below 3 μm. Thus, it is possible, with the help of a microscope, to investigate the 3-D microstructure of the object from a hologram, e. g. in medical diagnostics and archaeological documentation and analysis.

1 Holographic Contact Recording in True Color

Three laser beams (red, green, blue) are combined with mirrors and beam-splitters, and then focussed into one pinhole used as a spatial filter. During holographic recording the expanded, white laser beam illuminates the object through the holographic material (Denisyuk geometry) (Fig. 1).

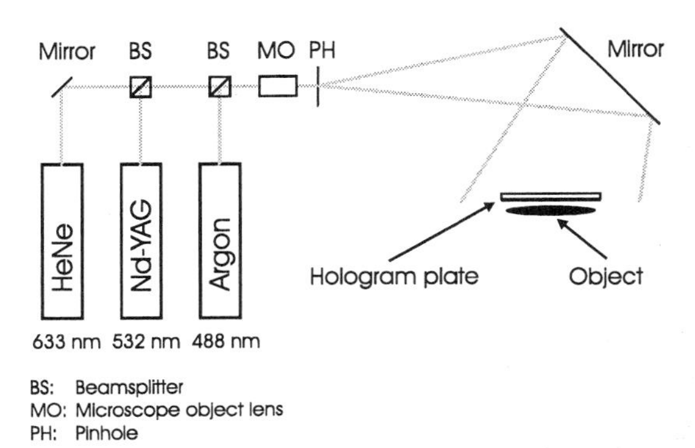

Fig. 1. Holographic contact recording with three lasers (red, green, blue) to achieve high resolution color holograms in only one holographic emulsion [1].

2 Lateral Resolution in The Holographic Image

The lateral resolution of the holographic image is determined with an USAF-resolution test chart. For reconstruction of a hologram recorded with the above described set up, a halogen spot light is used. Fig. 2 shows such a reconstructed holographic image. The smallest resolved element is No 4 out of group 7 defining a lateral resolution of 2.8 μm.

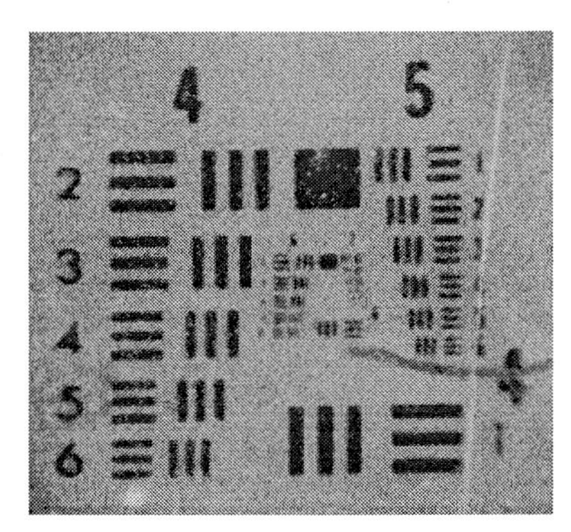

Fig. 2. White-light reconstruction of a color hologram. Shown is a USAF resolution test chart. The achieved lateral resolution in the holographic image is 2.8 μm (group 7, element 4).

3 Holographic Biopsy

High-resolution holography allows microscopic investigations of holographically recorded tissue without taking samples from the body as is usual in conventional biopsy. Fig. 3 shows human mastocarcinoma cells. Microscopic images of the original cells (on the left) and of a holographic recording of these cells (on the right) is presented for comparison.

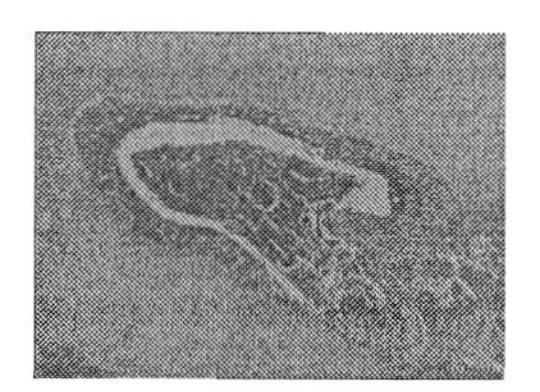

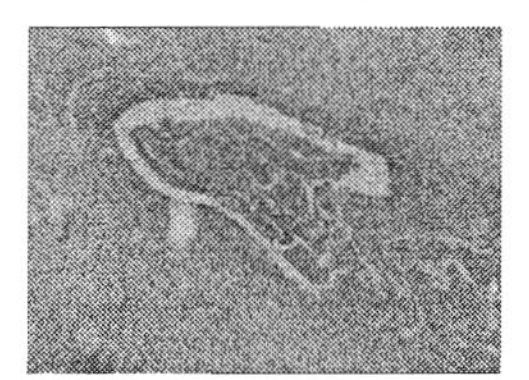

Fig. 3. Microscopic recording of human mastocarcinoma cells: (left) original, (right) hologram.

Human epithelial cells from the mucous membrane of the mouth were recorded holographically to demonstrate the lateral resolution of low contrast objects (Fig. 4). Here, both the original cells and their holographic image are magnified by a microscope, picked up by a video-camera and displayed on a monitor. A combination of endoscopy with this holographic technique is expected to extend conventional biopsy by microscopic analysis of holograms of vitally stained tissue.

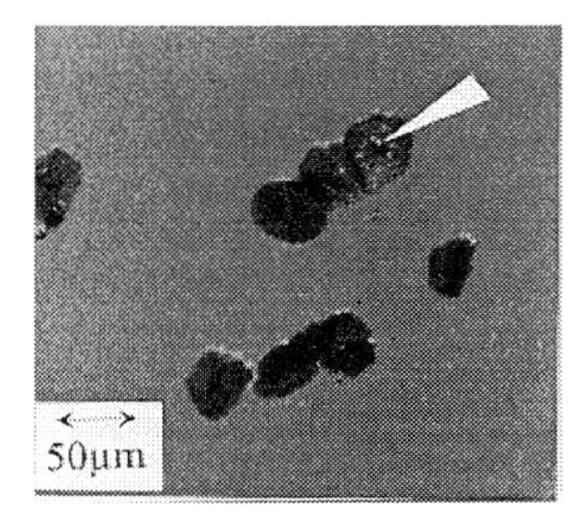

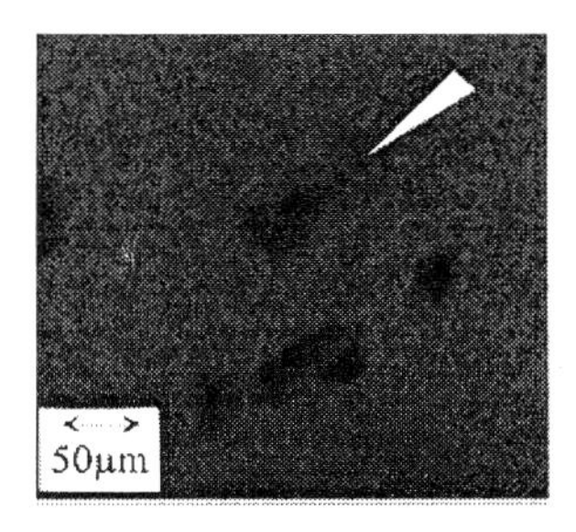

Fig. 4. Microscopic recording of human epithelial cells from the mucous membrane of the mouth: (left) original, (right) hologram. The arrow points to the nucleus of a cell (diameter of the nucleus approx. 10 μm) demonstrating resolution of subcellular structures.

4 Archaeoholography

The study of ancient civilizations is particularly difficult due to the fragility of the excavated objects and to their distribution throughout various museums. Only in rare cases can an archaeologist have all these objects at his disposal at the same time. A solution could be offered by high-resolution color-holography (practised at the Laboratory of Biophysics, University of Muenster [2]). Thus, the study of archaeological items could be improved with color-holograms and, as with the original object, they can be studied down to the finest structures. In Fig. 5 the 3-D representation of a hologram is demonstrated (the original hologram is of a true-color type). It is possible to look at the fragmented cuneiform tablet from different views, as can be recognized by changing position of the shadows. Even fragments can be recombined holographically (Fig. 6, left), showing the 3-D shape and details of the complete original. With this holographic technique a new type of a world archive of master holograms could be established, which may serve as a source for hologram copies. This way, the information of those valuable cultural objects can be secured, and every specialist could build up an individual library. Museums could make use of this technique to complete their collections without having to buy or borrow expensive original artefacts. Entire exhibits could be transported without the usual concerns of space and costs. Private collectors could even establish their own "holographic museum".

5 Transportable Holographic Camera System

The high-resolution holography technique is made mobile by a compact transportable holographic camera system. The dimensions of the holographic recording system are 45 x 60 x 30 cm^3 so that it fits into a standard sized suitcase. For recording the holograms, only a stable ground to place the camera is required. Thus, holographic recordings can be realized outside a special

Fig. 5. One hologram - different views. The photos show images reconstructed from the same hologram of an original Sumerian cuneiform-tablet from the Hilprecht-Collection, Jena, Germany (it shows a contract between merchants, 2100 BC).

holographic laboratory e. g. in museums or even at excavation sites. Fig. 6 shows reconstructed images of holograms recorded with the holographic camera system. On the left, the technique of holographic recombination is demonstrated. Two fragments of one cuneiform tablet from the Pergamon Museum in Berlin were recorded separately with the transportable camera system. Later, at the Laboratory of Biophysics, University of Muenster, these two holograms were copied into only one hologram (in the center). On the right, Fig. 6 shows an undeciphered Minoan tablet with inscriptions (Cyprus Museum, Nicosia).

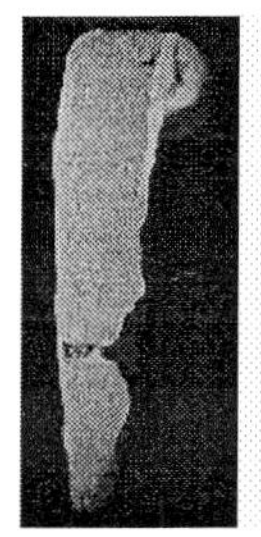

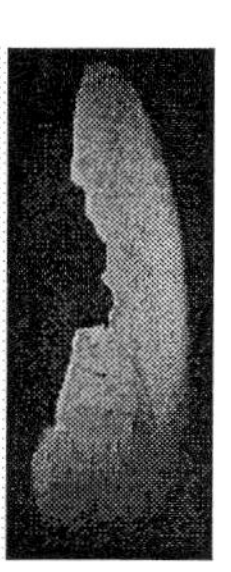

Fig. 6. : Holograms recorded with the transportable holographic camera system: (left) holographic recombination of a cuneiform tablet, fragments recorded at the Pergamon Museum, Berlin; (right) tablet with undeciphered Minoan inscriptions from the Cyprus-Museum in Nicosia.

References

1. Dreesen, F.; von Bally, G.: Color holography in a single layer for documentation and analysis of cultural heritage, series on Optics Within Life Sciences, Springer Verlag, Berlin, 4 (1997) 79-82.
2. Sommerfeld, W.; von Bally, G.; Dreesen, F.; Roshop, A.: Holography: A new technology in cuneiform research, series on Optics Within Life Sciences, Springer Verlag, Berlin, 4 (1997) 42-51.

Financial support by the Ministry of Education, Research, Science and Technology of the Federal Republic of Germany (Project-No 03VB9MU18) is gratefully acknowledged.

Protoporphyrin IX Kinetics in Rat Peritoneal Organs and Tumor After Systemic 5-Aminolevulinic Acid Administration

Maurice C.G. Aalders[1], Henricus J.C.M. Sterenborg[2], Fiona A. Stewart[3], Nine van der Vange[4]
1 Laser Center, Academic Medical Center, Amsterdam, The Netherlands.
2 Department of Radiotherapy, Daniel den Hoed Cancer Center, Rotterdam , The Netherlands.
3 Department of Experimental Therapy, Antoni van Leeuwenhoek hospital, Amsterdam , The Netherlands.
4 Department of Gynecology, Antoni van Leeuwenhoek hospital, Amsterdam, The Netherlands.

Abstract: This work was performed within the framework of a study investigating the possibility of using photodetection for localization of small volume, macroscopically non-visible peritoneal metastasis of ovarian tumors using 5-AminoLevulinic Acid (ALA) induced protoporphyrin IX. This technique is based on differences in protoporphyrin concentration between tumor tissue and normal tissue. We investigated the fluorescence kinetics of protoporphyrin IX (PpIX) in rat tumor and abdominal organs following intravenous administration of 100 and 25 mg/kg ALA. The highest level of fluorescence was achieved in the liver after 4 h of 100 mg/kg ALA administration. Peak fluorescence in the liver was about 5 times higher than maximum tumor fluorescence. Maximum fluorescence contrast between tumor and intestines was measured at 2.5 hours after administration of 25 and 100 mg/kg ALA.

1 Introduction

In the majority of patients with ovarian cancer the diagnosis is established at an advanced stage when spread has already occurred into the abdominal cavity. In case of small volume macroscopically non-visible metastasis, the risk of understaging and subsequent undertreatment is high. Currently staging in ovarian cancer without overt spread is done by taking multiple blind biopsies at random from peritoneal surfaces for microscopic examination. Photodetection using 5-Aminolevulinic acid (ALA) induced Protoporphyrin IX (PpIX) as a fluorescent tumorlocalizing agent may improve detection of superficial malignant lesions during staging and second look surgery by enabling directed non-blind biopsies. Photodetection is based on differences in concentration of fluorescent dye in tumor and normal tissue. We investigated the pharmacokinetics of protoporphyrin IX in the rat abdominal cavity after intravenous administration of ALA. Fluorescence kinetics were followed for up to 20 hours after intravenous administration of ALA on the liver, intestines, skin, peritoneum and tumor tissue.

2 Materials & Methods

The setup that was used for the fluorescence measurements is depicted in figure 1. A 100-watt mercury lamp was used to deliver 405 nm light for excitation of PpIX and an optical multichannel analyzer was used for detection of the fluorescence light. The ratio between the induced fluorescence, detected at a photosensitizer emission peak, and the fluorescence at an autofluorescence emission wavelength area was calculated from the fluorescence spectra.. This ratio is independent of fluctuations in excitation light and setup geometry.

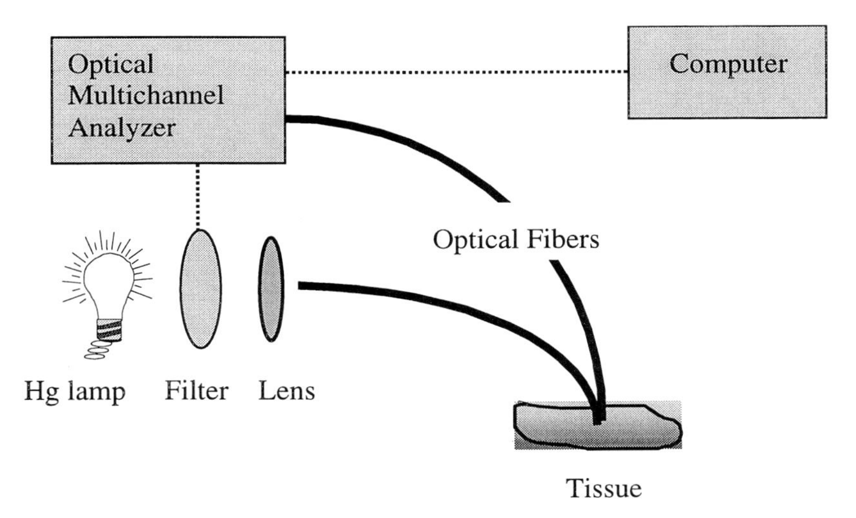

Figure 1: The fluorescence setup. A 100-watt mercury lamp is filtered to deliver 405 nm. Optical fibers deliver the light to and from the tissue. The fluorescence signals are analyzed using an Optical Multichannel Analyzer and a personal computer.

Four weeks prior to the procedure, female rats were injected intra-peritoneally with $1*10^6$ CC531 tumor cells. This resulted in tumor nodules with an average diameter of about 1 mm which were distributed diffusely throughout the abdominal cavity.

ALA was dissolved in PBS and injected intravenously in the tail vein of female rats at doses of 25 or 100 mg/kg. At different times after administration of ALA, the abdominal cavity was opened and fluorescence kinetics was followed for 4 hours.

3 Results

The fluorescence kinetics curves were normalized to a base-line value measured before administration of ALA. All organs had a significant rise in fluorescent level after ALA. Peak fluorescence in the tumor after administration of 25 and 100 mg/kg was reached at 2.5 hours after administration, followed by a rapid decline, reaching normal level at 5 hours. The highest level of fluorescence was measured in the liver 4 hours after administration of 100 mg/kg ALA. Peak fluorescence in the liver was about 5 times higher than maximum tumor fluorescence and was maintained for two hours. Fluorescence in the small intestines, cecum and skin was significantly lower than tumor fluorescence at 2.5 hours after 100 mg/kg ALA administration, and at 2.5-4.5 hours after 25 mg/kg ALA (small intestines and cecum only) (figures 2 and 3).

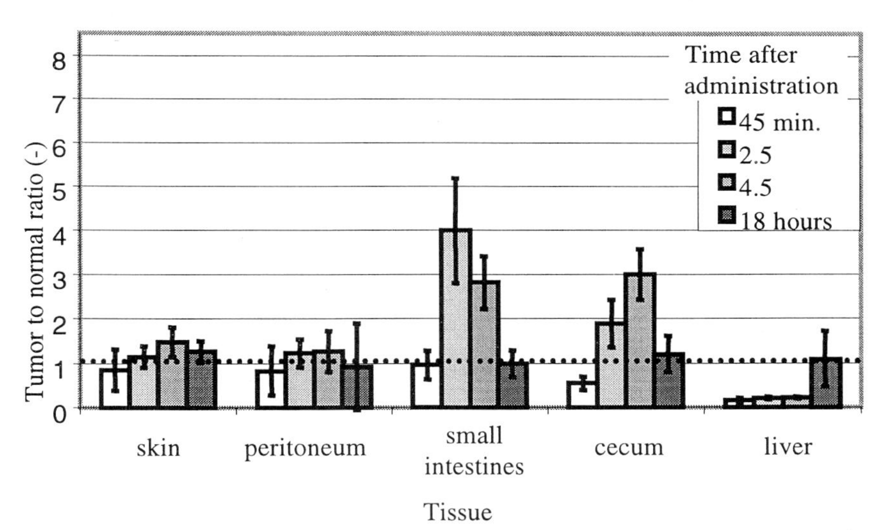

Figure 2: The tumor- to normal tissue fluorescence ratios at different time intervals after administration of 25 mg/kg ALA. All fluorescence kinetics curves were normalized on a base-line value, measured before administration. The values from the tumor curve were divided by the tissue values at different time intervals.

Porphyrin kinetics in the normal tissue was dependent on the drug dose. Peak fluorescence level was reached more quickly after administration of 25 mg/kg than 100 mg/kg.

4 Conclusions

Maximum tumor to normal fluorescence contrast is reached at 2.5-4.5 hours after administration of 25 or 100 mg/kg ALA.

Fluorescence kinetics are dependent on drug dose and tissue type.

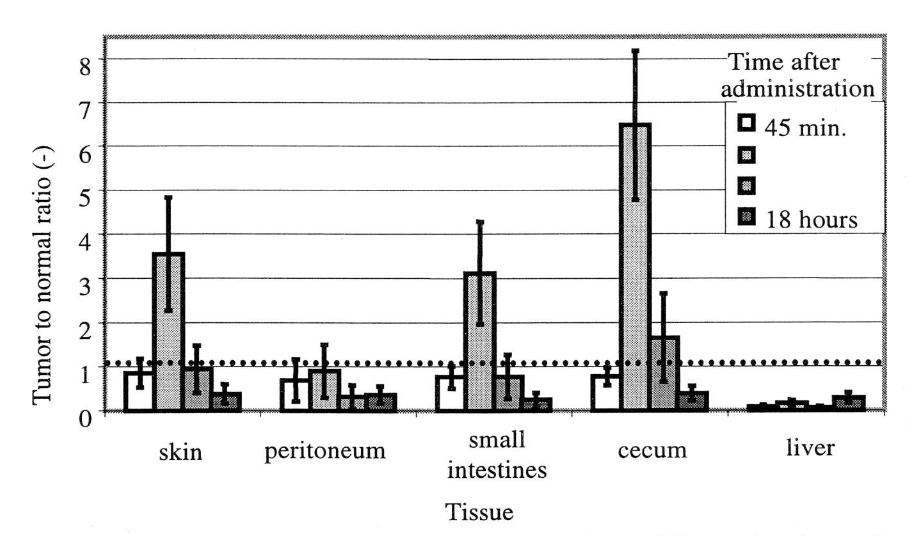

Figure 3: The tumor- to normal tissue fluorescence ratios at different time intervals after administration of 100 mg/kg ALA

Detection of small volume tumor is possible when localized near the intestines or the abdominal wall. Strong fluorescence from the liver prevents detection of tumor tissue near the liver.

Photodetection using ALA may improve the detection of small volume metastases in the abdominal cavity.

Druck: Strauss Offsetdruck, Mörlenbach
Verarbeitung: Schäffer, Grünstadt